WILFRIED BRUTSAERT

Cornell University

Evaporation into the Atmosphere

Theory, History, and Applications

D. REIDEL PUBLISHING COMPANY

DORDRECHT: HOLLAND/BOSTON: U.S.A.

LONDON: ENGLAND

Library of Congress Cataloging in Publication Data

CIP

Brutsaert, Wilfried, 1934–
Evaporation into the atmosphere.

(Environmental fluid mechanics ; 1)
Bibliography: p.
Includes index.
1. Evaporation (Meteorology) 2. Atmosphere. I. Title. II. Series.
QC915.6.B78 551.57′2 81-12186
ISBN 90-277-1274-6 AACR2

Published by D. Reidel Publishing Company,
P.O. Box 17, 3300 AA Dordrecht, Holland

Sold and distributed in the U.S.A. and Canada
by Kluwer Boston Inc.,
190 Old Derby Street, Hingham, MA 02043, U.S.A.

In all other countries, sold and distributed
by Kluwer Academic Publishers Group,
P.O. Box 322, 3300 AH Dordrecht, Holland

D. Reidel Publishing Company is a member of the Kluwer Group

Printed in The Netherlands

Table of Contents

Foreword

The phenomenon of evaporation in the natural environment is of interest in various diverse disciplines. This book is an attempt to present a coherent and organized introduction to theoretical concepts and relationships useful in analyzing this phenomenon, and to give an outline of their history and their application. The main objective is to provide a better understanding of evaporation, and to connect some of the approaches and paradigms, that have been developed in different disciplines concerned with this phenomenon.

The book is intended for professional scientists and engineers, who are active in hydrology, meteorology, agronomy, oceanography, climatology and related environmental fields, and who wish to study prevailing concepts on evaporation. At the same time, I hope that the book will be useful to workers in fluid dynamics, who want to become acquainted with applications to an important and interesting natural phenomenon.

As suggested in its subtitle, the book consists of three major parts. The first, consisting of Chapters 1 and 2, gives a general ouline of the problem and a history of the theories of evaporation from ancient times through the end of the nineteenth century. This history is far from exhaustive, but it sketches the background and the ideas that led directly to the scientific revolution in Europe and, ultimately, to our present-day knowledge. The second and central part of the book is covered in Chapters 3 through 7; it deals with the conceptualization and the mathematical formulation of water vapor transport in the lower atmosphere from or to natural surfaces. Some basics in the physics of the lower atmosphere are treated in Chapter 3, in order to support the details of the descriptions in later chapters and to make the book more self-contained. In Chapter 4, I have tried to relate the different concepts and formulations for atmospheric vapor transport over a statistically uniform surface, by considering them within the framework of advances in atmospheric boundary-layer theory. The parameterization of the turbulent transport characteristics of different types of surfaces is elaborated on in Chapter 5. Some elementary aspects of the energy budget at the earth's surface are covered in Chapter 6. Different ways of describing local advection effects in boundary layers disturbed by a step change in surface conditions are dealt with in Chapter 7. Finally, the third part, in Chapters 8 though 11, provides an overview of currently available techniques for measuring or calculating the rate of evaporation. These different methods are arranged according to their conceptual basis with reference to the principles described in the second part. The choice of any given method depends on the problem under consideration, and it is governed by the available data or the available instrumentation.

I have not tried to cover all possible angles and points of view of the subject matter. Rather, I have followed a line of thought, which over the years I have found to be productive in conveying an understanding of the phenomenon. Similarly, no attempt has been made to compile a complete bibliography. Nevertheless, the references that are listed contain references to other work, so that it should be possible to trace back the more important developments. Except in a few cases, when no other sources were at hand, I have avoided listing references to research and progress reports and other types of more informal publications, which are less generally accessible.

The material for this book developed from the preparation of lectures for my courses in hydrology and micrometeorology in the School of Civil and Environmental Engineering, at Cornell University. In addition, major parts of the book were presented in courses, which I gave as visiting professor at the Agricultural University (LH) in Wageningen, the Netherlands, and at the Federal Institute of Technology (ETH) in Zürich, Switzerland.

I owe gratitude to my former graduate students and also to my professional friends in different parts of the world, with whom I have had the good fortune of discussing problems related to evaporation. Their ideas and their insight have stimulated me to clarify my own thinking and thus contributed to the completion of this book.

WILFRIED BRUTSAERT
Ithaca, N.Y.,
Spring 1980

CHAPTER 1

Introduction

1.1. DEFINITIONS

The main concern of this book is the evaporation of water in the natural environment. In general, *evaporation* is the phenomenon by which a substance is converted from the liquid or solid state into vapor. In the case of a solid substance, the phenomenon is often referred to as *sublimation*. The vaporization of water through the stomata of living plants is called *transpiration*. Over land transpiration from vegetation and direct evaporation from the soil and small water surfaces are difficult to separate in computations; therefore these two terms are often combined in the term *evapotranspiration*. All these distinctions are useful at times; however, the term evaporation is usually adequate to cover all processes of vaporization, unless specified otherwise.

1.2. PRACTICAL SCOPE

a. The Water Budget

Evaporation of water in the natural environment, be it from free water surfaces or from land surfaces covered by vegetation, is one of the main phases of the hydrological cycle. This hydrological, or water, cycle consists of the perpetual transfer of water from the atmosphere to the earth's surface by precipitation, whence it runs off to rivers, to lakes and to the seas, either through infiltrated underground seepage, or directly as surface flow. The cycle is closed as the water vaporizes back into the atmosphere.

Water entering the evaporation phase of the hydrological cycle becomes unavailable and cannot be recovered for further use. This is an important consideration in the planning and management of water resources. In many parts of the world the available water resources are presently being tapped close to the limit, so that an accurate knowledge of the consumptive use through evaporation is indispensable. Evapotranspiration from land surfaces, together with precipitation, governs the amount of runoff that is available from a watershed or river basin. It also determines, to a large extent, the response characteristics of a watershed, to produce storm runoff and flooding as a result of heavy precipitation. Potential evaporation, which can be defined loosely as the evaporation that would occur if water were plentiful, is often used as the required water supply in the design of proposed irrigation schemes. The amount and rate of evaporation from water surfaces is information which is required to design storage reservoirs or to assess the value of natural lakes for such purposes as municipal and industrial water supply, irrigation of agricultural lands, condenser cooling water, hydroelectric power, navigation and, in some cases, even recreation.

Unfortunately, however, evaporation from land surfaces covered by vegetation and evaporation from free water surfaces are still among the less understood aspects of the hydrological cycle, and it is still rather difficult to estimate these quantities on a regional basis. The regional estimation of the other phases in the cycle, such as precipitation or streamflow, involves formidable sampling problems. But in the case of evaporation beside sampling, there is also the problem of simply determining it at a point location.

The water budget equation, which expresses the conservation of mass in a lumped or averaged hydrological system, can be written as follows

$$(P - E)A + Q_i - Q_o = dS/dt, \tag{1.1}$$

where P is the mean rate of precipitation on the system, E the rate of evaporation, A the surface area, Q_i the surface and ground water inflow rate, Q_o the surface and ground water outflow rate and S the water volume stored in the system. A rather obvious way of determining the evaporation rate E is to take it as the rest term in (1.1), when the other terms are known. But even if the other terms are known, this method is not generally practical. For relatively small but unavoidable errors in measuring precipitation and runoff can often produce large absolute errors in the resulting evaporation. Furthermore, this method would be inapplicable to predict evaporation in the design of planned water storage or irrigation engineering projects. This explains why it is usually necessary to determine evaporation, independently from the water budget, on the basis of meteorological data.

b. The Energy Budget

In any given system at the earth's surface, evaporation is the connecting link between the water budget and the energy budget. For a simple lumped system, when effects of unsteadiness, ice melt, photosynthesis and lateral advection can be neglected, the energy budget is

$$R_n = L_e E + H + G \tag{1.2}$$

where R_n is the specific flux of net incoming radiation, L_e the latent heat of evaporation, E the rate of evaporation, H the specific flux of sensible heat into the atmosphere and G the specific flux of heat conducted into the earth. The major portion of the incoming total radiation is absorbed near the surface of the earth, and it is transformed into internal energy. The subsequent partition of this internal energy into long wave back radiation, upward thermal conduction and convection of sensible heat, evaporation of water, and downward conduction of heat into the earth, is one of the main processes driving the atmosphere. The global pattern of heating forces the circulation of the planetary atmosphere. As a result of the large latent heat of vaporization of water, evaporation involves the transfer and redistribution of large amounts of energy under nearly isothermal conditions. Because, even at saturation, air can contain only relatively small amounts of water vapor, which can easily be condensed at higher levels, the air can readily be dried out; this release of energy through condensation and precipitation is the largest single heat source for the atmosphere. In other words, evaporation as a latent heat flux plays a crucial role in governing the weather and the climate.

Also, the availability, or the degree of shortage of water is a useful measure in portraying the climate. In this context actual evaporation is often compared with the potential evaporation to characterize the aridity of a region.

On a more local scale, electrical power generation and other industrial processes commonly produce waste energy. The disposal of this energy in coastal waters, lakes and rivers is a potential cause for large disruptions in the natural energy budget and the ecology of such water bodies. An understanding of the latent and sensible heat flux processes is essential in the design of suitable energy disposal systems to avoid any undesirable impact on the human environment.

As shown in the surface energy budget (1.2), the net radiation flux is also disposed of in the form of H, the sensible heat flux into the atmosphere. Actually, this sensible heat, which near the surface may conveniently be expressed as c_pT, where c_p is the specific heat of air at constant pressure and T the temperature, can be considered as a scalar admixture of the air just like water vapor. Consequently, the atmospheric transport mechanisms of sensible heat are quite similar to those of water vapor. Moreover, in many situations it is practically impossible to deal with E, without also considering H in the analysis, or vice versa. Therefore, the sensible heat flux H is generally treated together with the rate of evaporation E. The ratio of these two flux terms, called the Bowen ratio,

$$Bo = H/L_eE \tag{1.3}$$

is a useful meteorological and climatological parameter.

1.3. GLOBAL CLIMATOLOGY

Numerous studies have been conducted to estimate the magnitude of the most important components of the water and energy budget equations on a global scale. Because the available data base required for this purpose is far from adequate, several of the methods used in these estimates may be open to criticism. Nevertheless, there is a fair agreement among some of the recently-calculated values and, within certain limits, they provide a useful idea of the long-term average evaporation in different climatic regions of the world.

TABLE 1.1
Estimates of world water balance since 1970 (m y^{-1})

Reference	Land (1.49×10^8 km^2)		Oceans (3.61×10^8 km^2)		Global
	P	E	P	E	$P = E$
Budyko (1970, 1974)	0.73	0.42	1.14	1.26	1.02
Lvovitch (1970)	0.73	0.47	1.14	1.24	1.02
Lvovitch (1973)	0.83	0.54	–	–	–
Baumgartner and Reichel (1975)	0.75	0.48	1.07	1.18	0.97
Korzun *et al.* (1978)	0.80	0.485	1.27	1.40	1.13

As shown in Table 1.1, the average annual evaporation for the entire earth is of the order of 1 m. The evaporation from the land surfaces is around 60 to 65 percent of the precipitation. Under steady conditions, that is for long time periods, the remainder can be considered to be runoff, $q_r = (Q_o - Q_i)/A$, or

$$q_r = P - E \tag{1.4}$$

which is therefore around 35 to 40 percent of the precipitation averaged over all continents. Except for South America and Antarctica (See Table 1.2), the values of the individual continents are not very different from the global values.

TABLE 1.2
Some estimates of the mean evaporation (and precipitation) for the continents (m y^{-1})

Reference	Europe	Asia	Africa	N. America	S. America	Australia & Oceania	Antarctica
Lvovitch (1973)	0.415	0.433	0.547	0.383	1.065	0.510	–
	(0.734)	(0.726)	(0.686)	(0.670)	(1.648)	(0.736)	–
Baumgartner and Reichel	0.375	0.420	0.582	0.403	0.946	0.534	0.028
(1975)	(0.657)	(0.696)	(0.696)	(0.645)	(1.564)	(0.803)	(0.169)
Korzoun *et al.*	0.507	0.416	0.587	0.418	0.910	0.511	0
(1977)	(0.790)	(0.740)	(0.740)	(0.756)	(1.60)	(0.791)	(0.165)

Estimates of the average distribution of water in different forms expressed as depth of water covering the globe, assumed to be a perfect sphere, are given in Table 1.3. These indicate that the 1 m of evaporation is relatively large as compared to the fresh water on earth, which is not stored in permanent ice and deep ground water. The soils, lakes and rivers of the world appear to store much less than 1 m, and the water in the atmosphere amounts to only about 2 to 3 cm. of condensed liquid. In other

TABLE 1.3
Estimates of different forms of global water storage (as depth in m per unit area of entire earth surface)

Source of Data	Lvovitch (1970)	Baumgartner and Reichel (1975)	Korzun *et al.* (1978)
Oceans	2686	2643	2624
Ice caps and glaciers	47.1	54.7	47.2
Total Ground Water	117.6	15.73	45.9 (excluding Antarctica)
(Active Ground Water)	(7.84)	(6.98)	–
Soil Water	0.161	0.120	0.0323
Lakes	0.451	0.248	0.346
Rivers	0.00235	0.00212	0.00416
Atmosphere	0.0274	0.0255	0.0253

words, the turnover in the active part of the hydrological cycle is rather fast. For example, a global evaporation rate of 1 m y^{-1} with 0.025 m of storage in the atmosphere gives a mean residence time of the order of 9 days. A continental runoff rate of 0.30 m y^{-1} (Table 1.1) and a storage in the rivers of (0.003/0.29) m of water on the 29 percent of the world occupied by land, gives a mean residence time of the order of 13 days. These are very short residence times. Moreover, as the oceans occupy about 71 percent of the earth surface, the active fresh water in the hydrological cycle is continually being distilled anew through ocean evaporation.

Maps depicting the approximate distribution of evaporation, and other components of the water balance in different parts of the world have been presented by Lvovitch (1973), Budyko (1974), Baumgartner and Reichel (1975), Bunker and Worthington (1976), Korzoun *et al.* (1977), and Hastenrath and Lamb (1978). Figure 1.1 gives a sketch of average annual values in dm based on an approximate consensus of these recent estimates. It can be seen that evaporation in nature can vary widely from one location to another, with the sharpest contrasts in arid regions. Figure 1.1 shows that the largest rate of evaporation occurs in the North-Western Atlantic where it exceeds 320 cm y^{-1}. Actually, Bunker and Worthington (1976) indicate a maximum mean value of 373 cm y^{-1} for that area. Clearly this large rate of evaporation is not only due to the high local net radiation but also to the advection of energy by the Gulf Stream. A similar advective situation was described by Assaf and Kessler (1976) who concluded that evaporation from the Gulf of Aqaba, which is surrounded by desert and which receives large quantities of warm water from the Red Sea, is of the order of 3.65 m y^{-1}; however, they added that there are indications, from the oceanographic point of view, that the upper limit for the Gulf of Aqaba may be as high as 5 m y^{-1}.

Irrigation engineers, when lacking better information, sometimes use the rule of thumb that the duty of water for a well-irrigated crop ranges from about 1.0 to 1.5 l s^{-1} ha^{-1}. This corresponds to a range between about 3.2 and 4.7 m y^{-1}. Irrigation efficiencies are commonly of the order of 25 to 40 percent so that this rough estimate is consistent with the climatic values given here. Farmers in the north-eastern United States tell us that they like a weekly rainfall of about one inch, that is 2.5 cm, to maintain field crops in good condition during their active growing period. For a growing period of about six months, this practical estimate of the evaporative requirement of agricultural fields is in good agreement with the data in Table 1.2.

TABLE 1.4
Estimates of mean global heat budget at the earth surface in kcal/ (cm^2 y)

Reference	Land			Oceans			Global		
	R_n	L_eE	H	R_n	L_eE	H	R_n	L_eE	H
Budyko (1974)	49	25	24	82	74	8	72	60	12
Baumgartner and Reichel (1975)	50	28	22	81	69	12	72	57	15
Korzun *et al.* (1978)	49	27	22	91	82	9	79	67	12

The estimates presented in Table 1.4 provide an idea of the order of magnitude of the main components of the surface energy budget on a global scale. These data

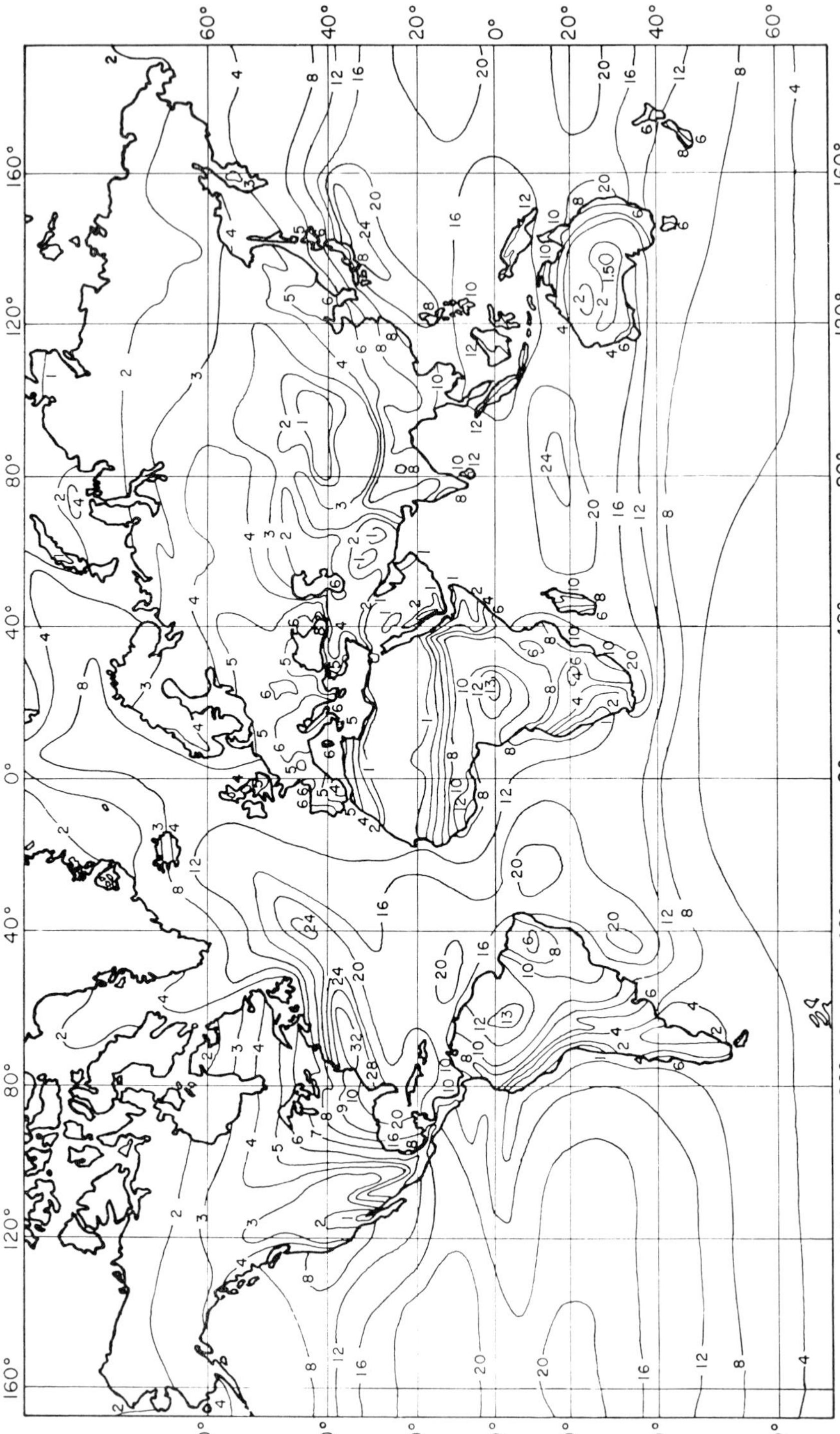

Fig. 1.1. Sketch of global distribution of average annual rate of evaporation in dm.

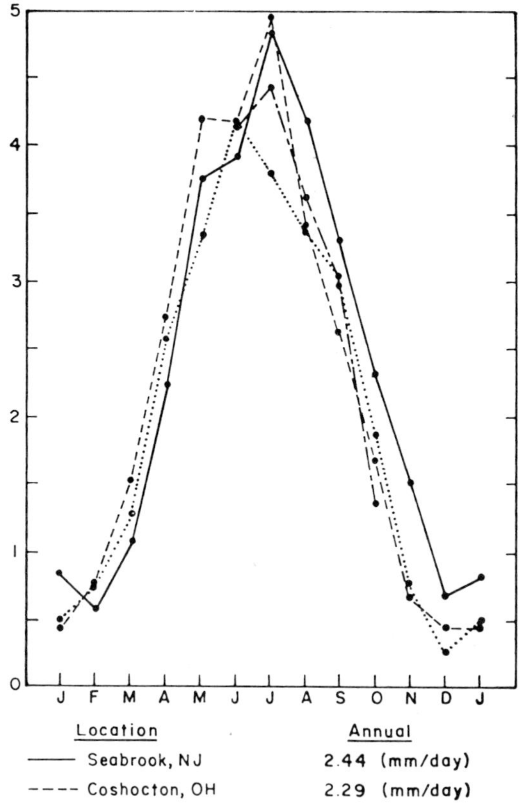

Fig. 1.2. Measured monthly evapotranspiration rate in mm/day for meadow covers on lysimeters at four locations in the Eastern United States. The data at Seabrook, Waynesville and Raleigh represent maximal or near-maximal values, but the Coshocton data are actual values, possibly affected by moisture deficits in the profile (adapted from Van Bavel, 1961).

show that the net energy is mainly disposed as evaporation. Over the oceans the latent heat flux L_eE is, on average, larger than approximately 90 percent of the net radiation. Over the land surfaces of the Earth, L_eE is, on averaged, slightly larger than 50 percent of R_n. Naturally, this is subject to large local and regional variations. The references listed in Table 1.4 also contain estimates of the terms in the energy budget for zones of 10° latitude. These results, which are not reproduced here, show that between approximately 20 and 40° latitude over land H is, on average, larger than L_eE. This is not surprising since these zones are occupied by extensive arid lands and the large deserts of the world. The relative magnitude of the evaporation term in the global energy budget underscores again its importance as a linkage term with the water budget. Thus changes in the surface energy budget are related not only to climatic changes but also to changes in water budget.

At any given location, and at any given time, the actual evaporation is usually quite different from the climatological mean. The deviations from this mean can

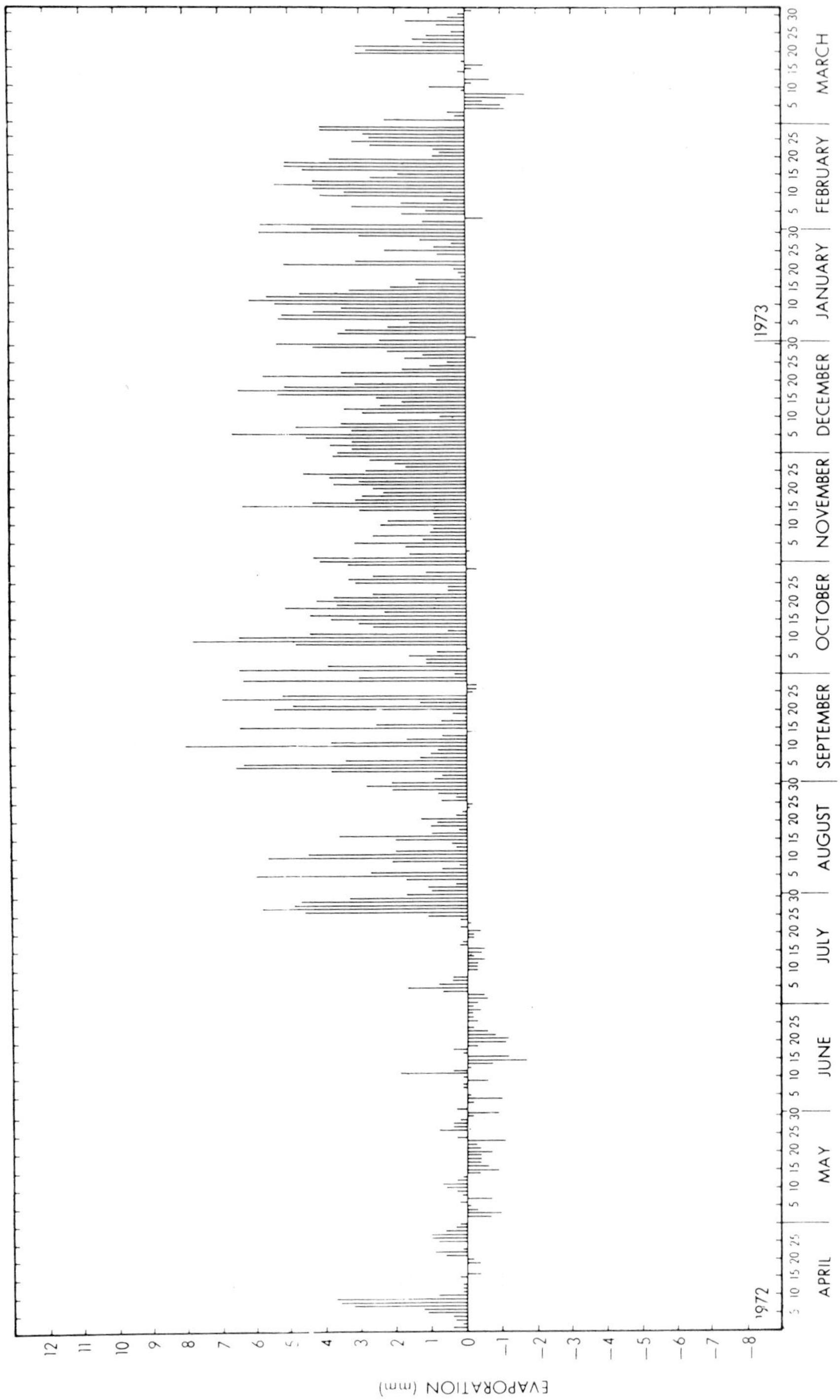

Fig. 1.3. Daily evaporation in mm for Lake Ontario during 1972–1973 computed by a mass transfer method (from Phillips, 1978).

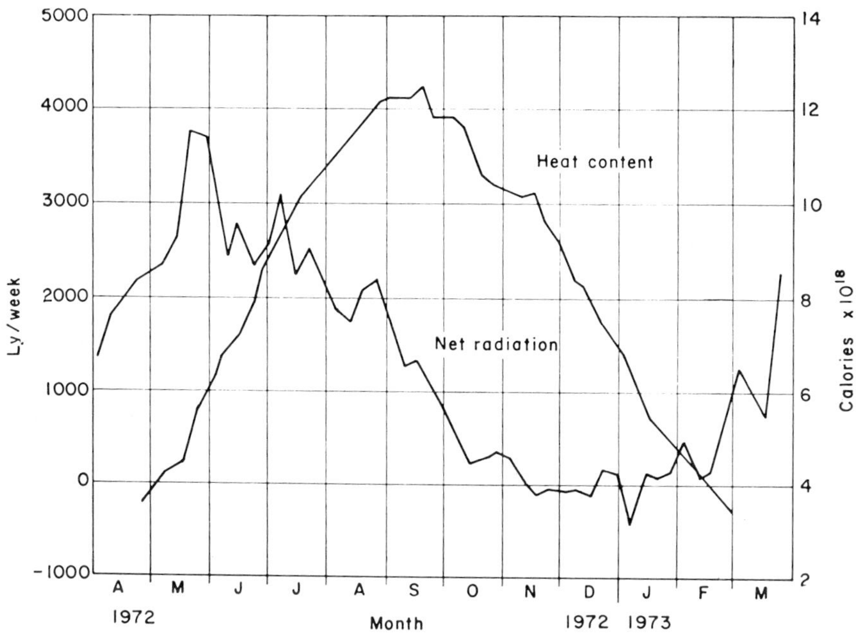

Fig. 1.4. Weekly net radiation and heat content of Lake Ontario during 1972–1973 (adapted from Pinsak and Rogers, 1974).

be characterized by a cyclic or periodic behavior, namely with a daily and with a seasonal time scale. In the extreme case of an arid, warm climate, with a pronounced dry and a wet season, the seasonal evaporation cycle is similar to the rainfall cycle. In a humid climate, or over water, the seasonal march of the evaporation rate follows closely the cycle of the energy available for evaporation. In most climates over land the seasonal evaporation cycle is affected both by the available water and by the available energy. As an example, in Figure 1.2 the monthly mean evaporation rates are shown for several locations in the eastern United States. The evaporation rate is maximal in summer and minimal in winter. Thus, the cyclic behavior here is similar to that of the solar radiation input and to that of the air temperature. The same holds true for shallow water bodies. However, for deep water bodies, the evaporation cycle does not coincide with the solar summer-winter cycle. In contrast to a land surface, a water body can store and release large amounts of heat and thus acts like a fly wheel; as a result the cycle of the available energy for evaporation may lag several months behind the solar input cycle. For example, as can be seen in Figure 1.3, the rate of evaporation from Lake Ontario is maximal in late fall and early winter, and minimal in late spring and early summer; the corresponding net radiation and heat storage are shown in Figure 1.4.

The daily evaporation cycle is usually less pronounced over water than over land. Over land, where much less heat is conducted below the surface, the daily cycle generally follows the daily march of the solar radiation. Illustrations of the daily cycles of evapotranspiration from vegetated surfaces and from water surfaces are shown in Figures 6.1–6.4 together with other components of the energy budget. In

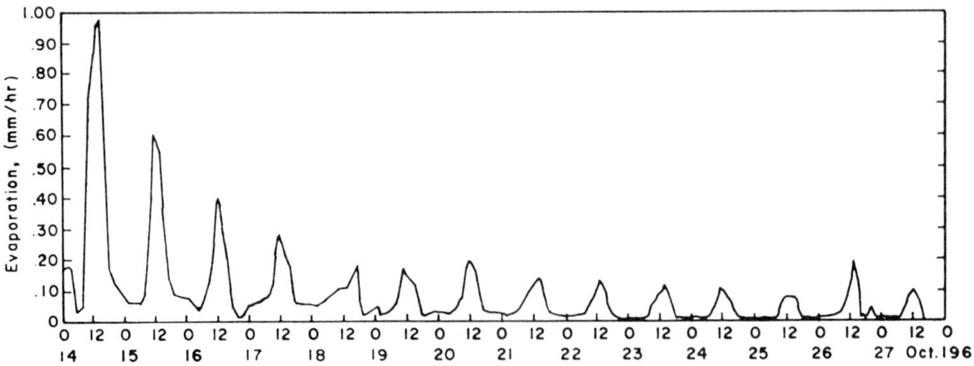

Fig. 1.5. Rate of evaporation from a bare soil during a drying cycle, measured with a weighable lysimeter in Arizona (from Van Bavel and Reginato, 1965).

Figure 1.5 an example is shown of the daily cycle of evaporation from bare soil. This figure also illustrates the general behavior of evaporation after a rainfall or after irrigation, when the available water stored in the soil profile is being depleted. Because the experiment took place during a drying period, the daily cyclic behavior is superimposed on a trend of decreasing daily mean evaporation. A similar trend in daily mean evaporation after rainfall is also illustrated in Figure 10.1 for pasture and in Figure 11.7 for bare soil.

The daily and seasonal cycle are actually only one feature of the general stochastic behavior of evaporation in nature. A few attempts have been made (e.g., Yu and Brutsaert, 1969a, b; Pruitt *et al.*, 1972; Shahane *et al.*, 1977; Magyar *et al.*, 1978) to study stochastic and statistical aspects of evaporation, but still very little is known. Progress will be possible by the acquisition of better data and longer time series.

1.4. THE TRANSFER OF OTHER ADMIXTURES AT THE EARTH-ATMOSPHERE INTERFACE

In addition to evaporation of water and the sensible heat flux, there are transport phenomena of other admixtures or constituents of the atmosphere which are of physical and biological consequence in the environment. For instance, the transfer of oxygen through the water surface is one of the main mechanisms for maintaining or restoring the water quality of lakes and rivers. Carbon dioxide is another constituent of the air that is essential for the biological metabolism. However, it is also one of the major waste products resulting primarily from the combustion of various fuels; in the past few decades its concentration in the air has been increasing. The consumption of CO_2 by vegetation and the transfer of CO_2 through the surface of water bodies govern to a large extent its removal from the air. Together with sunlight and nutrients, the surface transfer of CO_2 is a factor governing the eutrophication rate of lakes. Beside CO_2 many other gaseous combustion products are being discharged into the earth's atmosphere. Dry deposition of gaseous pollutants on the earth's surface is a major cleansing mechanism of the atmosphere, in addition to precipitation scavenging or wet deposition. The volatilization of certain hydrocarbons from lakes and rivers is also increasingly a cause for serious concern among environmental engineers.

This book does not deal with any of the problems just mentioned. However, several of the phenomena treated in connection with the transport of water vapor and sensible heat, are equally pertinent for the analysis of transport of other admixtures. The similarity is, of course, perfect in the case of evaporation of pure liquids or solids. But the similarity also holds when the gas has a low vapor pressure, when it has a high solubility in the surface material or when it has a rapid chemical reaction with it. For example, the transfer of gases such as NH_3, SO_2, SO_3 and HCl at a water surface is probably quite analogous with surface transfer of water vapor (e.g., Hicks and Liss, 1976). Actually, most gases do not fall in this category, and their transfer through an air-water interface is usually controlled by transport mechanisms in the water. Nevertheless, away from the immediate proximity of the surface, in the fully turbulent boundary layer all passive admixtures are transported in a similar way. For example, above a vegetation, the transport of CO_2 (Shawcroft *et al.*, 1974) O_3 (Wesely *et al.*, 1978), or NH_3 (Denmead *et al.*, 1978) may be treated by methods which are the same as some of those used to determine evaporation.

CHAPTER 2

History of the Theories of Evaporation – A Chronological Sketch

Since time immemorial human beings have observed evaporation of water in their surroundings, and they undoubtedly have speculated on the nature of this phenomenon. For a better understanding of the discovery of our present knowledge, it is appropriate to review briefly some of the concepts of the past and their evolution.

2.1. GREEK ANTIQUITY

Among the peoples of antiquity, the Greeks are renowned for the large effort their natural philosophers made to arrive at a rational explanation of the physical world in which they lived (e.g., Burnet, 1930; Freeman, 1953). Not much is left of their original writings; moreover, whatever is left is not easy to interpret, not only because major parts of the early theories are deduced from secondary sources, but also because the meanings of even the most elementary concepts have evolved in the mean time. Still, inspection of these works and theories shows that evaporation must have occupied a central position in the cosmology of the ancient Greeks (cf., Gilbert, 1907; p. 439 ff.). Indeed, telluric exhalations from water and earth formed the basis for their understanding of all meteorological phenomena. These exhalations were referred to as anathymiasis (*ἀναθυμίασις*) or more specifically in the case of vaporizing or vaporized moisture as atmis (*ἀτμίς*); they provided the connection and the interaction between the lower elements, earth and water, and the upper elements, air and fire.

Already in pre-philosophic times, Hesiod in the eighth century B.C. commented on the formation of mist. In a passage with advice to farmers to get dressed warmly and to finish the work in time, Hesiod (1928; 1978; vv. 547–553) wrote:

> For the morning is cold when the north wind bears down; in the morning from the starry sky over the earth a fertilizing mist spreads over the cultivations of the fortunate; this [mist], drawn from ever flowing rivers, and lifted high above the earth by a storm wind, sometimes falls as rain toward evening, or sometimes blows as wind, while the Thracian Boreas chases the clouds.

This passage probably reflects the intuition of most natural peoples. It contains two interesting features; there is a hint of the atmospheric phase of the hydrological cycle, and it is implied that evaporation may be both a cause and a result of the wind.

The formal inquiry into the reality behind the changes in the universe is usually assumed to have started with Thales of Miletos in Ionia, who flourished around 585 B.C. He probably did not commit his ideas to writing, and no actual quotations of him have survived; but there is evidence that he attached some significance to evaporation. He posited that water is the main principle of everything, and he is said (Aetius in Diels, 1879; p. 276) to have given as one of the three reasons for this "that even the fire of the sun and of the stars, and even the kosmos itself, are fed by the evaporation of the waters."

The earliest known Greek philosophical writings are those of his younger associate Anaximander of Miletos; he was born ca. 610 B.C. and he must have been in his prime ca. 565 B.C. Concerning phenomena related to evaporation, Anaximander's views were summarized by Hippolytus in his doxography as follows (Diels, 1934; p. 84, I, 6, 7):

> Winds are generated when the finest vapors [atmos] of the air are separated off and when they are put into motion by being assembled together; rains are generated from the evaporation [atmis] that is sent up from the earth toward under the sun.

On this topic, Aetius (Diels, 1934, p. 87, III, 7, 1) described Anaximander's view as "Wind is a stream of air, of the finest in it and of the moistest, which are moved or dissolved by the sun." Aetius was a doxographer who lived probably in the second century A.D.; he obtained his information indirectly from a lost work of Theophrastos (Diels, 1879; Burnet, 1930, pp. 33–35). Also Hippolytus, who died in 235 A.D., derived his information from indirect sources, but almost certainly independently from Aetius. Both passages are in the main quite consistent. Thus, although some uncertainties remain (e.g., Diels, 1934, p. 84; Gilbert, 1907, p. 512), they probably provide a fair representation of Anaximander's philosophy. They indicate that Anaximander considered a moist exhalation, which is caused by the sun, and which transforms into air, whose motion is wind. Evaporation is a cause rather than a result of the wind.

Xenophanes of Colophon [ca. 570–460 B.C.] was probably in his prime ca. 530 B.C. According to Aetius (in Diels, 1934, p. 125, III, 4, 4) Xenophanes said that

> . . . what happens in the sky is caused by the heat of the sun; for, when the moisture is drawn up out of the sea, the sweet part, which is distinguished by its fine texture, forms a cloud, and drips out as rain by compression like that of felt, and the winds spread it around.

And he wrote emphatically [in verse, Diels, 1934, p. 136]

> The sea is the source of the water, the source of the wind. For in the clouds, neither would the force of the wind, which blows outward, originate without the great sea, nor the flowing of streams, nor the rainwater from the sky; but the great sea is the originator of the clouds, winds and streams. . .

Diogenes Laertius (1925, p. 427, IX, 2, 19) gave a similar description of Xenophanes' views in the following brief sentence (Hicks's translation is not adhered to):"Clouds are formed when the vapor is carried upwards by the sun and lifts them into the surrounding air. . ." All this shows that Xenophanes had some idea of the hydrological cycle, which may have been more complete than Hesiod's and Anaximander's, because he also included streams. But evaporation, which is driven by the sun, plays a central role. He may also have suggested some kind of dual exhalation, because he is said, again by way of Theophrastos (Aetius, II, 20, 3 in Diels, 1934, p. 124), to have thought that ". . .the sun is made up from the fiery particles from the moist exhalation".

The concept of two exhalations was apparently first introduced by Herakleitos of Ephesos who flourished around 500 B.C. Diogenes Laertius (1925, IX, 1, 9–11, p. 416; Diels, 1934, p. 141) described his views as follows (Hicks's translation is not adhered to):

For by becoming denser fire exudes moisture, this by condensing becomes water, and by congealing water changes to earth; and this is the downward path. Then again earth is liquified, from which water originates and from water the remainder; he reduces nearly everything to exhalation from the sea. This is the upward path.

In other words, Herakleitos allowed the elements to transform into one another, and he specified that evaporation is the most important process. However, at the same time he considered another type of exhalation:

But exhalations arise also from the earth as well as from the sea; the former are bright and pure, the latter are dark; fire is favored by the bright ones, moisture by the others. . . Day and night, months, seasons, years, rains, winds and similar things originate by the various exhalations. For the bright exhalation, ignited in the circle of the sun, causes day, and when the opposite predominates, it produces night; and the increasing heat from the bright exhalation produces summer, whereas the preponderance of the moisture from the dark exhalation produces winter. And he attributes causes to other things accordingly.

This passage suggests that the sun is the principle governing the preponderance of the bright or the moist exhalation, and that it drives the transformations of the elements. To Herakleitos the exhalations were not merely physical phenomena but the soul of everything, for (Aetius IV, 3, 12, in Diels, 1934, p. 147)

The soul [psyche] of the kosmos originates from the exhalation of its waters, and the soul in the living beings originates from the outside exhalation and from that taking place in themselves.

and elsewhere (Arius Didymus, by Eusebius XV, 20, in Diels, 1934, p. 154): "But also the souls evaporate out of the moist things."

Hippocrates of Cos [ca. 460–370 B.C.], the best known of the physicians of antiquity, also had some ideas on evaporation. Because his main objective was not philosophy, his views are less esoteric, and they probably represent more closely the general opinion of his contemporaries. Writing on the properties and quality of rainwater (Hippocrates, 1923, VIII) he elaborated as follows:

To begin with the sun raises and draws up the finest and lightest part of water, as is proved by the formation of salt. . . the finest part, owing to its lightness, is drawn up by the sun. Not only from pools does the sun raise this part, but also from the sea and from whatever has moisture in it – and there is moisture in everything. Even from humans it raises the finest and lightest part of their juices. The plainest evidence thereof is that when a man walks or sits in the sun wearing a cloak, the parts of his skin reached by the sun will not sweat, for it draws up each layer of sweat as it appears. But those parts sweat which are covered by his cloak or by anything else.

The development of Greek natural philosophy culminated with Aristotle [ca. 384–332 B.C.]. Although he criticized many of the views of his predecessors (cf., Cherniss, 1964), he was strongly influenced by the earlier ways of thinking. In his book 'Meteorology' Aristotle further developed Herakleitos's concept of the dual exhalation and made it a central point for his physical theories. The following quotations illustrate his views on evaporation (Aristotle, 1952, 340b 3):

For vapor [atmis] is a separation* of water;... [346b 24] The earth is at rest, and the moisture about it is evaporated by the sun's rays and the other heat from above and rises upwards: but when the heat which caused it to rise leaves it, some being dispersed into the upper region, some being quenched by rising so high into the air above the earth, the vapor [atmis] cools and condenses again as a result of the loss of heat and the height and air turns into water: and the generated water falls again into the earth. [Lee's translation is not adhered to.] The exhalation from water is vapor, that from air to water is cloud;...

This evaporation involved a vague notion of latent heat (347a 22):

Dew tends to form rather in fair weather and mild districts; hoar frost, as said, under opposite conditions. For it is obvious that vapor [atmis] is warmer than water, as it still contains the fire that caused it to rise, and so needs more cold to freeze it.

In addition, there is also a second type of exhalation (360a 6):

Now there is in the earth a large amount of fire and heat, and the sun not only draws up the moisture on the earth's surface, but also heats and so dries the earth itself; and this must produce exhalations which are of the two kinds we have described, namely vaporous and smoky. The exhalation containing the greater amount of moisture is, as we have said before, the origin of rain: the dry exhalation is the origin and natural substance of winds... Since the two exhalations differ in kind, it is clear that the substance of wind and of rain water also differ and are not the same, as some maintain: for they say that the same substance, air, is wind when in motion, water when condensed again. Yet it is absurd to suppose that the air which surrounds us becomes wind simply by being in motion, and will be wind whatever the source of its motion; for we do not call a volume of water, however large, a river whatever its flow but only if it flows from a source, and the same thing is true of the winds,...

Clearly Aristotle, like many of his predecessors, realized that the moist exhalation requires solar radiation or some other heat; however, he denied any direct connection between evaporation and the wind except that both, as separate exhalations, are caused by the sun. It was impossible for Aristotle to allow any cause-and-effect relation between wind and evaporation, because he disagreed so vehemently with the view that wind is merely air in motion. Since this view had already been proposed at least some 200 years earlier by Anaximander, Aristotle's theory represented a set-back.

It may be that Herakleitos, who initiated the concept of the two exhalations, and perhaps also Xenophanes, already had the same ideas concerning the nature of wind as Aristotle; but this cannot be ascertained from the available evidence.

The possibility has been raised (Needham, 1959, p. 637) that the concept of the two exhalations may be of an older Mesopotamian origin, which may also have influenced ancient Chinese philosophy. This may be so, but at present satisfactory evidence for this is totally lacking. Certainly, the idea that moist vapor is related with rain and smoke with ashes, minerals and thunder, could very well have originated independently in Greece and in China. But early Chinese concepts on wind do not appear to have been related to any dry exhalation. For example, in a naturalist work 'Chi Ni Tzu', probably of the late fourth century B.C. (Needham, 1959, p. 467) wind and rain are dealt with as follows:

Wind is the chhi [spirit, mind] of heaven, and the rain is the chhi of earth. Wind blows according to the seasons and rain falls in response to wind. We can say that the chhi of the heavens comes down and the chhi of the earth goes upwards.

Evidently, rather than a dry telluric exhalation in the Aristotelian sense, the wind is represented here as an 'exhalation' from above.

---

*Lee translated *διάκρισις* as evaporation, but this alters Aristotle's statement.

Theophrastos [ca. 372–287 B.C.] was Aristotle's successor in directing the Peripatetic School at the Lyceum in Athens. Among his works is one 'On Winds' in which he made a number of important contributions both to the description and the genesis of winds. As noted by Coutant and Eichenlaub (1974), these contributions have often been overlooked in the past mainly because of his close association and parallelism in other matters with Aristotle. Although he was Aristotle's student, Theophrastos (1975, 15, 22, 29) did not attach great importance to the dry exhalation in his wind theory, and he went back to the older idea that wind is air in motion. He introduced his views as follows (Section 15): "If the generation of all winds is the same and caused by the same factors (by taking on some material), the sun is the agent." And then, as if to appease those holding Aristotle's view, "Perhaps this is not correct taken universally, but rather the exhalation is the cause, while the sun assists." However, he continued, "But the sun, by rising, seems to set the winds in motion and to halt them." In a later passage he drastically departed from Aristotle (Section 22):

> If air were self-moving, being cold and vaporous by nature, it would move downwards; if it were moved by heat, it would move upwards. For the motion of fire is naturally upwards; in fact, the motion is in a sense a mixture of both because neither prevails.

And finally he asserted (Section 29): "But the movement of air is wind." Because he assumed that wind is not due exclusively to some dry and smoky exhalation, Theophrastos (Section 22) could also explain the observation that the wind may be cold and vaporous.

These innovations allowed Theophrastos to see a more correct relationship between the wind and evaporation. Like his predecessors, he took the sun as the most important agent (Section 24, 48), but this is not the only factor (Theophrastos, 1975, 60).

> The reason that winds which are cold dry more quickly than the sun, which is warm, and the coldest winds most of all, must be that they create a vapor and remove it, especially the coldest winds, while the sun leaves the vapor.

This is a remarkable statement, for it is perhaps the first instance on record in Greek science that the possibility was raised that, distinctly from the sun, wind has a drying effect and creates vapor. It should be recalled that both Anaximander, who also considered wind as moving air some 200 years earlier, and Xenophanes had implied that wind and evaporation, caused by the sun, were very intimately related. Thus it would have been impossible for them to consider the wind and the sun as separate agents.

It is difficult to ascertain whether the Aristotelian or the Theophrastian view subsequently prevailed at the Peripatetic School. From the present vantage point at least, it would seem that Aristotle's 'Meteorologica' was better known and probably better accepted; it was an essential part of the Aristotelian corpus which came to the Arabs and later to Western Europe. However, in Book 26 of 'Problems', attributed to Aristotle (1938b, 26, 28; 26, 34) there are several passages which throw a curious light on this issue. For example (Hett's translation is not adhered to),

> Why do the winds which are cold have a drying effect? Is it because the colder winds cause evaporation? Why do they cause more evaporation than the sun? Is it because they drive off the vapour, but the sun leaves it? So it moistens more and dries less?

And again, "Why as the sun is rising do the winds rise and fall? Is it because wind is a movement either of air or of rising moisture?..."

Clearly, these passages are so alien to the two-exhalation theory that they could not possibly have been written by Aristotle. And indeed, 'Problems' is now generally accepted to be spurious: it was probably compiled later in the Peripatetic School, and it may have continued to evolve to its present form until as late as the fifth century A.D. (Hett, in Aristotle, 1938). All this suggests that the Peripatetic School did not always adhere to Aristotelian orthodoxy, and that the ideas of Theophrastos on wind were taken seriously; so seriously, in fact, that later some of them became associated with the name of the old master himself.

2.2. THE ROMAN PERIOD AND THE MIDDLE AGES

The Romans are less known for their natural philosophers than for their engineering works and their contributions in law and administration. As their views on scientific matters were heavily influenced by the Greeks, their writings are often dismissed as merely reviews and commentaries of the Greek works. This is an over-simplification. Because the Romans were generally more practically oriented, they tended to rely also on observation rather than on speculation only, arriving at interesting insights in some cases.

This is illustrated by the explanation of Lucretius [ca. 99–55 B.C.] in his work 'On Nature'. On why the level of the sea does not increase, Lucretius (1924, VI, 617 ff) wrote:

> Besides the sun by his heat draws off a great portion. For certainly we do see that clothes soaking with wet are dried up by the sun with his burning rays. But we see that the seas are many and stretching wide beneath; therefore although the sun may sip but a small portion from the surface in any given place, yet over so great an expanse he will take away from the waves in abundance. Then further the winds also can lift a goodly portion of moisture by sweeping the surface, since under the winds we see very often the roads grow dry in one night, and the soft mud massing together in crusts. Besides I have shown that the clouds also lift a great deal of moisture from the great surface of the main, which they sprinkle everywhere over the whole world when it rains on earth and the winds carry the clouds along. Lastly since the earth is a porous body...

A similar account is given in V, 264 ff. Lucretius derived his philosophy primarily from Epikuros who adhered to the atomic theory of Demokritos and Leukippos, and his views on evaporation seem to reflect this. Unfortunately, nothing is left that might provide some clue as to what the Greek atomists themselves thought about this problem. Lucretius's description of evaporation is definitely superior to that given by Theophrastos. It is noteworthy that he referred to concrete observation, giving two examples to support his statements on the effects of sun and wind.

The views of Seneca [ca. 4 B.C.–A.D. 65], born in Cordoba, and teacher and later advisor of Emperor Nero, provide a different example of the general status of natural philosophy during the Roman period. His book 'Natural Questions' contains some new interpretations, but it is strongly influenced by the earlier Greek theories, and it is replete with moralizing conclusions and analogies. Seneca quoted nearly 40 references, among whom five Latin authors, and the remainder Greek. As regards evaporation, he explained (Seneca, 1972; V, 8, 1) that the sun is nourished by the exhalations from the marshes and the rivers; the sun draws up the fresh water from the sea because it is the lightest (IV, 2, 24); the wind is flowing air in one direction (V, 1, 1); evaporation from land and water is sometimes the only cause of the wind (V, 5, 1; V, 8, 1; V, 9, 1), but in addition the atmosphere also has an inherent capacity for movement by itself without external agent (V, 5, 1). A brief account (II, 12, 4) was also given of Aristotle's two-exhalation theory, but in connection with the origin of thunder. In summary, it can be noted that Seneca's views were related to those of Aristotle and Theophrastos; although he admitted that also the moist exhalation may produce wind, the possibility that wind affects evaporation was not mentioned. Thus, his ideas on evaporation appear less sound than those of Lucretius.

Pliny the Elder [ca. A.D. 23–79] was a contemporary of Seneca; in spite of an active career in public life he kept up continuous study allowing him to write a gigantic and encyclopedic 'Natural History' in 37 volumes. In the preface Pliny (1938, 17) stated that it deals with "20,000 noteworthy facts obtained from one hundred authors that we have explored." When discussing the phenomenon of evaporation in general terms, Pliny (1938; II, 42, 111–44, 114) gave a synthesis of the earlier Greek theories. He mentioned the two exhalations, the effect of the sun, and also the wind as air in motion; but he did not take sides, carefully couching his statements in terms of "I would not deny..." and "Similarly, I am not prepared to deny..." Pliny took a pragmatic attitude, and he was willing to accept several possibilities as not necessarily exclusive mechanisms. This attitude is again illustrated by his comments on evaporation in colder regions. In an interesting description of the Chauci, a people who lived along the North Sea between the Ems and the Elbe, and whom he must have observed while he served as a cavalry officer in Germania, he wrote (Pliny, 1945; XVI, 1, 4): "...they scoop up mud in their hands and dry it by the wind more than by sunshine, and with earth as fuel warm their food and so their own bodies, frozen by the north wind." Thus Pliny inferred that because of the often cloudy conditions in North-Western Europe, the sun could not possibly be the only agent, and that the wind had to play a more important role.

The end of the Roman era witnessed the rise of Christianity. The writings of the early leaders, or fathers, of the church reflected an eclecticism between biblical interpretation and pagan philosophies. An example of this is the set of homilies 'On the Hexaemeron', written by Basileios of Cappadocia [St Basil, ca. 330–379]. Basileios had been educated at Caesarea, Constantinople and Athens, so naturally in this work he drew freely from the classical Greek philosphers, among whom were Herodotos, Plato, Aristotle, and Theophrastos. He considered the sun the only cause of evaporation (Basil, 1963; 4, 6).

...and it [the sea] is good because, being the receptacle of rivers, it receives the streams from all sides into itself but remains within its own limits. It is good also because it is a certain origin and source for aerial waters. Warmed by the rays of the sun, it gives forth through vapors a refined form of water, which, drawn to the upper regions then chilled because it is higher than the reflection of the sun's rays from the ground and also because the shadow from the cloud increases the cooling, becomes rain and enriches the earth.

To support his contention that the heat of the sun is the cause of evaporation he then made the analogy with a kettle which boils until it is empty. The same view is also given in (3, 7). A decade or two later, around 389, Ambrosius [St Ambrose, ca.

333–397] also wrote a 'Hexameron' which was partly inspired by that of Basileios. Ambrosius was then bishop of Milan, but he had become a Christian only in 374, and his early education in Rome had been in the classical Latin tradition. His descriptions (Ambrose, 1961; 2, 13; 2, 14; 3, 22) of evaporation are very similar to those of Basileios.

It is of some interest to note, incidentally, that both Basileios and Ambrosius wrote on evaporation in connection with the problem why the sea does not overflow even though all rivers flow into it. Already Aristotle (1952; 355b 22) had dealt with it, and he had called it an 'ancient difficulty'. As seen above, Lucretius had thought about it. The problem was also of concern in ancient China. In the third century B.C. Lü Shihh Chhun Chhiu wrote (Needham, 1959; p. 467):

> The waters flow eastwards from their sources, resting neither by day nor by night. Down they come inexhaustibly, yet the deeps are never full. The small (streams) become large and the heavy (waters in the sea) become light (and mount to the clouds). This is (part of) the Rotation of the Tao.

However, the preoccupation of Basileios and Ambrosius with this problem stemmed directly from (1, 7) in Ecclesiastes [ca. fourth to third century B.C.] as follows (Oxford Study Edition, 1976): "All streams run into the sea, yet the sea never overflows; back to the place from which the streams ran they return to run again." This preoccupation was shared by most later Christian writers, and it was to endure well into the Middle Ages. But the theme has kept recurring: Dobson (1777) contended that his data supported the wisdom in this biblical passage, and, as recently as 1877, Huxley (1900, p. 74) used the passage in his description of the hydrological cycle.

Just like the Roman Period the Middle Ages are also often dismissed as a period of relapse in the long development of physical science. Again, this is a convenient oversimplification (e.g., Lear, 1936; Pernoud, 1977) and, as will be seen below, several concepts of medieval meteorology were at least as advanced as those of the Renaissance and even later. Although the divine was considered to be the ultimate principle of everything – but also the Greeks assumed this – the fundamental concept of science, as Thales had initiated it, was retained. In other words, the Greek tradition of searching for an explanation of the physical world within that same world, without animistic or direct divine intervention, was continued. The main difference was that this scientific knowledge had to provide a first step and an aid for the transmission of the Christian doctrine.

This can be seen, for example, in the work 'On Nature' written around 613 by Isidorus Hispalensis of Sevilla [ca. 560–636] for the benefit of Sisebut, king of Visigothic Spain at Toledo. The title of the work is the same as that of Lucretius; its outline is in some places similar to those of Aristotle, Lucretius and Pliny, but it is very close to that of Aëtius (cf. Fontaine, in Isidore, 1960). So to organize his material, Isidore must have had at his disposal some doxographic treatise, or at least a monastery school manual or summary of it. Beside a few pre-Christian references, most of the explicit references are to patristic writers. But these were largely products of the classical tradition; hence, both in form and contents, Isidore's work was an outgrowth of the same.

On the nature of wind Isidore (1960; 36, 1) wrote:

> Wind is moved and agitated air, as proved by Lucretius: "For wind is generated where the air is

brought in motion by agitation'. Which can be shown to be true even in a place which is most calm and protected from all winds, by means of a small fan, with which we move the air, as we chase the flies, and we feel a blowing.

And on why the sea does not increase (41, 1),

Bishop Clemens said that it is because the naturally salty water consumes the flow of fresh water which it receives, in such a way that, however large the masses of the waters it receives, this salty element of the sea nevertheless does not absorb them totally. Add to this also what the winds take away, and what the evaporation and the heat of the sun absorbs. Finally, we see lakes and many ponds being consumed in a short time by the blowing of the wind and the glowing of the sun. On the other hand, Solomon* said: the streams return to where they come from.

Finally, Isidore (1960; 42, 1) explaining why the sea has bitter water, continued,

It is again the learned Ambrosius who says in his teachings: the reason why, so say the ancients, the sea has salty and bitter waters is that whatever flows in it from the different streams, is absorbed by the heat of the sun and the flowing of the winds, and as much is consumed by the diurnal evaporation as is brought in by all the river flows during each single day. It is said that this happens by virtue of the sun, which takes up what is pure and light, but leaves what is heavy or earthy, because it is also bitter and not drinkable.

These views of Isidore on evaporation, with a suggestion of a hydrological cycle, are very similar to those of Lucretius, and they are superior to those of Aristotle and even Theophrastos. Just like Lucretius, Isidore used practical examples to support his statements. The reference to Ambrosius probably exaggerates his importance, for Ambrosius (1961; Hex. 2, 14; 3, 22) did not mention any wind effect. In a later work 'Etymologies' completed around 620, Isidore gave similar descriptions of the wind (Isidorus, 1911; 13, 11) and of why the sea does not increase (13, 14), "...or because the clouds attract a large amount of water; or because the winds partly carry it up, and the sun partly dries it out."

Isidore's work had a vast impact during the early middle ages. About a century later, Bede [ca. 673–735] a Benedictine monk at Jarrow in England, known mainly for his historical writings, also wrote a book 'On Nature', in which he followed Isidore's rather closely. In connection with meteorological phenomena, Bede summarized Isidore's statements and, in some instances, he rendered them almost verbatim. For example, Sections 26 and 40 of Bede (Beda, 1843) regarding evaporation, correspond very closely to Sections 36, 41 and 42 of Isidore (1960), which are quoted above. A book 'On the Times of the Year' written around 993 by Aelfric (1942, 1961), under inspiration of Bede's work, shows that some of Isidore's ideas had even found their way into the Anglo-Saxon vernacular. Isidore's influence is also evident in the work of Hrabanus Maurus of Mainz [ca. 776–856]. Entitled variously 'On Nature or 'On the Universe', it was written around 844; it was designed as an aid in preparing sermons, and it contains numerous Christian explanations, allegories, and biblical references. Hrabanus appears as a well-read author, but his primary source was clearly Isidore. His descriptions (Rabanus Maurus, 1852; 9, 25; 11, 2) of the wind and of why the sea does not increase, were taken nearly literally from Sections 13, 11 and 13, 14 of Isidore (Isidorus, 1911) quoted above.

*It has in the past often, apparently mistakenly, been assumed that Solomon was the author of Ecclesiastes.

These examples illustrate that by the early Middle Ages several concepts of Greek and Roman science had been fairly well propagated in Western Europe through Isidore's writings. If Isidore deserves a place in the history of Western thought, it is not by virtue of the originality or correctness – by today's standards – (e.g., Lear, 1936) of his overall view of the world. As regards evaporation, however, he was part of a tradition that has scientific merit. Isidore's descriptions of evaporation were inspired indirectly by those of Lucretius, and they are thus related to the views of the earlier atomists Demokritos and Leukippos, rather than those of Aristotle. The most notable feature of the evaporation theory of the early Middle Ages is that both the solar heat and the wind were taken as active but distinct agents.

A further development in this same direction were the explanations given in the 'Dialogue on Physical Substances'; this book in the form of questions and answers was written by Vuilhelmus [William or Guillaume] of Conches in Normandy, who flourished between ca. 1120 and 1155 as a teacher at the School of Chartres. Vuilhelmus (1567; p. 159) described wind as air flowing in one direction, which is reminiscent of Seneca (1972; V, 1); but his description of the causes of the wind (p. 160) involved not only an exhalation but also specifically atomistic features, which albeit different were explicitly inspired by ideas of Demokritos. On the drying effects of wind he wrote perceptively (p. 168):

> Duke: If a certain wind is moist, as you state, how come that all winds, provided they do not bring rain, dry out the water which is in the earth's surface, or wet clothes, or herbs or trees? Philosopher: the wind, which is humid, such as the southerly wind, is also warm, and the heat actually causes the dryness. But the wind which is cold, is also dry; hence it is no surprise if it has a drying effect.

General descriptions of the evaporation process remained roughly the same until the beginning of the thirteenth century. Thomas Cantimpratensis [1201 – 1270] of St Pieters-Leeuw in Brabant was a notable writer of that period. In his 'Book on Nature', completed prior to 1244, he was still of the opinion that wind is only a flow of air (Thomas Cantimpratensis, 1973; 18, 4, 2); and in the context of the familiar problem why the ocean is not increased by the influx of the streams, he gave as one of the mechanisms that fresh waters (Section 19, 4, 14) "...are taken up by the wind or by the glowing of the sun..." The book of Thomas, sometimes mistakenly attributed to Albertus Magnus [ca. 1193–1280] of Cologne, had a considerable influence. For example, it served as one of the major sources for the 'Natural Mirror' of his fellow-Dominican Vincent of Beauvais [ca. 1190–1264] in France; around 1267 Van Maerlant (1878) of Damme in Flanders made a paraphrased translation of it in Middle-Dutch verse, entitled 'The Flower of Nature', unfortunately without Chapters 18 and 19, which are of concern here; and it was used around 1350 by Conrad von Megenberg (1897) when he wrote his 'Book of Nature' in Middle-German. Other works in which it was used as a source have been listed by Kaufmann (1899; p. 36).

But the description by Thomas was probably one of the last instances that evaporation was dealt with in this way. Around that same time, Aristotle's philosophical works were becoming known in Western Europe. The Latin translations of these works were derived from Greek originals, as a result of intensified contacts with Constantinople during the crusades, and from Arabic translations mostly in Moorish Spain (e.g., Jourdain, 1960; Peters, 1968).

The Arab world became acquainted with the works of the ancient Greeks through the translations of Syrian scholars at the end of the eighth and the beginning of the ninth centuries. One of the earlier figures among these was Ayyub al-Ruhawi al-Abras [Job of Edessa, ca. 760–835], a Nestorian, who translated both into Syriac and into Arabic. He also wrote his own 'Book of Treasures', an encyclopedia containing all knowledge of interest in Baghdad about A.D. 817. Ayyub apparently had a thorough knowledge of Aristotle's meteorological philosophies, but he did not accept them all uncritically. Actually, he categorically rejected Aristotle's wind theory (Job of Edessa, 1935; IV, 14, 190–192) using the following refutation of its dry-vapor origin:

> Since there are sometimes winds and sometimes not, while that vapor constantly rises from the earth, in a great or small quantity, it does not seem likely that it constitutes the origin of the winds or the measure (of their strength).

He considered wind as air in motion, and he also had an idea of the effect of advection by the wind on cloud movement and precipitation. It is surprising therefore that he considered only the sun and other sources of heat as causes of evaporation. On evaporation he wrote (IV, 17, 193).,

> Further, water found in a southern country is generally brackish, on account of the nearness of the sun. This happens because the sun sucks from it the expanded and thin part.

and again (V, 1, 194),

> The heat which in the winter is inside the earth heats the water which is in it, and this water rises upwards in the form of vapors, and mounts up in the air, in the same way as the water which is in a pot, when heated by fire, leaves it and rises upwards in the air.

Ayyub failed to mention any wind effect, and unlike some others such as Anaximander or Theophrastos, who took wind as air in motion, he did not even entertain the possibility that moist evaporation might be a cause of wind. His concept of evaporation was clearly Aristotelian.

Once his theories became accessible in translation, Aristotle was apparently held in high esteem among the Arabs. This is witnessed by the fact (cf. Mieli, 1966; pp. 95, 102) that the famous philosophers Al-Farabi [d. 950] from Turkestan, and the Iranian Ibn-Sina ['Avicenna', 980–1037] have also been called the second and third master, respectively, after Aristotle.

Available descriptions of evaporation and related phenomena in the Arabic literature strongly suggest that Aristotle's two-exhalation theory exerted a powerful influence. An example of this is found in the large encyclopedia 'Rasail' written in the tenth century by the Brothers of Purity (Ikhwan al-Safa, 1861; IV, pp. 76–81) of Basra in Iraq. Their treatment of the wind reveals their opinion on evaporation.

> The wind is nothing but the surging back and forth of the air, when it is moved to the six directions, just like the surges of the sea are nothing but the motion of the water, in which the parts push one another toward the four directions. For the water and the air are two standing seas, but the parts of the water are dense and they are heavy in motion, whereas the air is fine and it is light in its motion.
>
> A cause of the motion of the air is the rising of vapors from the sea and of smoke from the earth. For when the sun shines on the surface of a sea, of a field or of a desert, it stirs up fine moist vapors from the sea, but a dry smoke from the dry earth surfaces. Its heat makes both rise into the air; then one part of the air pushes the other toward different directions, in order to create space for the two

kinds of rising vapors. When there is a lot of dry smoke, the winds originate from it, for as these parts arrive at the upper edge of the windy region, they become cold, and the cold of the icy region prevents them from rising any higher. They then return continually and push the air to the four directions, and from this the different winds originate.

The description of surging back and forth suggests that the Ikhwan were dealing with the land and sea breezes which they observed at Basra. The analogy of the wind with tides appears to be directed against Aristotle's (1952; 360a 6) argument, quoted earlier, that wind cannot possibly be air in motion since it has no source. Their reasoning was that if tides can be water in motion, there is no reason why wind cannot be air in motion. This difference in outlook was undoubtedly due to the fact that the tides in the Persian Gulf are considerably larger than in the Mediterranean at Athens. At any rate, whatever the nature of the wind, the Ikhwan clearly indicated that it is caused by the dry exhalation, and the role of the moist vapor seems to be secondary. Evaporation is caused solely by the sun. They also described it as part of the hydrological cycle in similar terms.

The sun dissolves the seas, swamps and ponds in small parts, and raises them from the upper surface as vapor; from this originate mist and clouds, which are driven to certain places by the winds, just like it happened the year before. This is the way it keeps on happening, and this is the designation by the glorious, knowing God. So consider then this divine, all-encompassing care and the magnificant, wise guidance; ...

Thus winds have an advective effect on the clouds, but they are not directly related to evaporation. Interestingly, even the distinction of the different causes involved in vaporization and precipitation was stated in Aristotelian terms (Ikhwan al-Safa, 1861; p. 86).

The material cause for the clouds, the rain and their consequences are, as we have described above, the two rising vapor streams. The efficient cause for these are the sun and the stars, because they cast their rays, as we have mentioned before.

Although the details of their descriptions do not correspond with those of the 'Meteorologica', the Ikhwan al-Safa used the theory of the two exhalations as the basis of their concepts. But their description of evaporation, just like that of Ayyub, was practically identical with Aristotle's. Equally significant in this respect, as measures of Aristotle's influence on Arabic science, are the commentaries on the 'Meteorologica' by Al-Farabi (1969; III, 9, 110–111) and by Ibn-Rushd ['Averroes', 1126–1198] of Cordoba in Spain (Aristoteles, 1574; pp. 416, 432).

The first three books of the 'Meteorologica' had been translated from Arabic by Gerardus Cremonensis, who died in 1187, and the fourth directly from Greek by Henricus Aristippus, who died in 1162 (cf. Grabmann, 1916). As a result, during the first half of the thirteenth century, copies of these Latin translations were appearing in Western Europe. One unmistakable indication that the two-exhalation theory was beginning to make its presence felt is a fleeting comment that wind originates from 'dry vapor'; it appeared in a versified work 'On the Praises of Divine Wisdom' written by Alexander Neckam [ca. 1157–1217] (1863; 4, 116) of St Albans, toward the end of his life. But the full penetration of Aristotle's theories did not take place easily. During the initial period, in 1210 and 1215, the teaching of Aristotle's works in natural philosophy became the subject of official prohibitions in Paris, which remained in effect actually until at least 1241 and officially until 1255 (Van Steenberghen,

1955; pp. 98, 109, 164). This was extended later to Toulouse, but it never applied to Oxford. As Thomas Cantimpratensis had been a student at Paris, Aristotle's prohibition there probably explains why, around 1240, he still described evaporation in Isidorian terms. At any rate, the writers after Thomas gradually adopted Aristotle's theory.

Among the Latins, one of the first specific and formal accounts of Aristotle's two-exhalation theory was given by Vincent of Beauvais (Vincentius Bellovacensis, 1964; 4, 27) in the context of the origin of wind. His physical encyclopedia, which was probably written within a decade after the publication of the book of Thomas Cantimpratensis, contained a detailed account of several other opinions on the cause of wind, namely those of Pliny, Vitruvius, Demokritos, and Guilelmus de Conchis. Vincent, however, asserted that wind is moving air, referring to Seneca and Thomas C. (Section 4, 26), and he felt that nobody among mortals could know the real cause of the winds. In a later passage (Section 5, 8) he gave Isidore's view on why the sea does not increase. In addition, Vincent dealt extensively with vaporization, distillation and related phenomena of oils and other substances beside water.

Bartholomaeus Anglicus [ca. 1190], an English Franciscan, wrote a similar encyclopedia, 'On the Properties of Things'; he started it after 1231 in Magdeburg in Germany, so that it was probably published around the same time as Vincent's; there is no evidence that they were aware of each other's work. Although Bartholomaeus (1601; 11, 2) summarized the views of Beda, Aristotle, and unnamed others on the nature and the causes of the wind, he gave preference to the two-exhalation mechanism as the first cause without, however, dismissing the notion of the wind as moving air. On why the sea does not increase (13, 21), he quoted Isidore's views, involving evaporation due to sun and wind; but then he explained the salinity of the sea by means of Aristotle's theory, giving the Sun as the only cause of evaporation, which leaves the salt behind. Also this book had an enormous success in Europe; in the course of the fourteenth century it was translated into French, Spanish, Dutch, and English, and later several editions of it appeared in print.

Aristotle's theories rapidly penetrated into other languages and the vernacular. Around 1273 an unknown poet from Ghent in Flanders (sometimes alleged to be Gheraert Van Lienhout) published a didactic poem 'Physics of the Universe' in Middle-Dutch verse. In it (Jansen-Sieben, 1968; Section 819–872; 1717–1726) the simultaneous moist and dry exhalations of Aristotle are repeatedly mentioned in connection with wind genesis, precipitation and clouds. Aristotle's influence is also clearly evident in 'The Gate of Heaven', written in Hebrew in the second half of the thirteenth century by Gershon Ben Shlomoh (1953), a rabbi at Arles in Province: there are two kinds of vapors, one is moist and the material of the rains, snow, frost, hail, and the other is dry and warm and the material of the winds (Section I, 49, 50, 99); the Sun is the only cause of evaporation (Section I, 91). Similarly, passages in the 'Book of Nature', published around 1350 by Conrad von Megenberg (1897; 2, 15; 2, 31) from near Schweinfurt, indicate the general acceptance of Aristotle's views by that time. The cause of the wind was explained as smoke evaporated from the Earth. Although he cited the book of Thomas Cantimpratensis as his main source, Conrad explained that the evaporation, which prevents the sea from increasing, is due to the sun and other stars; but unlike Thomas, he made no mention of any wind effect.

Aristotle's monopoly continued for the next three centuries and at the height of the Renaissance European literature was fully imbued with his theory of the two exhalations. This theory served not merely as a physical explanation, but it was used as a source of metaphors and poetic imagery. For example, Heninger (1960) has given an extensive discussion of its omnipresence in the works of such English writers of the Renaissance as Spenser, Marlowe, Jonson, Chapman, Donne and most of all, Shakespeare.

The completeness of Aristotle's philosophical system undoubtedly provided a great stimulus from the thirteenth century onward for the development of European thought, which ultimately led to the scientific revolution. However, from the point of view of evaporation theory, his concept of the two exhalations represented for the second time a setback, which had even more serious ramifications than when it was first developed in antiquity. As noted, even at present very little is known about the meteorological theories of the Greek atomists Demokritos and Leukippos. Many of their physical concepts were closer to present day concepts than those of Aristotle. One can only speculate about the different development science might have undergone if, instead of Aristotle's, their works had been rediscovered in the late Middle Ages.

2.3. THE SEVENTEENTH AND EIGHTEENTH CENTURIES: INITIAL MEASUREMENTS AND EXPERIMENTATION

Descartes (1637) was one of the first natural philosophers to break away from Aristotle's concepts. In his book 'The Meteors' he started out with the hypothesis that all bodies in the environment are made up of small particles, and that the spaces between these particles are not empty but filled with a subtle matter, through which the action of light is communicated. He specified (Disc. 1);

> ... that those, of which water is composed, are long, united and slippery, just like little eels, which although they join and interlace each other, do not knot or get hooked, so that they can easily be separated; and on the contrary, that almost all those, as well of the earth as of the air, and of most of the other bodies, have very irregular and unequal shapes, such that they cannot be interlaced a little without becoming hooked and bound to each other just like the various branches of the shrubs which cross together in a hedge; ...

The sensation of heat or cold he attributed to the intensity of agitation of these little particles. But he pointed out that he did not conceive of these particles as indivisible atoms, but rather he believed that they are of the same matter and that each could be redivided in an infinity of ways.

With these preliminaries it is easy to follow Descartes' description of evaporation (Disc. 2):

> Consider that the subtle matter present in the pores of the earthly bodies, which is being more strongly agitated one time than another, either by the presence of the sun or by some other cause, also agitates more strongly the little particles of these bodies; you will then easily understand that this [matter] must effect that those [particles], which are the smallest and which have such shapes or are in such a situation that they can easily separate from thier neighbors, part here and there from one another and rise into the air; not because they might have any inclination to rise, nor because the sun might have a force which attracts them, but solely because they find no other space in which they would easily continue their movement; this is similar to the dust of the countryside which rises when it is only pushed and agitated by the feet of some passer-by.

Since Aristotle's theory had enjoyed such a universal acceptance, it was unavoidable (see also Gilson, 1920, 1921) that some elements of the two-exhalation theory were retained in the concepts proposed by Descartes. He indicated that most of the particles which rise into the air have the same shape as those of water, because they can easily separate, and he referred to these as vapors. And those with more irregular shapes he called exhalations for want of a more suitable name. However, he attached very little importance to these [dry] exhalations, among which he included particles of earth, spirits, volatile salts, burned oils, and smokes. This is also illustrated in his treatment of the winds, which he defined as follows (Disc. 4): "Any sensible agitation of air is called wind, and any invisible and impalpable body is called air." The wind is caused by the dilatation and expansion of the vapors rising from water surfaces and from the humid earth, snow and clouds. Descartes proved this by referring to the artificial wind which is created by the aeolipile, an apparatus which was invented by Hero of Alexandria [first century A.D.] and which has been called the first steam engine. Thus the vapors

> ... take along or chase the whole of the air and all the [dry] exhalations which it contains; in such a way that, although the winds are caused nearly only by the vapors, they are not composed only of vapors; the dilatation and condensation of these [dry] exhalations of this air may aid in the production of these winds, but this is so little in comparison with the dilatation and condensation of the vapors that they should hardly be taken into account.

In summary, Descartes attempted to explain evaporation and wind by postulating the existence of small particles. Evaporation is caused by the heat of the sun; heat is equivalent to agitation of the particles. Wind is air in motion, but it is the result of evaporation, rather than one of its causes. Seen against the long domination of the Aristotelian theories since the thirteenth century, the proposals of Descartes are radical. However, they were still the result of speculation, and except for the aeolipile, Descartes had little concrete evidence to back up his views.

During this period, partly as a result of the writings of Descartes, the general approach to science had started to change, and gradually experimentation became an essential part of it.

One of the earliest evaporation experiments on record was run by Perrault (1733). During the cold winter of 1669–1670, "... having exposed 7 pounds of [frozen] water to the cold air, found them diminished in 18 days by nearly one pound; which is an astonishing evaporation for this season." He also considered evaporation from various kinds of oils. This experimental result led him to theorize (Perrault, 1674, p. 239 ff.), "Although Aristotle and all the other philosophers give only one cause for the evaporation of water, namely the heat, I would be able to find two more, one the cold, its contrary, and the other the movement of the particles of air." After having described the above experiment, he continued, showing some hesitation as regards the effect of the cold.

> As it is very certain that the heat causes evaporation, and that one could doubt that the cold, its contrary, might produce a similar effect, I do not see any difficulty in attributing evaporation to the action of air particles, ... What makes me use this idea is that I see that evaporation occurs even without the aid of heat or of cold. ... The effect of this evaporation, whether it be caused by the heat or by the cold, or only by the agitation of the air particles, is always similar; I mean that the evaporated water remains always what it was; the evaporated water is always water, and its evaporation being only a separation of its parts, it will not fail to become water again as soon as this separation ceases ...

Apart from the hesitant suggestion of the effect of the cold, these ideas already appear rather modern; yet it must be pointed out that very similar notions had been put forth already by writers from Lucretius onward until Vuilhelmus of Conches or perhaps even Thomas Cantimpratensis in the early Middle Ages. As a matter of fact, in the twelfth century Vuilhelmus (1567, p. 168) had already dealt with the problem of the cold as a cause of evaporation in a much more satisfactory manner. But what was definitely new here was Perrault's reliance on experiment.

A few years later (1687) Halley also presented experimental data. From weight changes during evaporation of water from a small pan he deduced that, on warm days, evaporation amounted to approximately 0.1 inches in 12 hours, which he estimated "...will be abundantly sufficient to serve for all the rains, springs and dews..." He attributed this evaporation mainly to the effect of the sun "To estimate the quantity of vapour out of the sea, I think I ought to consider it only for the time the sun is up, for that the dews return in the night, as much if not more, vapours are then emitted." He realized, in addition, that the wind has an effect.

And this quantity of vapour, though very great, is as little as can be concluded from the experiment produced: and yet there remains another cause, which cannot be reduced to rule. I mean the winds, whereby the surface of the water is lick'd up sometimes faster than it exhales by the heat of the sun; as is well known to those that have considered those drying winds which blow sometimes.

In a second paper, Halley (1691) elaborated on a kind of particle, quite different from that of Descartes.

I have formerly attempted to explain the manner of rising of vapour by warmth, by shewing that if an atom of water were expanded into a shell or bubble so as to be ten times as big in diameter as when it was water, such an atom would become specifically lighter than air, and rise so long as that flatus or warm spirit that first separated it from the mass of water shall continue to distend it to the same degree; and that warmth declining ... Yet I undertake not that this is the only principal of the rise of vapors, ... But whatever is the true cause, it is in fact certain, that warmth does separate the particles of water and emit them with a greater and greater velocity as the heat is more and more intense, as is evident in the steam of a boiling cauldron.

But he also compared the process of vaporization with that of solution of salt.

... I take it, that it would follow that the air of itself would imbibe a certain quantity of aqueous vapours and retain them like salts dissolved in water; that the sun warming the air and raising a more plentiful vapour from the water in the daytime, the air would sustain a greater proportion of vapour, as warm water will hold more dissolved salts, which upon the absence of the sun in the nights would be all again discharged in dews, analogous to the precipitation of salts on the cooling of liquors; ...

In a third paper, Halley (1694) indicated that he knew of the work of Perrault; he further elaborated on the sun and the wind as the main causes of evaporation, and he presented additional experimental data.

Thus, experimentation had become an integral part of the scientific method. This new spirit is reflected in the introduction of the paper on evaporation which Sedileau (1730a, 1733a) presented to the Académie Royale in 1692:

There are certain fundamental experiments on which all of physics is based, and which one must necessarily make, however annoying they may be, if one wants to reason correctly in this science: otherwise all the reasonings which are made on natural things are speculations in the air.

The motivation of Sedileau's study was the design of the reservoirs needed to maintain the artificial fountains and water jets of the Parc of Versailles of Louis XIV; Colbert, superintendent of the king's buildings, and later his successor de Louvois, had asked the Académie, and in particular Sedileau, to investigate how much water the rains on the plains around Versailles could furnish, and how much of it would be lost by evaporation. Accordingly, Sedileau carried out measurements of precipitation and evaporation during three years. He used two basins of tin; one, 2′ by 1.5′ and 1.5′ deep, was used to measure precipitation, and the other, 3′ by 2′ and 2′ deep, was used to observe evaporation. Both basins were placed on the terrace of the Royal Observatory. He found that from June 1688 until December 1690 the average annual precipitation was about 19 inches, in agreement with Perrault's (1674) finding, but somewhat more than the 17 inches observed by Mariotte at Dijon (1 French inch or 'pouce' equals 2.707 cm). The average annual evaporation was found to be 32.5 inches. He also concluded that evaporation from a smaller pan is larger than from a large pan, all other conditions being the same. In addition, he made some observations on snow and ice. In a paper presented in 1693 (Sedileau, 1730b, 1733b) dealing with the validity of the concept of the hydrological cycle put forth by Perrault and Mariotte, he also attempted to explain why evaporation was almost twice the precipitation. He reasoned that part of the rain water seeps into the ground and is preserved there with little evaporation; the remainder runs off to lower places where it collects into a large volume with little exposed area. It should be noted here that Sedileau's result is not surprising; pan evaporation is usually much larger than regional evapotranspiration; also, even today rainfall is notoriously difficult to measure, and it is very likely that Sedileau's result was an underestimate.

The same question concerning the hydrological cycle and the origin of rivers was the incentive for the experimental study by de LaHire (1703). While this experiment did not deal directly with evaporation theory, it is of interest because it was a forerunner of the lysimeter. It was described as follows:

> I chose a place on the lower terrace of the Observatory, and in 1688 I had a leaden basin with a surface area of 4 feet installed in the ground at a depth of 8 feet. This basin had side walls of 6 inch height, and it was slightly inclined toward one of its corners, where I had a 12 foot long leaden tube soldered, which had a considerable slope and which entered in a small excavation at the other end. This basin was kept far from the wall of the excavation, in order that it would be surrounded by a greater quantity of soil similar to that which was on top, and that it would not dry out by the proximity of the wall.

This set-up had serious shortcomings: evidently, the basin side walls did not extend to the soil surface so that percolating rain water could move away laterally. With present understanding of the flow in a partly saturated soil, it is no wonder that de LaHire had to report that "not a single drop of water has come out through the tube in 15 years". He also conducted some experiments with a basin at more shallow depths and under conditions of minimal evaporation, but here some water would only be collected after heavy rainfall and large snowmelt. From these percolation experiments, he deduced that rain water cannot penetrate the earth very deeply. He then proceeded with an experiment to determine the evaporative loss from two individual fig leaves inserted in water, and this led him to infer that rain alone is not sufficient to support vegetation in summer, let alone to feed rivers. De LaHire concluded that the theory of Perrault and Mariotte is not generally valid and that other causes have to be found to explain the origin of springs. The experiments of de LaHire were a failure, but then, this is understandable, since it took another two hundred years (e.g. Buckingam, 1907) before developments in soil physics would allow a satisfactory explanation of his findings.

Inevitably experimentation stimulated thought, which resulted in various hypotheses and theoretical models to explain the phenomenon of evaporation. Often these explanations, which involved heavy debate, were reflections of developments taking place in other branches of the physical sciences. Interestingly, the main points of the debate, which was to stay around into the nineteenth century, were already contained in Halley's (1691) paper quoted above. The debate revolved around the following questions. Is vaporization a solution process, like that of salt in water, implying that without air there cannot be any vaporization, or is it merely a breaking up into particles? Then, do these particles maintain themselves because they are vesicules, or because they are repelling each other as a result of heat? Since the vapor particles and the nature of heat (cf., Fox, 1971) were ill-defined and poorly understood, many additional points of uncertainty arose. In some cases, several possibilities were admitted to coexist.

At the time, the particle separation theory was probably considered an older and more established view; it had elements of the concepts of Descartes and even of the Greek atomists. Adherents of various versions of this theory were among others 's Gravesande (1742; 1747, p. 2535), Desaguliers (1729; 1744, p. 313), and Van Musschenbroek (1739, par. 1495, p. 735; 1732). The latter also reported on evaporation measurements taken from a square basin placed in his garden in Utrecht; over a ten-year period the average yearly evaporation was approximately 29 Rhenish inches (one Rhenish inch – 'Rijnlandse duim' or 'pouce Rhénan' – equals 2.618 cm).

As mentioned, the particle separation mechanism was usually assumed to involve some form of heat or fire. Later, as another possibility, Desaguliers (1744, p. 333) invoked also electrostatic effects to explain evaporation: small particles of water jump toward particles of air, whose specific gravity is somewhat larger, and they adhere to them. However, the air in motion then repels these water particles as soon as they have become electrical; these electrical particles repel one another and also the particles of air. Water vapor is less dense so that it rises. This same idea was expressed by Van Musschenbroek (1769, p. 272) who specified that evaporation is the result not only of fire, but also of electricity, which causes the particles to repel each other. The wind has a double effect: first, it takes away vapors which have already some tendency to rise; and second, especially when it is dry, it contains a large amount of fluid electricity and enhances the separation of the particles. But the electrical effect was never accepted very widely; for instance, within a few years it was already argued against by de Saussure (1783), who otherwise had views similar to those of Desaguliers and Van Musschenbroek.

The solution theory of evaporation, which had no antecedents, must have appeared more innovative and modern. Among its adherents Bouillet (1742) explained that air absorbs and 'drinks' the water particles at the water surface with which it is in contact. In this way water particles, which are continually being detached, unite with the air by being stored and supported in the interstices, so that these particles follow all the movements of the air. He likened the phenomenon to the dissolution of copper or silver in *aqua fortis* (nitric acid); similarly Bouillet viewed the absorption of air in water as a sort of evaporation, taking place in an analogous way. The salt solution analogy was also supported by LeRoy (1751) and by Franklin (1765). Hamilton (1765) reviewed earlier opinions, which all involved some rarefaction by one or other form of heat; but he felt that this is not the real mechanism, because water in a closed room does not evaporate faster than when exposed in a colder place with a current of air, and because evaporation is carried on even after water is condensed into ice, that is in the absence of heat. Similarly, Dobson (1777) and Achard (1780a) supported the solution theory on the basis of their experiments showing that evaporation of water under a vacuum is slower than in the outside air. Monge (1790) gave as proof that water in air, just like salt in water, preserves its transparency, and that as the dissolved amount increases, the solving power decreases; moreover, at a high temperature saturated air contains more water than at a lower temperature, and as saturated air is cooled it precipitates the water.

Incidentally, Monge's (1790) explanation is just one indication of the fact that by then temperature was a standard measurement in the physical sciences. As a result several concepts were being put forward which were important in the development of evaporation theory. For example, LeRoy (1751) introduced the concept of 'degree of saturation' of air, which corresponds to the present-day dew point temperature, in an attempt to characterize the moisture content of the air. He found that this degree increases with increasing heat content of the air, and that it depends on the strength of the wind and its direction. In another temperature related development, it had become commonly known that evaporation causes some cooling. Around 1757 Franklin (1887) reported on this cooling effect, which is indicated by the fact that "wetting the thermometer with spirits brought the mercury down by five or six degrees". Apparently, as noted by Lavoisier (1777), similar qualitative observations had already been made earlier by Richmann in 1748, de Mairan in 1749, and several others. These observations undoubtedly contributed to the discovery of the latent heat concept by Black (1803) around 1760, after which the cooling effect could be studied in a more quantitative way.

The theory, that evaporation requires the presence of air to dissolve the water, evidently had many proponents in the eighteenth century. But toward the end of the century, its foundations became rather tenuous, mainly as a result of the work by De Luc (1787, 1792). From his close acquaintance with the work of Watt and from his own experiments, he arrived at the following conclusions. When water evaporates, an expansible fluid is produced; it may be called steam, and it is composed of water and fire ('free fire' or the 'cause of heat'). The pressure exerted by this fluid has a constant maximum at a given temperature, which increases with temperature. This fluid, regardless of the presence of air, affects the manometer by pressure and the hygrometer by moisture; when this fluid and air are mixed, they act on the manometer or the barometer according to their respective 'power'.

De Luc's findings clearly contain the essence of the law of partial pressures in gas mixtures, which is now generally associated with the name of Dalton. Although he proved it perhaps more clearly and convincingly, Dalton (1801, 1802a) arrived at the same conclusion. Thus, the pressure or density of a gas is independent of the amount of other gases or vapors present, and each one of them presses separately as if it were the only elastic fluid constituting the atmosphere. This follows from the fact that the 'force of vapor' produced by a liquid depends only on temperature, and that it is the same in an 'exhausted receiver' as it is in the atmosphere. At any rate, the proof and acceptance of the law of partial pressures made the controversy, whether or not air is required to dissolve water vapor, virtually a dead issue. The way was now open for more quantitative theories of evaporation.

2.4. FOUNDATIONS OF PRESENT THEORIES IN THE NINETEENTH CENTURY

Undoubtedly the publication of Dalton's paper in 1802 was one of the major events in the development of evaporation theory. In it he first recapitulated his views on gas mixtures, and he gave a table of the saturated vapor pressure as a function of temperature. Then, in the essay on evaporation, Dalton (1802a, p. 576) summarized the consensus among scientists at the end of the eighteenth century, as follows:

> The following positions have been established by others, and need therefore only to be mentioned here.
> (1) Some fluids evaporate much more quickly than others.
> (2) The quantity evaporated is in direct proportion to the surface exposed, all other circumstances alike.
> (3) An increase of temperature in the liquid is attended with an increase of evaporation, not directly proportionable.
> (4) Evaporation is greater where there is a stream of air than where the air is stagnant.
> (5) Evaporation from water is greater the less the humidity previously existing in the atmosphere, all other circumstances the same.

A more definite formulation of positions 1, 2 and 4 eluded Dalton and several generations after him. In fact, it is now known that, as already observed by Sedileau (1730a), position 2 is not quite correct. And positions 1 and 4 have only been elucidated relatively recently. Dalton's contribution was the quantification of positions 3 and 5. From his experiments he concluded that "The quantity of any liquid evaporated in the open air is directly as the force of steam from such liquid at its temperature, all other circumstances being the same." And he explained his findings, "...in short, the evaporating force must be universally equal to that of the water, diminished by that already existing in the atmosphere." When the evaporating force is the same in different cases, the different rates of evaporation are "...regulated solely by the force of the wind." From these comments, it follows that Dalton's result can be written in present-day notation as

$$E = f_D(\bar{u})(e_s^* - e_a) \tag{2.1}$$

where E is the rate of evaporation as height of water per unit time, e_s^* the saturation vapor pressure at the temperature of the water surface, e_a the vapor pressure in the air and $f_D(\bar{u})$ is a function of the mean wind speed $\bar{u}$. (Note that Dalton did not give this equation.)

It is of some interest to elaborate on Dalton's consideration of the effect of the wind. For practical use, Dalton presented a table of the rate of evaporation in dry air as a function of the water temperature and for three different categories of wind speed; he described these categories as evaporation in calm air in the middle of a room with closed doors and windows, evaporation in the room with open windows and strong wind outside resulting in a large draft, and evaporation in the open air exposed to high winds. Each category was characterized by the rate of evaporation at boiling, and the values in the table were calculated by multiplying the evaporation at boiling by the fraction (e_s^*/p) in which p is the pressure of the air. Dalton found good agreement between the values in this table and numerous experiments run under different conditions of draft and exposure.

Soldner (1804) pointed out that Dalton had failed to include the effect of the atmospheric pressure in his theory, and that evaporation does not solely depend on the water temperature and the wind as implied in his table. Dalton had concluded that at their boiling point all liquids evaporate equally quickly at a rate E_b which depends only on the wind speed. Consequently, Soldner felt that a more appropriate description of Dalton's findings than his table, would be the following:

$$E = E_b \frac{(e_s^* - e_a)}{p} \tag{2.2}$$

where E_b is the evaporation rate at boiling into dry air of atmospheric pressure p, that is the saturation vapor pressure at boiling. To apply (2.2) Soldner (1804) also fitted a mathematical expression to Dalton's experimental results on the saturation vapor pressure; he proposed,

$$e^* = p \exp\left[-(250 + T_b - T)(T_b - T)/6976\right] \tag{2.3}$$

where T is the temperature (K) and T_b the boiling temperature at atmospheric pressure p. Around the same time, Laplace proposed a similar equation; however, it fitted Dalton's data slightly less closely than (2.3), (e.g., Soldner, 1807).

Except for some sceptics (e.g., Parrot, 1804) most of Dalton's contemporaries quickly recognized his contribution, but they also realized its limitations. Soldner's (1807) comments illustrate this:

> His [Dalton's] laws on evaporation are as such very correct; but the practical application will hardly ever succeed. I have considered his observations [Dalton, 1802b] of evaporation, temperature, dew point and barometric pressure; but nothing agrees, except for the month of August, which Dalton himself has chosen as example. (Why doesn't he say anything about the others? This would have been really something valid for Mr. Parrot.) But when one considers how irregular the wind is, how quickly and considerably it changes its temperature through clouds and through the terrain, over which it has come from its momentary direction, this is not surprising ... I am convinced that as such the matter is correct, and that it agrees in a room where the air is completely quiet; but in the open it will never work out.

These perceptive remarks notwithstanding, during the next half century, apparently little progress was made as regards the effect of the air stream. For instance, the best Schmid (1860, p. 598) could muster on the subject was a table with Schübler's data obtained during 1826 at Tübingen; these merely showed that evaporation of a water surface exposed to wind was 1.7 times larger than that of a sheltered surface in summer, and 4 times larger in winter. Such qualitative findings did not really prove anything new. But then within two decades several significant contributions followed in succession.

By means of a newly designed 'evaporameter', Tate (1862) arrived at the conclusion that the rate of evaporation is nearly proportional to the velocity of the wind. He did not specify the proportionality, but his statement was unambiguous. Note that he also concluded that the rate of evaporation is nearly inversely proportional to the atmospheric pressure, and that it is nearly proportional to the difference of the temperatures indicated by the wet- and dry-bulb thermometers. The former conclusion is reminiscent of Soldner's interpretation of Dalton's results, and the latter represents a setback relative to Dalton's (2.1), since it dismisses any direct effect of the water temperature.

Further insight was provided by Weilenmann's (1877a, b) work. By considering the air in contact with the water surface, he expressed the evaporation rate as a linear function of the mean wind speed, $\bar{u}$. However, he also reasoned that it can be taken to be proportional to the saturation deficit of the air. He tested the following equation with experimental data:

$$E = (A_w + B_w\bar{u})(e_a^* - e_a) \tag{2.4}$$

where A_w and B_w are constants and e_a^* is the saturation vapor pressure at the temperature of the air. Evidently, Weilenmann (1877a, p. 11) had been inspired by Tate's conclusions in the derivation of his equation.

In view of Dalton's law (2.1), the use of the saturation deficit in (2.4) is tantamount to assuming that the temperatures of the water surface and the air are the same. No wonder, therefore, that (2.4) yielded better results in the shade than in open sunshine. To some extent Weilenmann was aware of this limitation, for he asserted that the saturation deficit should actually be calculated by using the temperature at the air-water interface; since he did not have any water temperatures available, he felt justified in using air temperatures as an approximation. As a further extenuating circumstance, it should be pointed out that with his experimental data this approximation was not very critical: most of the evaporation measurements he had available had been taken in some type of shelter. As a matter of fact, even Dalton (1802a) in the calculation of some examples with his tables, had tacitly made the same assumption that air and water temperature are the same. Dalton clearly intended (2.1) in his general theoretical discussion; still, his own rather ambiguous interpretation of (2.1) in practical application may have contributed later to the confusion exemplified in the studies of Tate (1862) and Weilenmann (1877a, b).

The matter was finally presented correctly by Stelling (1882). Combining Weilenmann's wind function with Dalton's (2.1), he was probably the first to formulate the following:

$$E = (A_s + B_s\bar{u})(e_s^* - e_a) \tag{2.5}$$

where A_s and B_s are empirical constants; by means of evaporation data obtained with a Wild-evaporometer (See Figure 11.16) exposed at 1 m. above the surface at Nukus in Uzbekistan, he tested this equation extensively and concluded that A_s = 0.0702 and B_s = 0.00319 when E is in mm (2 h)$^{-1}$, e in mm Hg and $\bar{u}$ at 7.5 m above the ground in km h^{-1}. Stelling's equation, with various values of A_s and B_s, soon became very popular. It was applied by Fitzgerald (1886) in Massachusetts and by Carpenter (1889, 1891) in Colorado. It is still widely used in engineering practice today.

Equation (2.5) was arrived at in an empirical way. Countless different values of A_s and B_s have appeared in the literature, and even now attempts are still being made to determine their optimal values for various conditions. However, as sometimes happens when a formula is moderately successful in describing experimental data for practical purposes, it stifles and impedes further progress in the more fundamental aspects of the phenomenon. As such, therefore, Stelling's equation represented a dead end from the theoretical point of view.

Subsequent progress in evaporation theory has come out of different directions, and it has followed generally developments in fluid mechanics and transport phenomena in turbulent flow. However, this association did not come about very rapidly. A crucial contribution to the understanding of mass transfer in fluids was made by Fick (1855). He found experimentally that the local specific flux of an admixture of an undisturbed fluid, as a result of molecular action only, is proportional to the gradient of its concentration. This was but a confirmation of his inkling at first that this phenomenon should proceed (Fick, 1855, p. 65) "... according to the same laws, which Fourier has laid down for the propagation of heat in a conductor, and which Ohm has already extended with such splendid successes to the propagation of electricity ..." In fact, Fick's law was also analogous in this respect to Newton's law for viscous shear. This law of viscous shear was extended to turbulent flow by Boussinesq (1877). He hypothesized that the shear stresses in a turbulent flow are proportional to the velocity gradients, and specified that the coefficient of proportionality (Boussinesq, 1877, p. 46) "... must depend at each point, not only on the temperature and perhaps the pressure p, but also and especially on the intensity of the mean agitation which happens to be produced there."

Reynolds (1874), in the context of transfer equations for heat and resistance similar to (2.5), supposing that the respective A's and B's should be proportional, implied that the transport mechanisms of heat and momentum in turbulent flow might be similar. It would be a small step to extend this 'Reynolds analogy', as it is now called, to water vapor as well. Thus, with interest developing in the description of the vertical change of the wind velocity over the earth's surface as a power of elevation (e.g. Stevenson, 1880; Archibald, 1883), the stage was set for a more fundamental approach. The seminal ideas of Fick, Boussinesq, and Reynolds finally came to full fruition in Schmidt's (1917) work. His proposal of the same 'exchange coefficient' for momentum and any other admixture signaled the recognition of evaporation as a regular problem of turbulent flow; also, it led directly to the development of present-day similarity theories for the turbulent transport of water vapor and other scalars in the lower atmosphere.

Little has been said in this review so far about the development of the energetic aspects of evaporation in the nineteenth century. As noted above, since about the middle of the eighteenth century it was known that the evaporation from a moistened bulb of a thermometer would lower its temperature. Thus, it was generally understood that evaporation causes cooling and that it requires heat. Although the nature of heat was unknown and the subject of much controversy (e.g., Fox, 1971), by the end of the century it had become possible to express heat in quantitative terms. The latent heat of vaporization had been discovered by Black around 1760 (cf., McKie and Heathcote, 1935). Two years later Black (1803, p. 157) had made his first experiments and found that the latent heat of water at boiling is equivalent to about 810 degrees, that is the amount which would be required to warm the water by 810° F. This is about 810 × 5/9 = 450 cal g^{-1}, which is amazingly close to the presently-accepted value of 539.1 cal g^{-1}. Shortly thereafter, but probably before 1765, this estimate had been further improved to between 900° and 950° by Watt, an associate of Black at Glasgow, in the course of his studies on the steam engine (Black, 1803, p. 174). Also, by the turn of the century, the major heat transfer phenomena, viz. conduction, convection, and radiation, had been identified. However, the prevailing uncertainty was still sufficiently large that it would take a long time before all these concepts could be related to the evaporation taking place in the outdoor atmosphere.

That there is a strong relationship between solar radiation and evaporation had, of course, been known since prehistoric days. It is scarcely surprising, therefore, that at first investigations were aimed at the discovery of this relationship. For example, Heller (1800), observing that for the same temperature evaporation increases with sunlight, concluded that, beside temperature, radiation measurements would be useful. Probably the first quantitative study, albeit on a global scale, was made by Daubrée (1847). He estimated from available data that the average annual rainfall over the globe amounts to approximately 1.379 m.; this is not very different from the current consensus of about 1.0 m. Some ten years earlier, Pouillet (1838; 1847, p. 680) had already estimated by means of a 'pyrhéliomètre' that solar radiation at the upper limit of the atmosphere is about 1.7633 cal cm^{-2} min^{-1}, also remarkably close to the present value of 1.98; to give a better idea of the significance of this quantity, Pouillet had added that in one year this heat would be capable of melting a layer of ice with a thickness of 31 m uniformly covering the earth. The average annual rainfall is also the average annual evaporation. Thus on the basis of Pouillet's estimate, Daubrée (1847) concluded that evaporation consumes approximately one-third of the solar energy received at the outer boundary of the atmosphere, or the equivalent of approximately 10.7 m of ice melt uniformly over the globe. Interestingly, he also observed that the annual combustion of fuels in France at that time was approximately equivalent to melting a crust of 0.0017 m of ice covering the whole country, which corresponded to 1.6 × 10^{-4} times the average heat used in evaporation. The same theme, based on Pouillet's ice-melt equivalent of solar radiation, was taken up by Maury (1861, p. 107) when he estimated that the energy used up in evaporation over the Mississippi Basin is 5/6 of "as much as would be set free by the combustion of 30 000 tons of coal multiplied 6 540 000 times".

More importantly, however, in his description of the distribution of the incoming solar radiation at the earth surface, Maury (1861, p. 422) introduced the energy budget concept in terms acceptable even today.

> As the heat comes from the sun, part of it is absorbed by the atmosphere, but the largest portion of it is impressed upon the land and water. From these a portion passes off into the atmosphere by conduction, while another portion is radiated directly off into the realm of space. . . . The remainder of this heat, . . . is absorbed in the process of evaporation. It is then delivered to the atmosphere, latent in the vesicles of vapor, to be set free in the cloud region, rendered sensible and imparted to the upper air, whence it is sent off by radiation into the "emptiness of space". Thus the air, with its actinometry, presents itself in the light of a thermal adjustment, by which the land and sea are prevented from becoming seething hot, . . .

The relationship between evaporation, solar radiation and other heat flux components in an energy budget context was also implicit in the works of such contemporaries as Wollny (1877) in agronomy and Woeikoff (1887) in climatology. Clearly, the energy budget concept was gradually being accepted at that time.

However, the first quantitative and detailed analysis of the energy budget at the earth's surface was probably carried out by Homén (1897). He determined daily soil heat flux from measurements of soil temperature at 10 cm. intervals down to 60 cm. and from the heat capacity of the soil, and he measured radiation by adapting the method of K. Ångström published in 1893; evaporation was determined from weight changes of a metal cylinder, 35 cm in diameter and 30 cm deep, filled with soil and kept flush with the surroundings. The sensible heat flux into the atmosphere was taken as the only unknown rest term in the budget equation. Homén's contribution came at a time when rapid advances were being made in the understanding of radiation as a result of the discoveries by Stefan (1879) and Boltzmann (1884). Thus the groundwork had been laid for the further developments of energy budget procedures through the work of Schmidt (1915), Bowen (1926) and others to the present.

CHAPTER 3

The Lower Atmosphere

3.1. MOIST AIR

a. Some Parameter Definitions

For many practical purposes, the air of the lower atmosphere can be considered as a mixture of perfect gases; in the present context these may conveniently be assumed to be dry air of constant composition and water vapor. The water vapor content of the air can be expressed in terms of the mixing ratio, defined as the mass of water vapor per unit mass of dry air,

$$m = \rho_v/\rho_d \tag{3.1}$$

where ρ_v is the density of the water vapor and ρ_d the density of the air without the water vapor. The specific humidity is defined as the mass of water vapor per unit mass of moist air,

$$q = \rho_v/\rho \tag{3.2}$$

where $\rho = \rho_v + \rho_d$. The relative humidity is the ratio of the actual mixing ratio and the mixing ratio in water vapor saturated air at the same temperature and pressure,

$$r = m/m^*. \tag{3.3}$$

This is nearly equal to (e/e^*), the ratio of the actual vapor pressure and the equilibrium vapor pressure at saturation.

According to Dalton's law, the total pressure in a mixture of perfect gases equals the sum of the partial pressures, and each of the component gases obeys its own equation of state. Thus, the density of the dry air component is

$$\rho_d = \frac{p - e}{R_d T} \tag{3.4}$$

where p is the total pressure in the air, e the partial pressure of the water vapor, T the ('absolute') temperature, and R_d, which is given in Table 3.1, is the specific gas constant for dry air. Similarly, the density of water vapor is

$$\rho_v = \frac{0.622e}{R_d T} \tag{3.5}$$

where $0.622 = (18.016/28.966)$ is the ratio of the molecular weights of water and dry air.

The density of moist air is from (3.4) and (3.5)

TABLE 3.1
Some physical constants

Dry Air
Molecular weight: 28.966 g mol^{-1}
Gas constant: R_d = 287.04 J kg^{-1} K^{-1}
Specific heat: c_{pd} = 1005 J kg^{-1} K^{-1}
c_{vd} = 716 J kg^{-1} K^{-1}
Density: ρ = 1.2923 kg m^{-3}
(p = 1013.25 mb, T = 273.16 K)
Water Vapor
Molecular weight: 18.016 mol^{-1}
Gas constant: R_w = 461.5 J kg^{-1} K^{-1}
Specific heat: c_{pw} = 1846 J kg^{-1} K^{-1}
c_{vw} = 1386 J kg^{-1} K^{-1}

Note: The values listed in Tables 3.1, 3.4 through 3.6 are adapted from the Smithsonian Meteorological Tables (List, 1971), where the original references are cited.

$$\rho = \frac{p}{R_d T}\left(1 - \frac{0.378e}{p}\right) \tag{3.6}$$

showing that it is smaller than that of dry air at pressure p. This means that water vapor stratification plays a role in determining the stability of the atmosphere. The equation of state of moist air can be obtained by eliminating e from (3.4) and (3.5)

$$p = \rho T R_d(1 + 0.61\ q). \tag{3.7}$$

This indicates that the mixture behaves as a perfect gas provided it has a specific gas constant $R_m = R_d(1 + 0.61\ q)$ that is a function of the water vapor content. Therefore, (3.7) is often expressed as

$$p = R_d \rho T_V \tag{3.8}$$

where T_V is the virtual temperature defined by

$$T_V = (1 + 0.61\ q)T. \tag{3.9}$$

It is the temperature dry air should have in order to have the same density as moist air with given q, T and p.

For convenient reference, some common units and conversion factors are listed in Tables 3.2 and 3.3.

b. Useful Forms of the First Law of Thermodynamics

This law states that the heat added to a system equals the sum of the change in internal energy and the work done by the system (e.g., Fermi, 1956)

$$\delta H = \delta U + p\ \delta V \tag{3.10a}$$

or, per unit mass and in differential form

$$\mathrm{d}h = \mathrm{d}u + p\ \mathrm{d}\alpha \tag{3.10b}$$

TABLE 3.2
Common units

	SI (mks)	cgs
Length	meter	centimeter
	m	cm
Mass	kilogram	gram
	kg	g
Time	second	second
	s	s
Force	newton	dyne
	N = kg m s^{-2}	dyn = g cm s^{-2}
Pressure	pascal	microbar
	Pa = N m^{-2}	μ bar = dyn cm^{-2}
Energy	joule	erg
	J = N m	erg = dyn cm
Power	watt	erg
	W = J s^{-1}	erg s^{-1}

TABLE 3.3
Conversion factors

Pressure	millibar	1 mb = 10^3 μb = 10^2 Pa = 10^3 dyn cm^{-2}
	millimeter mercury	1 mm Hg = 1.333 224 mb
	atmosphere	1 atm = 1.01325 10^5 Pa
Energy	calorie (IT)	1 cal = 4.1868 J = 4.1868 10^7 erg

where V is the volume and $\alpha = \rho^{-1}$, the specific volume. The equation of state of any gas relates the three variables α, T and p, so that the state can be defined by any two of the three. If α and T are chosen as independent variables, (3.10) becomes

$$\mathrm{d}h = \left(\frac{\partial u}{\partial T}\right)_\alpha \mathrm{d}T + \left[\left(\frac{\partial u}{\partial \alpha}\right) + p\right]\mathrm{d}\alpha. \tag{3.10c}$$

Since, by definition, the specific heat capacity for constant volume is

$$c_v = \left(\frac{\partial h}{\partial T}\right)_\alpha \tag{3.11}$$

and since it can be shown experimentally, as done by Joule, and from kinetic gas theory that $(\partial u/\partial \alpha) = 0$, the first law can also be written as

$$\mathrm{d}h = c_v\,\mathrm{d}T + p\,\mathrm{d}\alpha \tag{3.10d}$$

or, by virtue of the equation of state, as

$$\mathrm{d}h = (c_v + R)\,\mathrm{d}T - \alpha\,\mathrm{d}p. \tag{3.10e}$$

This shows that

$$R = c_p - c_v \tag{3.12}$$

where by definition

$$c_p = \left(\frac{\partial h}{\partial T}\right)_p \tag{3.13}$$

is the specific heat for constant pressure.

c. Saturation Vapor Pressure

The saturation vapor pressure e^* is the vapor pressure at which a change in phase can occur at constant temperature. It is an important variable in the study of evaporation, so that it deserves some elaboration. Consider (e.g., Fermi, 1956) a liquid-vapor system of total mass M in which an amount δM of the substance passes isothermally from the liquid to the vapor state, with a corresponding change δV in total volume and δU in internal energy. When liquid and vapor are in equilibrium, the pressure and the densities depend only on temperature. If M_l and M_v are the masses of liquid and vapor, α_l and α_v their specific volumes, the total volume before the isothermal transformation is

$$V = M_l\alpha_l(T) + M_v\alpha_v(T).$$

After the transformation this becomes

$$V + \delta V = (M_l - \delta M)\alpha_l + (M_v + \delta M)\alpha_v$$

so that

$$\delta V = [\alpha_v(T) - \alpha_l(T)]\,\delta M. \tag{3.14}$$

A similar argument with the internal energy produces

$$\delta U = [u_v(T) - u_l(T)]\,\delta M \tag{3.15}$$

where u_l and u_v are the specific internal energies of liquid and vapor, respectively.

By definition, the heat added per unit mass of phase change is the latent heat of vaporization

$$L_e = \delta H/\delta M. \tag{3.16}$$

Combining this with (3.10a), (3.14) and (3.15) one obtains

$$L_e = u_v - u_l + p(\alpha_v - \alpha_l). \tag{3.17}$$

Equations (3.14) and (3.15) also show that

$$\left(\frac{\partial U}{\partial V}\right)_T = \frac{u_v - u_l}{\alpha_v - \alpha_l}$$

so that with (3.17)

$$\left(\frac{\partial u}{\partial V}\right)_T = \frac{L_e}{\alpha_v - \alpha_l} - p. \tag{3.18}$$

It is convenient here to introduce the definition of entropy

$$\mathrm{d}S = \mathrm{d}H/T. \tag{3.19}$$

Combined with the first law (3.10c) this can be written as

$$\mathrm{d}S = \frac{1}{T}\left(\frac{\partial U}{\partial T}\right)\mathrm{d}T + \frac{1}{T}\left(\frac{\partial u}{\partial V} + p\right)\mathrm{d}V.$$

In order that this expression be a perfect differential, as its left-hand side indicates, the following must be satisfied

$$\frac{\partial}{\partial V}\left(\frac{1}{T}\frac{\partial U}{\partial T}\right) = \frac{\partial}{\partial T}\left[\frac{1}{T}\left(\frac{\partial U}{\partial V} + p\right)\right]$$

in which it is understood that V and T are the independent variables. This yields

$$\left(\frac{\partial U}{\partial V}\right)_T = T\left(\frac{\partial p}{\partial T}\right)_V - p. \tag{3.20}$$

The pressure in the system under consideration is a function of T only, and it is the pressure exerted by the vapor in equilibrium with the liquid. Thus combination of (3.18) with (3.20) produces

$$\frac{\mathrm{d}e^*}{\mathrm{d}T} = \frac{L_e}{T(\alpha_v - \alpha_l)} \tag{3.21}$$

which is the well-known Clausius–Clapeyron equation. As α_1 is usually negligible as compared to α_v, (3.21) may be written as

$$\frac{\mathrm{d}e^*}{\mathrm{d}T} = \frac{0.622\, L_e e^*}{R_d T^2} \tag{3.22}$$

in which use is made of (3.5) for saturation.

Equation (3.22) can be integrated when a suitable expression is available for $L_e(T)$. One of the better integrals of the Clausius–Clapeyron equation is the formulation of Goff and Gratch (1946), who considered the deviations from a perfect gas and also available experimental data; this formulation is widely adopted as a standard, and it has been the basis for published tables (e.g., List, 1971). Some values of e^*, $\mathrm{d}e^*/\mathrm{d}T$ and the saturation vapor pressure over ice e_i^* are presented in Tables 3.4 and 3.5, respectively.

Fore easy computation, numerous simpler expressions have been developed. Among the earliest is (2.3) of Soldner based on Dalton's data. Lowe (1977), who has also compared other currently used expressions for e^*, has presented polynomials

TABLE 3.4
Some properties of water

Temperature (°C)	c_w (J kg^{-1} K^{-1})	L_e (10^6 J kg^{-1})	e^* (mb)	$\mathrm{d}e^*/\mathrm{d}T$ (mb K^{-1})
−20	4354	2.549	1.2540	0.1081
−10	4271	2.525	2.8627	0.2262
0	4218	2.501	6.1078	0.4438
5	4202	2.489	8.7192	0.6082
10	4192	2.477	12.272	0.8222
15	4186	2.466	17.044	1.098
20	4182	2.453	23.373	1.448
25	4180	2.442	31.671	1.888
30	4178	2.430	42.430	2.435
35	4178	2.418	56.236	3.110
40	4178	2.406	73.777	3.933

c_w: specific heat; L_e: latent heat of vaporization; e^*: saturation vapor pressure.

TABLE 3.5
Some properties of ice

Temperature (°C)	c_i (J kg^{-1} K^{-1})	L_{fu} (10^6 J kg^{-1})	L_s (10^6 J kg^{-1})	e_i^* (mb)	de_i^*/dT (mb K^{-1})
−20	1959	0.2889	2.838	1.032	0.09905
−15	—	—	—	1.652	0.1524
−10	2031	0.3119	2.837	2.597	0.2306
− 5	—	—	—	4.015	0.3432
0	2106	0.3337	2.834	6.107	0.5029

c_i: specific heat; L_{fu}: latent heat of fusion; L_s: latent heat of sublimation; e_i^*: saturation vapor pressure over ice.

for e^*, de^*/dT, e_i^* and de_i^*/dT, which are quite accurate and suitable for rapid computation. For computational speed these polynomials should be used in nested form; the representation for e^* takes the form

$$e^* = a_0 + T(a_1 + T(a_2 + T(a_3 + T(a_4 + T(a_5 + a_6 T))))) \tag{3.23}$$

where the polynomial coefficients are as follows when T is in K, $a_0 = 6984.505\,294$, $a_1 = -188.903\,931\,0$, $a_2 = 2.133\,357\,675$, $a_3 = -1.288\,580\,973 \times 10^{-2}$, $a_4 = 4.393\,587\,233 \times 10^{-5}$, $a_5 = -8.023\,923\,082 \times 10^{-8}$, $a_6 = 6.136\,820\,929 \times 10^{-11}$. Richards (1971) has proposed an expression which is perhaps more suitable for desk calculations, as follows

$$e^* = 1013.25 \exp\,(13.3185\, t_R - 1.9760\, t_R^2 - 0.6445\, t_R^3 - 0.1299 t_R^4) \tag{3.24a}$$

where $t_R = 1-(373.15/T)$ in which T is the temperature in K. Although even in nested form (3.24a) requires approximately three times the computational time of Lowe's polynomial, its accuracy is practically the same and of the order of 10^{-2} percent, compared to the Goff–Gratch standard. An advantage of (3.24a) is that it yields the variation of saturation vapor pressure with temperature, which is of the same form as the Clausius–Clapeyron Equation (3.22)

$$\frac{de^*}{dT} = \frac{373.15\, e^*}{T^2}(13.3185 - 3.952\, t_R - 1.9335\, t_R^2 - 0.5196\, t_R^3). \tag{3.24b}$$

3.2. HYDROSTATIC STABILITY OF PARTLY SATURATED ATMOSPHERE

a. Small Adiabatic Displacements

The criterion for the stability of an atmosphere at rest can be obtained by considering a small vertical displacement of a small parcel of air without mixing with the surrounding body of air. The displacement is sufficiently small and fast, so that the pressure of the particle adjusts to its new environment in an adiabatic fashion, that is as a reversible process without heat transfer. Moreover, the moisture content of the particle remains constant, whereas that of the surrounding atmosphere changes as $\partial q/\partial z$.

The vertical acceleration of the parcel is given by the following equation of motion

$$\ddot{z} = -g - \frac{1}{\rho_1}\frac{\partial p}{\partial z} \tag{3.25}$$

where ρ_1 is the density of the moist air of the particle. The pressure gradient in the surrounding atmosphere at rest, acting on the parcel, is

$$\frac{\partial p}{\partial z} = -\rho g \tag{3.26}$$

where ρ is the density of the surrounding air. Substitution of (3.26) and (3.8) for ρ and ρ_1 in (3.25) yields

$$\ddot{z} = -g\left(\frac{T_V - T_{V1}}{T_V}\right) \tag{3.27}$$

where T_{V1} and T_V are the virtual temperatures of the parcel and of the surrounding air, respectively. If it is assumed that the small displacement z originates at a reference level $z = 0$, where the parcel has the same density and thus virtual temperature T_{V0} as the surrounding air, one can write $T_{V1} = T_{V0} + z\partial T_{V1}/\partial z$, and a similar equation for T_V, so that (3.27) becomes

$$\ddot{z} = -\frac{gz}{T_V}\left(\frac{\partial T_V}{\partial z} - \frac{\partial T_{V1}}{\partial z}\right). \tag{3.28}$$

In view of (3.9), the virtual temperature gradient in the surrounding air can also be written as

$$\frac{\partial T_V}{\partial z} = (1 + 0.61q)\frac{\partial T}{\partial z} + 0.61T\frac{\partial q}{\partial z}. \tag{3.29}$$

The rate of change of virtual temperature with elevation of the parcel during its displacement can be obtained by considering that it does not exchange moisture or heat with the surroundings. Thus $\partial q_1/\partial z = 0$ and $(\partial T_1/\partial z)$ is that of an adiabatic process. Equation (3.10e) gives for an adiabatic process applied to the parcel

$$c_p\frac{\partial T_1}{\partial z} - \frac{1}{\rho_1}\frac{\partial p}{\partial z} = 0. \tag{3.30}$$

The specific heat at constant pressure of moist air is the weighted sum of the specific heats of the dry air and the water vapor component, namely $c_p = qc_{pw} + (1 - q)c_{pd}$; with the values given in Table 3.1 this is

$$c_p = c_{pd}(1 + 0.84q) \tag{3.31}$$

which shows that the effect of the water vapor is nearly negligible. Substituting in (3.30) the pressure gradient given by (3.26), and the densities ρ_1 and ρ given by (3.7), one can write (3.30) as

$$\frac{\partial T_1}{\partial z} = -\frac{g}{c_p}\frac{T_1}{T}. \tag{3.32}$$

The vertical rate of change of virtual temperature of the parcel is, on account of (3.9) (3.31), and (3.32), since $\partial q_1/\partial z = 0$,

$$\frac{\partial T_{V1}}{\partial z} = -\frac{g}{c_{pd}}\frac{T_1}{T}(1 - 0.23q)$$

or, since (T_1/T) and $(1 - 0.23q)$ are very close to unity,

$$\frac{\partial T_{V1}}{\partial z} = -\Gamma_d \tag{3.33}$$

where $\Gamma_d = (g/c_{pd})$, is known as the *dry adiabatic lapse rate*, which is of the order of 9.8 K km^{-1}. It is in fact the vertical rate of temperature decrease, which the (positively) displaced parcel would undergo, if it were completely dry. Substitution of (3.33) in (3.28) finally produces the acceleration of the displaced parcel

$$\ddot{z} = \frac{gz}{T_V}\left(-\frac{\partial T_V}{\partial z} - \Gamma_d\right). \tag{3.34}$$

Equation (3.34) shows how the dry adiabatic lapse rate can serve as a stability criterion for a partly saturated atmosphere at rest. If the lapse rate of the virtual temperature, that is the rate of virtual temperature decrease with height, in such an atmosphere equals Γ_d, the acceleration of the slightly displaced particle is zero; its density is the same as that of the surrounding air, and it has no tendency to increase or decrease the displacement. The atmosphere is in statically-neutral equilibrium. On the other hand, if the lapse rate of the virtual temperature is superadiabatic, that is larger than g/c_p, the solution of (3.34) yields a displacement z which increases exponentially with time; the parcel tends to move even further away from its original position. The atmosphere is now statically unstable. Conversely, when $-(\partial T_V/\partial z) < \Gamma_d$, that is subadiabatic, $\ddot{z} < 0$, the parcel decelerates, and it tends to return to its original position. The atmosphere is statically stable. In the simplified case described by (3.34) the motion is oscillatory around $z = 0$, but in reality friction would damp the motion.

b. Potential Temperature

The potential temperature is the temperature which would result if air were brought adiabatically to a standard pressure level $p_0 = 1000$ mb. For an adiabatic process (3.10e) with (3.8) can be integrated to yield Poisson's equation, which allows the definition of the potential temperature θ as follows

$$\theta = T\left(\frac{p_0}{p}\right)^{\kappa} \tag{3.35}$$

where p is in mb for $p_0 = 1{,}000$ and where for moist air $\kappa = R_d(1 - 0.23q)/c_{pd}$, approximately. Differentiation of (3.35) yields

$$\frac{\partial \theta}{\partial z} = \frac{\theta}{T}\left(\frac{\partial T}{\partial z} + \Gamma_d\right) \tag{3.36}$$

if the effect of q is neglected. The potential temperature is conserved during an adiabatic displacement. Comparison of (3.36) with (3.34) shows that it can serve as a stability criterion for an atmosphere which is dry or which has a uniform moisture content q; when θ decreases with elevation, the atmosphere is unstable, and *vice-versa.* As a stability criterion for an atmosphere, in which $\partial q/\partial z$ is not zero, it is also useful to define, after Montgomery and Spilhaus (1941) the *virtual potential temperature.* It is the virtual temperature a substance would have if changed adiabatically from its actual state to the standard pressure p_0. Thus, equating (3.7) and (3.8) for $p = p_0$, one has

$$p_0 = R_d(1 + 0.61q)\,\rho_0\theta = R_d\rho_0\theta_V$$

where ρ_0 is the density after the adiabatic change, or

$$\theta_V = (1 + 0.61q)\,\theta. \tag{3.37a}$$

For a constant specific heat, on the basis of (3.9) and (3.35) this can also be written as

$$\theta_V = T_V\left(\frac{p_0}{p}\right)^{\kappa}. \tag{3.37b}$$

If κ denotes R_d/c_{pd}, instead of $R_d(1 - 0.23q)/c_{pd}$, (3.37b) yields the *potential virtual temperature*; it is the potential temperature of dry air at the same pressure and density. Clearly, for many practical purposes, it is unnecessary to distinguish between the virtual potential temperature and the potential virtual temperature. Differentiation of (3.37b) yields, to a very close approximation

$$\frac{\partial\theta_V}{\partial z} = \frac{\theta_V}{T_V}\left(\frac{\partial T_V}{\partial z} + \Gamma_d\right). \tag{3.37c}$$

Comparison with (3.34) shows that, when θ_V is constant, the atmosphere is statically neutral. When it decreases with elevation the atmosphere is statically unstable, and *vice versa.*

3.3. ATMOSPHERIC TRANSPORT OF WATER VAPOR

a. Conservation of Water Vapor

In the absence of phase transitions water vapor in air is a conservative scalar admixture. Any conservative substance admixed in a moving fluid is transferred relative to a fixed coordinate system, first through convection with the fluid; and second, through molecular motion superimposed on the convective motion of the fluid. The total specific mass flux is

$$\mathbf{F} = \rho_v\mathbf{v} + \mathbf{F}_m \tag{3.38}$$

where $\mathbf{v} = \mathbf{i}u + \mathbf{j}v + \mathbf{k}w$ is the velocity of the air, $\mathbf{i}$, $\mathbf{j}$, $\mathbf{k}$ and u, v, w, the unit vectors and the velocity components in the x, y, z directions, respectively, and $\mathbf{F}_m$ is the specific mass flux due to molecular diffusion. The latter type of transfer may be taken to be proportional to the local gradient of the water vapor density, in accordance with Fick's (see Chapter 2) law,

$$\mathbf{F}_m = -\,\kappa_v\nabla\rho_v \tag{3.39}$$

where κ_v is the molecular diffusivity of water vapor in air; for air at 20°C and 1 atm it is of the order of 0.25 cm^2 s^{-1}. Its variation with temperature is given in Table 3.6.

The equation of continuity of water vapor is in the absence of sinks or sources

$$-\nabla\cdot\mathbf{F} = \frac{\partial\rho_v}{\partial t}. \tag{3.40}$$

Similarly, the equation of continuity for the moist air is

TABLE 3.6
Diffusivities in air in (10^{-5} m² s⁻¹) at 1013.25 mb

Temperature (°C)	Kinematic viscosity (ν)	Thermal (κ_h)	Water vapor κ_v
−20	1.158	1.628	1.944
−10	1.243	1.747	2.082
0	1.328	1.865	2.230
10	1.418	1.994	2.378
20	1.509	2.122	2.536
30	1.602	2.250	2.694
40	1.700	2.388	2.852

Note: Values from List (1971), p. 395, converted to 1 atm.

$$-\nabla\cdot(\rho\mathbf{v}) = \frac{\partial\rho}{\partial t}. \tag{3.41}$$

Upon combining (3.38), (3.40) and (3.41) one obtains

$$-(\mathbf{v}\cdot\nabla)q - \rho^{-1}\nabla\cdot\mathbf{F}_m = \frac{\partial q}{\partial t}. \tag{3.42}$$

If ρ and κ_v are assumed to be spatially constant, substitution of (3.39) in (3.42) yields

$$\frac{\partial q}{\partial t} + (\mathbf{v}\cdot\nabla)q = \kappa_v\nabla^2 q \tag{3.43}$$

which is the basic equation of conservation of water vapor. Note that (3.43) is sufficiently general, so that it may be used to describe the conservation of any conservative and passive admixture or property of the air by merely replacing q by the concentration, as mass per unit total (i.e., contaminated) mass of air, and κ_v by the molecular diffusivity of that admixture.

With suitable boundary conditions and a known velocity field $\mathbf{v}$, it might appear that it should be possible to solve (3.43) for q to study the transport of water vapor. As will be discussed below, among these boundary conditions one often has at or near the surface, $z = 0$, one of the following three: the value of q, the value of the specific flux of water vapor, or an energy budget equation relating the specific flux of water vapor to some other energy flux terms.

Unfortunately however, in practice (3.43) is not directly applicable, since the flow of the atmosphere is almost invariably turbulent. This means that the description of the velocity field and also the content of water vapor, or other admixtures, at any given point in time and space is practically impossible, and that it can only be accomplished in a statistical sense.

The simplest and probably most important statistic is the mean. The equation for the mean specific humidity, that is the first moment, can be obtained as follows. In accordance with the common convention, introduced and discussed by Reynolds (1894) for the velocity, the dependent variables are first decomposed into a mean and turbulent fluctuation, namely $u = \bar{u} + u'$, $v = \bar{v} + v'$, $w = \bar{w} + w'$ and $q = \bar{q} + q'$. After applying the customary time-averaging over a suitable time period and using the equation of continuity (3.48), one obtains from (3.43)

$$\frac{\partial \bar{q}}{\partial t} + \bar{u}\frac{\partial \bar{q}}{\partial x} + \bar{v}\frac{\partial \bar{q}}{\partial y} + \bar{w}\frac{\partial \bar{q}}{\partial z}$$
$$= -\left[\frac{\partial}{\partial x}(\overline{u'q'}) + \frac{\partial}{\partial y}(\overline{v'q'}) + \frac{\partial}{\partial z}(\overline{w'q'})\right] + \kappa_v \nabla^2 \bar{q}. \tag{3.44}$$

The terms on the left hand side represent the rate of change in mean specific humidity observed when following the mean motion of the air. The cross correlations on the right-hand side may be called Reynolds fluxes by analogy with the Reynolds stresses [cf., (3.62)], and they represent the diffusive flux components due to turbulent motion. The last term is the convergence of the transfer rate by molecular diffusion.

Equations of higher moments can also be useful in the study of atmospheric water vapor transport. These equations can be derived from the equation of the specific humidity fluctuations. The equation of the fluctuations, obtained by subtracting (3.44) from the decomposed form of (3.43), is

$$\frac{\partial q'}{\partial t} + \bar{\mathbf{v}}\cdot\nabla q' + \mathbf{v}'\cdot\nabla\bar{q} + \mathbf{v}'\cdot\nabla q' - \nabla\cdot(\overline{\mathbf{v}'q'}) = \kappa_v \nabla^2 q'. \tag{3.45}$$

Equations for various second moments can be obtained by multiplying (3.45) by u', v', q', etc., respectively, and then averaging. In the case of the mean square specific humidity fluctuation, $\bar{m}$, this is

$$\frac{\partial \bar{m}}{\partial t} + \bar{\mathbf{v}}\cdot\nabla\bar{m} = -\overline{q'\mathbf{v}'}\cdot\nabla\bar{q} - \overline{\mathbf{v}'\cdot\nabla m} + \kappa_v\overline{q'\nabla^2 q'} \tag{3.46}$$

where $m = (q')^2/2$. Making use of the equation of continuity for the velocity fluctuations (3.49), one can also write

$$\frac{\partial \bar{m}}{\partial t} + \bar{\mathbf{v}}\cdot\nabla\bar{m} = -\overline{q'\mathbf{v}'}\cdot\nabla\bar{q} - \nabla\cdot[\overline{\mathbf{v}'m} - \kappa_v\nabla\bar{m}] - \kappa_v\overline{\nabla q'\cdot\nabla q'}. \tag{3.47}$$

The left-hand side represents, again, the rate of change in m observed when following the motion of the air. The first term on the right represents the production of mean square humidity fluctuation by the specific humidity gradient. The second term represents the turbulent and molecular transport of m; this can be explained (e.g., Tennekes and Lumley, 1972) by the observation that, if (3.47) is integrated over a sufficiently large control volume so that the turbulence on the boundaries is essentially zero, this term which is a divergence, yields zero as the volume integral can be reduced to a surface integral; thus this term represents only redistribution of m within the control volume. The molecular diffusion transport term is usually neglected. The last term represents ε_q, the dissipation or smearing out of humidity fluctuations by molecular diffusion. Equation (3.47) is applied in the measurement of evaporation by means of the dissipation method (see Chapter 8).

As another example, the equation for the turbulent vapor flux component, $\overline{w'q'}$, can be obtained by multiplying (3.45) by w', similarly multiplying (3.63) for w' by q', adding the two equations and then averaging. Just like (3.47) the resulting equation contains production, transport and dissipation terms. This equation and its analog for sensible heat transfer $\overline{w'\theta'}$ have been discussed by, among others, Donaldson (1973), Launder (1975), and Warhaft (1976). Although it is useful in higher order modeling (see Chapter 7), its detailed coverage is beyond the scope of the present treatment.

b. Other Conservation Equations

Clearly, the equations of water vapor conservation, such as (3.44) and (3.45) or those derived from them, such as (3.47), do not suffice to determine the specific humidity or its transfer, unless information is available on the flow field. This requirement necessitates the introduction of additional equations; these are an equation of state, such as (3.8), and equations of conservation of bulk air mass, of momentum and of energy. For easy reference later on these equations are given in what follows.

At this point it is useful to introduce an approximation that is usually attributed to Boussinesq and that greatly simplifies the formulation of these equations. It involves the assumption that the compressibility of the fluid is negligible with respect to inertia effects, but that density changes with respect to gravity, affecting the buoyancy of the air, must be considered. In other words, density changes resulting from pressure changes are negligible but not those resulting from temperature and specific humidity changes, when multiplied by g the acceleration due to gravity.

Conservation of Bulk Mass

The equation of continuity in terms of the mean velocity of an incompressible fluid, such as air at velocities well below that of sound, is obtained by averaging the Reynolds-decomposed (3.41) for constant ρ;

$$\nabla \cdot \bar{\mathbf{v}} = 0. \tag{3.48}$$

Subtraction of (3.48) from the decomposed (3.41) for constant ρ yields the equation of continuity for the fluctuations

$$\nabla \cdot \mathbf{v}' = 0. \tag{3.49}$$

Conservation of Momentum

The equations of motion are the Navier–Stokes equations for an incompressible fluid with constant viscosity but with the inclusion of a term arising from the acceleration of the reference system due to the earth's rotation. If the z-axis is the vertical, one has

$$\frac{\partial \mathbf{v}}{\partial t} + (\mathbf{v} \cdot \nabla)\mathbf{v} = -g\mathbf{k} - \frac{1}{\rho}\nabla p + \nu \nabla^2 \mathbf{v} - 2\Omega \times \mathbf{v}. \tag{3.50}$$

The last term on the right represents the Coriolis acceleration resulting from the earth's rotation (e.g., Rossby, 1940); Ω is the angular frequency of rotation vector in a right-handed system, which may be written as

$$\Omega = \omega[(\cos\alpha \cos\phi)\mathbf{i} + (\sin\alpha \cos\phi)\mathbf{j} + (\sin\phi)\mathbf{k}]$$

if ϕ is the latitude or the angle the horizontal plane makes with the axis of rotation of the earth, and α the angle between the x-axis and north; ω is the angular speed of rotation.

Even in an atmosphere in hydrostatic equilibrium, the pressure, the density, the temperature, and the specific humidity vary with elevation; therefore it is convenient (Landau and Lifshitz, 1959) to decompose these variables in a static part (undisturbed or reference state) and a dynamic part as follows

$$p = p_S + p_D, \qquad \rho = \rho_S + \rho_D, \qquad T_V = T_{VS} + T_{VD} \tag{3.51}$$

where the D-subscript components are supposed to be small deviations from the static parts. The undisturbed state is horizontally uniform; it satisfies the hydrostatic law

$$-\nabla p_S = \rho_S g\mathbf{k} \tag{3.52}$$

the equation of state

$$p_S = \rho_S T_{VS} R_d \tag{3.53}$$

and its virtual temperature lapse rate is dry-adiabatic

$$-\frac{\partial T_{VS}}{\partial z} = \Gamma_d. \tag{3.54}$$

This is a linear profile

$$T_{VS} = T_{VSr}[1 - (z - z_r)\Gamma_d/T_{VSr}] \tag{3.55}$$

where T_{VSr} is the known T_{VS} at reference level z_r. On account of (3.54) the gradient of the dynamic part is approximately equal to that of θ_V given in (3.37c).

$$\frac{\partial T_{VD}}{\partial z} = \frac{\partial \theta_V}{\partial z}. \tag{3.56}$$

In terms of actual measurements the dynamic part of the virtual temperature is, by virtue of (3.9), (3.51), and (3.55),

$$T_{VD} = T(1 + 0.61q) + \Gamma_d z + \text{const} \tag{3.57}$$

where the constant, obtainable from (3.55), is zero when the reference level is at 1000 mb.

According to the Boussinesq assumption, the density does not depend on the pressure but only on temperature and humidity; hence since ρ_D and T_{VD} are small, one has $\rho = \rho_S + (\partial\rho/\partial T_V)(T_V - T_{VS})$, so that (3.8) the equation of state becomes to a close approximation,

$$\rho_D = -\rho_S T_{VD}/T_{VS}. \tag{3.58}$$

Also the pressure gradient in (3.50) can be approximated closely by

$$\frac{\nabla p}{\rho} = \frac{\nabla p_S}{\rho_S} + \frac{\nabla p_D}{\rho_S} - \frac{\rho_D}{\rho_S^2}\nabla p_S. \tag{3.59}$$

Thus, combining (3.59) with (3.52) and (3.58), and substituting in (3.50), one obtains the Navier–Stokes equations with Coriolis effect under the limitations of the Boussinesq assumption,

$$\frac{\partial \mathbf{v}}{\partial t} + (\mathbf{v} \cdot \nabla)\mathbf{v} = -\frac{1}{\rho_S} \nabla p_D + g\left(\frac{T_{VD}}{T_{VS}}\right)\mathbf{k} + \nu \nabla^2 \mathbf{v} - 2\Omega \times \mathbf{v}. \tag{3.60}$$

To describe the turbulent flow, the deviations from the reference state can be decomposed in a mean and a fluctuating part, as

$$p_D = \bar{p}_D + p'_D, \quad T_{VD} = \bar{T}_{VD} + \theta'_V \tag{3.61}$$

and as before $\mathbf{v} = \bar{\mathbf{v}} + \mathbf{v}'$. By virtue of (3.51), (3.56) and (3.61), it is clear that θ'_V is the fluctuation not only of T_{VD}, but also of T_V and θ_V. Thus decomposing (3.60) in this way and then averaging, one obtains the equations of the mean motion, i.e., the Reynolds equations to the Boussinesq approximation

$$\frac{\partial \bar{\mathbf{v}}}{\partial t} + (\bar{\mathbf{v}} \cdot \nabla)\bar{\mathbf{v}} + (\nabla \cdot \overline{\mathbf{v}')\mathbf{v}'} = -\frac{1}{\rho_S} \nabla \bar{p}_D + g\frac{\bar{T}_{VD}}{T_{VS}}\mathbf{k} + \nu \nabla^2 \bar{\mathbf{v}} - 2\Omega \times \mathbf{v}. \tag{3.62}$$

Subtraction of (3.62) from the decomposed (3.60) yields the equations for the fluctuating velocities

$$\frac{\partial \mathbf{v}'}{\partial t} + (\mathbf{v} \cdot \nabla)\mathbf{v}' + (\mathbf{v}' \cdot \nabla)\bar{\mathbf{v}} + (\mathbf{v}' \cdot \nabla)\mathbf{v}' - (\nabla \cdot \overline{\mathbf{v}')v'}$$
$$= -\frac{1}{\rho_S}\nabla p'_D + g\frac{\theta'_V}{T_{VS}}\mathbf{k} + \nu \nabla^2 \mathbf{v}' - 2\Omega \times \mathbf{v}'. \tag{3.63}$$

Just like (3.45), this equation is useful in the derivation of equations of higher moments.

The equation for the turbulent kinetic energy $\bar{e}_t$ of which (3.47) is the scalar analog, can be derived by dot multiplying (3.63) by $\mathbf{v}'$ and then averaging. Straightforward but somewhat lengthy operations finally yield

$$\frac{\partial \overline{e_t}}{\partial t} + \bar{\mathbf{v}} \cdot \nabla \overline{e_t} = -[\overline{\mathbf{v}'(\mathbf{v}'} \cdot \nabla)] \cdot \bar{\mathbf{v}} + \frac{g}{T_{VS}} \overline{w'\theta'_V}$$
$$- \nabla \cdot \left[\overline{\mathbf{v}'\left(e_t + \frac{p'_D}{\rho_S}\right)}\right] + \nu \overline{\mathbf{v}' \cdot (\nabla^2 \mathbf{v}')} \tag{3.64}$$

where $e_t = (u'^2 + v'^2 + w'^2)/2$. Note that the Coriolis effect has cancelled out in this derivation; this indicates that it does not contribute to the rate of change of turbulent energy. The first term on the right of (3.64) is a mechanical production term; it represents the energy transfer rate from the mean motion through the turbulent shear stresses. The second term represents the rate at which turbulent kinetic energy increases as a result of work by buoyancy forces; it is commonly referred to as the thermal production term. The third term is a transport term. It is the divergence of the turbulent fluxes of not just kinetic but total turbulent energy; as mentioned earlier for (3.47), since it is a divergence, this term represents the redistribution of energy from one place to another. The fourth term represents the effects of viscosity; it can readily (e.g., Hinze, 1959, p. 65) be shown that it can be decomposed as follows (tensor notation is used for convenience, repeated subscripts denoting summation over the indices $i, j = 1, 2, 3$)

$$\overline{\nu \mathbf{v}' \cdot \nabla^2 \mathbf{v}'} = \nu \frac{\partial}{\partial x_j} \overline{\left[u_i' \left(\frac{\partial u_i'}{\partial x_j} + \frac{\partial u_j'}{\partial x_i} \right) \right]} - \nu \overline{\frac{\partial u_i'}{\partial x_j} \left(\frac{\partial u_i'}{\partial x_j} + \frac{\partial u_j'}{\partial x_i} \right)}. \tag{3.65}$$

The first term on the right, which is a divergence, is a transport term of turbulent kinetic energy, resulting from viscous shear stresses of the turbulence; it is quite small and is usually neglected. The second term is ε the rate of dissipation per unit mass by the turbulence into internal heat.

Conservation of Energy

This principle applied to an incompressible fluid under the Boussinesq assumption can be shown to yield

$$\frac{\partial \theta}{\partial t} + (\mathbf{v} \cdot \nabla)\theta = \kappa_h \nabla^2 \theta - \frac{1}{\rho_s c_p} \nabla \cdot H_R \tag{3.66}$$

where κ_h is the thermal diffusivity (Table 3.6) and H_R is the radiative heat flux. The similarity of (3.64) with (3.43) shows that, except for the radiative transfer, the potential temperature can be considered a conservative property of the air, and $c_p\theta$ represents a measure of the sensible heat content of the air. The equation for the mean potential temperature $\bar{\theta}$ is obtained in the same way as (3.44), viz.

$$\frac{\partial \bar{\theta}}{\partial t} + (\bar{\mathbf{v}} \cdot \nabla)\bar{\theta} + (\nabla \cdot \overline{\mathbf{v}')\theta'} = \kappa_h \nabla^2 \bar{\theta} - \frac{1}{\rho_s c_p} \nabla \cdot H_R \tag{3.67}$$

where, again, the last term on the left represents the convergence of the Reynolds flux, which is the heat transport contributed by the turbulent motion.

c. Solution of the Transport Equations

The conservation equations presented above are not easy to solve. Firstly, it was noted that the equations for the mean quantities, i.e., the first moments, such as (3.44), (3.62) and (3.67), contain Reynolds fluxes, i.e., second moment terms. Similarly, the equations for second moments, such as (3.47) and (3.64), contain third moment terms, and so on. Thus, in general, any finite set of equations for moments of the turbulent fluctuations always has more unknowns than the number of equations, due to the presence of some higher order moments. This is the notorious problem of closure, which is inherent in all turbulent flow formulations based on Reynolds-decomposition, and which results from the nonlinearity of the equations. Secondly, daily experience shows that even the mean motion of the atmosphere involves phenomena of considerable complexity; hence, it seems an almost hopeless task to describe this mean motion and the distribution of the mean specific humidity by the solution of a number of partial differential equations, even if the closure problem did not exist.

Fortunately, however, it is possible to simplify the general problem formulated by the above conservation equations considerably, and still obtain some very meaningful results. This is accomplished, first, by considering the atmosphere nearest the surface as a steady boundary layer, and second, by the application of similarity principles to describe the turbulence. The boundary-layer concept, introduced by

Prandtl (1904), involves the assumption that the vertical scales of the problem are much smaller than the horizontal; this means that the vertical gradients are much larger than the horizontal. The application of similarity principles and semiempirical turbulence theory alleviates the closure problem, since it allows the substitution of second and higher order moments of the turbulent fluctuations by terms containing only mean and lower order variables, respectively. Probably the earliest attempt along this line was made by Boussinesq (1877) when he introduced the concept of eddy viscosity.

3.4. THE ATMOSPHERIC BOUNDARY LAYER

In the atmosphere the largest changes in wind, temperature and humidity usually take place in the vertical and very close to the surface. For this reason the air near the surface may be regarded as a boundary layer, a concept set forth by Prandtl (1904) for the momentum transport in the neighborhood of a solid wall. Accordingly, the horizontal scales of most problems are much larger than the vertical, so that the horizontal gradients and the vertical velocities are negligible as compared to the vertical gradients and the horizontal velocities.

The atmospheric boundary layer (ABL) can be defined as the lower part of the atmosphere where the nature and properties of the surface affect the turbulence directly. Under normal atmospheric conditions there are numerous factors that affect the major mass, momentum, and energy transport phenomena in the boundary layer. Nevertheless, for the present purpose, some useful results can be obtained by considering the ABL under the simplest conditions. These may be specified as steady motion, parallel to a uniform plane surface, intermediate between cyclonic and anticyclonic flow, that is driven by parallel equidistant straight isobars.

The governing equations can be simplified considerably for this situation. Equation (3.44), for the mean specific humidity, becomes simply

$$\kappa_v \frac{\partial^2 \bar{q}}{\partial z^2} - \frac{\partial}{\partial z}\overline{(w'q')} = 0. \tag{3.68}$$

The analogous equations of horizontal mean motion are obtained from the Reynolds Equations (3.62)

$$-\frac{1}{\rho}\frac{\partial \bar{p}}{\partial x} + f\bar{v} + \nu\frac{\partial^2 \bar{u}}{\partial z^2} - \frac{\partial}{\partial z}\overline{(w'u')} = 0, \tag{3.69}$$

$$-\frac{1}{\rho}\frac{\partial \bar{p}}{\partial y} - f\bar{u} + \nu\frac{\partial^2 v}{\partial z^2} - \frac{\partial}{\partial z}\overline{(w'v')} = 0 \tag{3.70}$$

where the gradients of the mean pressure are assumed constant in the horizontal; unless stipulated otherwise, the density ρ will from now on refer to the mean standard value in the boundary layer. The Coriolis parameter f represents the influence of the earth's rotation; it is defined as

$$f = 2\omega \sin \phi \tag{3.71}$$

where ω is the angular speed of rotation of the earth and ϕ is the latitude; at intermediate latitudes f is typically of the order of 10^{-4} s. The vertical equation of motion is often simplified to the hydrostatic equation. Finally, the analogous equation of heat transport in the stationary and horizontally homogeneous boundary layer is

$$\kappa_h \frac{\partial^2 \bar{\theta}}{\partial z^2} - \frac{\partial}{\partial z}(\overline{w'\theta'}) - \frac{1}{\rho c_p} \frac{\partial H_R}{\partial z} = 0. \tag{3.72}$$

In fully turbulent flow the molecular transport terms with viscosity ν and the molecular diffusivities κ_v and κ_h are orders of magnitude smaller than the Reynolds fluxes. From now on, unless specifically mentioned, it will be assumed that they are negligible.

To elucidate the structure of the boundary layer and to set the stage for the application of similarity concepts in the next chapter, some dynamic features are now briefly considered. It is assumed that outside the boundary layer lies the free atmosphere where the wind speed is that of a free stream, affected mainly by the pressure field and the rotation of the earth but very little by friction. In frictionless flow (3.69) and (3.70) become

$$v_g = \frac{1}{\rho f} \frac{\partial \bar{p}}{\partial x}, \qquad u_g = -\frac{1}{\rho f} \frac{\partial \bar{p}}{\partial y} \tag{3.73}$$

where by definition u_g and v_g are the x- and y-components of the geostrophic velocity **G**; it is the steady horizontal flow blowing along the isobars, and it results from the balance of horizontal pressure gradient and Coriolis force in the absence of frictional forces and centripetal and tangential accelerations. Although the assumptions leading to (3.73) for the free stream are rarely met, and the atmosphere is rarely barotropic, so that the pressure gradients may change with elevation, the geostrophic wind velocity is often considered a good approximation of the wind velocity outside the atmospheric boundary layer. The thickness of the boundary layer is of the order of 10^3 m, varying between approximately 500 and 2000 m (see Figure 3.1).

It is generally accepted (e.g., Csanady, 1967; Monin, 1970; Tennekes, 1973) that the structure of the atmospheric boundary layer is similar to the two-dimensional turbulent boundary layer, for example, as generated in a wind tunnel, in that both have a distinct inner and outer region. In the *outer region* or *defect layer* the flow is nearly independent of the nature of the surface and mainly determined by the free-stream velocity, whereas in the *inner region*, also called *wall*, *Prandtl* or *surface layer*, the flow is strongly affected by the nature of the surface. However, in the atmosphere the outer region is driven not only by pressure but also by the Coriolis force, as a result of the rotation of the earth. For this reason the atmospheric boundary layer is also often called the *Ekman layer*. Between the inner and outer region one can assume a region of overlap, sometimes called the *matched layer* or the *inertial sublayer*. It should be noted that this situation is fairly typical only for conditions which are not very different from neutral. Under strongly unstable conditions, the effects of pressure and Coriolis terms are small and the outer region is characterized by thermal convective turbulence which is often local and spasmodic. The outer region may then be referred to as *mixed* or *free convection* layer. Under unstable conditions the upper limit of the boundary layer typically indicated by an inversion, can be very variable, but it is on average higher than under neutral conditions. Under stable conditions

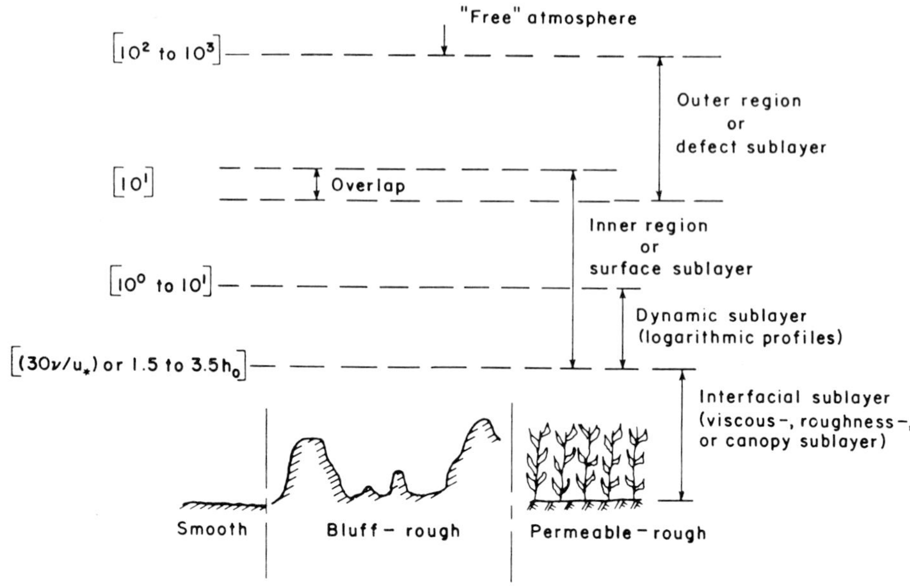

Fig. 3.1. Definition sketch showing orders of magnitude of the heights of the sublayers of the atmospheric boundary layer (ABL); h_0 is a typical height of the roughness obstacles; the (distorted) vertical scale is in meters.

the thickness of the boundary layer may range from only a few tens of meters to about 500 m, and in extreme cases the turbulence may be damped.

The surface sublayer may be defined as a fully turbulent region where the vertical turbulent fluxes do not change appreciably from their value at the surface. In the case of water vapor, (3.68) shows that in the absence of condensation the flux is indeed constant, namely

$$E = \rho \overline{w'q'} \tag{3.74}$$

where E is the evaporation rate at the surface. Similarly, for the sensible heat flux (3.72) shows that in the absence of radiative flux divergence,

$$H = \rho c_p \overline{w'\theta'} \tag{3.75}$$

where H is the surface heat flux. In the case of momentum transport the matter is not as simple. The surface sublayer is also defined sometimes as the layer near the ground in which the direction of the wind remains approximately constant with height, and in which the effects of the earth's rotation are negligible. Consequently, it is often assumed that (3.69) and (3.70) can be simplified to

$$-\overline{w'u'}\,[= (\tau_{xz}/\rho)] = u_*, \qquad -\overline{w'v'}\,[= (\tau_{yz}/\rho)] = 0 \tag{3.76}$$

where x is the direction of the mean wind near the surface, and where by definition

$$u_* = (\tau_0/\rho)^{1/2} \tag{3.77}$$

is the friction velocity and τ_0 is the shear stress at the surface. Actually, this assumption that the surface sublayer is a constant stress layer is not quite correct, as can be seen from (3.69) and (3.70). Since $\bar{v}$ is very small in the surface layer, from (3.69) with (3.73) it follows that

$$\left|\frac{\partial}{\partial z}(\tau_{xz}/\rho)\right| \simeq |f| v_g. \tag{3.78}$$

If H_c is the height at which the fractional deviation of the shear stress from its surface value is ε_τ, (3.78) with (3.77) yields approximately

$$\varepsilon_\tau \simeq \frac{|f| v_g H_c}{u_*^2} \tag{3.79}$$

Typical values are $|f| = 10^{-4}\, s^{-1}$ and $v_g/u_* = 12.5$ [$= (A/k)$ in (4.77)]. Hence, for $u_* = 0.3$ m s^{-1} and $H_c = 50$ m one obtains $\varepsilon_\tau = 20$ percent. In many situations, $\overline{u'w'}$ cannot be measured more accurately than 10 to 20 percent so that from the practical point of view, the assumption of a constant stress layer is sufficiently accurate for the lowest few tens of meters above the surface. From the theoretical point of view, however, it would seem that deviations of 20 percent at 50 m or 30 percent at 75 m are not negligible. In the development of those similarity models for the surface sublayer, which are based on mixing length- or K-theory, the requirement of a strictly constant shear, i.e. (3.76), is often included. Therefore, it has been puzzling that some of the results obtained with such K-theory models have been found to be valid up to elevations well above the, say, 1 percent or 5 percent constant stress layer. However, subsequently it has been found possible to formulate similarity hypotheses without the constant stress assumption, which yield the same results as K-theory. The idea of a constant stress layer is thus artificial, but it has been the starting point for some useful theoretical derivations. A detailed discussion (cf., Tennekes, 1973) would lead too far here, but let it be assumed that for practical purposes the thickness of the surface sublayer is of the order of approximately 50 to 100 m; this accords with the criterion $(z/\delta) < 0.10$ for the wall layer observed in non-rotating stratified flows over flat plates or in channels. In the steady horizontally homogeneous surface sublayer, in the absence of appreciable wind turning the equations for the mean square humidity fluctuations (3.47) can be written to a close approximation as,

$$\overline{q'w'}\frac{\partial \bar{q}}{\partial z} + \frac{\partial}{\partial z}\overline{(w'm)} + \varepsilon_q = 0 \tag{3.80}$$

in which ε_q is the last term of (3.47); similarily, the equation for the turbulent kinetic energy (3.64) can be written as

$$\overline{u'w'}\frac{\partial \bar{u}}{\partial z} - \frac{g}{T_{VS}}[\overline{w'\theta'} + 0.61\,\bar{T}\,\overline{w'q'}] + \frac{\partial}{\partial z}\left(\overline{w'e_t} + \frac{\overline{w'p'}}{\rho}\right) + \varepsilon = 0 \tag{3.81}$$

in which the reference virtual temperature T_{VS} and mean temperature $\bar{T}$ may usually be replaced by a reference temperature of the air T_a. The energy dissipation ε, given as the second term on the right of (3.65), can be closely approximated by

$$\varepsilon = \nu\overline{\frac{\partial u_i'}{\partial x_j}\frac{\partial u_i'}{\partial x_j}}. \tag{3.82}$$

The error can be shown to be $\partial^2(\overline{w'^2})/\partial z^2$ (cf., Hinze, 1959, p. 65) which is very small, since $\overline{w'^2}$ is approximately equal to u_*^2.

In general, under non-neutral conditions the air flow and the transfer of momentum are greatly affected by the transfer of sensible heat and water vapor, and *vice versa*. However, in the lower part of the surface sublayer it is found that water vapor and sensible heat may be considered as merely passive admixtures, and that the effects of density stratification resulting from humidity and temperature gradients are negligible. This lower region of the surface sublayer is referred to as the *dynamic sublayer*. Under neutral conditions, the whole surface sublayer behaves as a dynamic layer.

Finally, in the immediate vicinity of the surface, the turbulence is strongly affected by the structure of the roughness elements, or it is greatly damped by viscous effects; in most cases it is subjected to both effects, Thus the nature of the roughness elements must be considered, and the terms containing the viscosity ν and the molecular diffusivities κ_v and κ_h in the transfer equations are no longer negligible, as they are in the fully turbulent flow, higher up. The region nearest to the surface, where these effects are important, is referred to herein as the *interfacial* (*transfer*) *sublayer*. In the case of smooth flow it is often referred to as the *viscous sublayer*, and its thickness is of the order of 30 ν/u_*. Over a rough surface it may be referred to as a *roughness sublayer*, and its thickness is of the order of the mean height of the roughness obstacles h_0. When the roughness obstacles consist of vegetation which is more or less porous or permeable for the air stream, the interfacial sublayer may be referred to as the *canopy sublayer*. In air κ_v and κ_h are of the same order as ν (see Table 3.6) so the scaling lengths (and thicknesses of interfacial transfer sublayers) for momentum, water vapor and sensible heat are very similar.

CHAPTER 4

Mean Profiles and Similarity in a Stationary and Horizontally-Uniform ABL

In this chapter the so-called flux-profile relationships are presented for specific humidity, and other relevant quantities such as wind speed and temperature in the different sublayers of the ABL. Because of the above-mentioned difficulties of closure, these relationships are not derived by the solution of the transport equations; rather they are arrived at by invoking similarity, through the application of dimensional analysis. Thus, after the relevant physical quantities are identified from the governing equations or simply by inspection, they are organized into a reduced number of dimensionless quantities. Dimensional analysis only establishes the possible existence of a functional relationship between these dimensionless quantities; however, the function itself must usually be determined by experiment. Still, in some cases the functional form of the relationship may be inferred theoretically by means of a conceptual transport model or by applying a plausible closure assumption to the transport equations, and only some unknown constants need be determined experimentally. Especially in recent years numerous similarity models for the ABL have been proposed in the literature. This chapter does not present an exhaustive review but only the more important schemes that appear applicable in the determination of water vapor transport.

4.1. THE DYNAMIC SUBLAYER

The dynamic sublayer consists of the fully turbulent region, which is sufficiently close to the ground surface that the effects of the Coriolis and buoyancy forces due to density stratification are negligible; but it is far enough from the surface that both the viscosity of the air and the structure of the individual roughness elements also have no effect on the motion. Under diabatic conditions, i.e., with density stratification of the air, this layer may extend over only a few meters or less, whereas under conditions of neutral stability the dynamic sublayer occupies the entire surface sublayer.

a. The Logarithmic Profile

It has by now been well verified experimentally, and it is therefore almost accepted by definition, that in the dynamic sublayer the profiles of the mean wind speed, mean temperature, mean specific humidity and concentration of any other admixtures, provided they are released or absorbed uniformly at the surface, are all logarithmic functions of z. The logarithmic relationship was first established for the mean wind. Moreover, the analysis of the wind profile near the surface is prerequisite for the understanding of the turbulent transfer of water vapor as well. Therefore it is treated first.

Mean Wind Speed

The logarithmic wind profile law was developed in the late Twenties and it was introduced in meteorology by Prandtl (1932). One of the simplest derivations is that of Landau and Lifshitz (1959) which was first put forth in the 1944 edition of their book (Monin and Yaglom, 1971). The approach is based on dimensional analysis and it consists of noting that in plan-parallel flow an increase of the mean velocity in the z-direction, $(d\bar{u}/dz)$, is evidence of a downward momentum flux and a sink at the surface. Thus, the mean velocity gradient in a fluid of density, ρ, is determined by the shear stress at the wall, τ_0, and the distance from the wall, z. These variables can be combined into a single dimensionless quantity as follows

$$\frac{u_*}{z(d\bar{u}/dz)} = k \tag{4.1}$$

where u_* is defined in (3.77). Experimentally, this combination k is found to be approximately invariant and it is referred to as von Kármán's constant. Its value, which is of the order of 0.40 is still the subject of some uncertainty. In the literature, experimental values have been reported as low as 0.35 (e.g., Businger *et al.*, 1971; Högström, 1974) and as high as 0.47 (e.g., Pierce and Gold, 1971). Nevertheless, at present it appears (e.g., Hicks, 1976a; Yaglom, 1977) that there is still no compelling reason to abandon the consensus value $k = 0.4$.

The logarithmic wind profile equation follows immediately from (4.1), viz.

$$\bar{u}_2 - \bar{u}_1 = \frac{u_*}{k} \ln\left(\frac{z_2}{z_1}\right) \tag{4.2}$$

where the subscripts refer to two levels within the dynamic sublayer. Alternatively, one can write

$$\bar{u} = \frac{u_*}{k} \ln\left(\frac{z}{z_{0m}}\right) \quad \text{for } z \gg z_{0m} \tag{4.3}$$

where z_{0m} is an integration constant whose dimensions are length; herein it is referred to as the momentum roughness parameter. Its value depends on the conditions at the lower boundary of the region of validity of (4.1). Graphically, it may be visualized as the zero velocity intercept of the straight line resulting from a semi-logarithmic plot of mean velocity data versus elevation in the dynamic sublayer.

When the average height or size of the roughness elements of the surface is much larger than (ν/u_*), the surface is called dynamically rough. For a rough surface the momentum roughness is commonly written as

$$z_{0m} = z_0 \tag{4.4}$$

where z_0 is referred to as surface roughness length. Except for flexible obstacles or water waves, the value of z_0 is theoretically independent of the flow, and only a function of the nature of the surface, that is the geometry, the size and the arrangement of the roughnesses. Some methods to estimate it for natural surfaces are discussed in Chapter 5.

In the case of rough surfaces there is some uncertainty concerning the reference level $z = 0$ as used in (4.1). Clearly for very sparsely placed roughness elements this level can be taken at the base of the roughnesses; on the other hand, for extremely densely placed elements, $z = 0$ in (4.1) should refer to the level of the tops of the elements. Hence, in most situations the zero level reference should be located at a height somewhere between the tops and the bases of the roughness obstacles. To minimize this difficulty, it is common practice to define $z = 0$ as the level of the bases of the roughness elements, and to allow for a shift in reference level for the coordinate used in the similarity formulation; accordingly, instead of (4.1) one has

$$\frac{u_*}{(z - d_0)(\mathrm{d}\bar{u}/\mathrm{d}z)} = k \tag{4.5}$$

or, upon integration, instead of (4.3)

$$\bar{u} = \frac{u_*}{k} \ln\left(\frac{z - d_0}{z_{0m}}\right) \tag{4.6}$$

where d_0 is called the (zero-plane) displacement height. The idea of a displacement height, as given here, appears to have originated with Paeschke (1937).

Under conditions of low wind speeds over water, snow, ice, or salt flats, when (ν/u_*) is not very small as compared to the heights of the protuberances of the surface, the flow can no longer be considered fully rough. The effect of viscosity must be considered and the roughness Reynolds number

$$z_{0+} = \frac{u_* z_0}{\nu} \tag{4.7}$$

is then an important parameter. Experimental studies by Nikuradse (1933) have shown that in the approximate range $2 > z_{0+} > 0.13$, z_{0m} is not a constant; instead, one has a functional relationship

$$(z_{0m}/z_0) = f(z_{0+}) \tag{4.8}$$

which Nikuradse determined experimentally (e.g., Schlichting, 1960, Fig. 20.21; Monin and Yaglom, 1971; Fig. 28). Clearly, for $z_{0+} > 2$ it equals unity as required by (4.4).

At the lower extreme when $z_{0+} < 0.13$, the flow is independent of the nature of the surface, and it is referred to as hydrodynamically smooth. Extensive experiments, whose median is reproduced in Figure 4.1 have shown that (4.3) is valid for smooth flow with

$$z_{0m} = 0.135\, \nu/u_* \tag{4.9}$$

approximately, and provided $(u_* z/\nu) \equiv z_+ > 30$; Figure 4.1 shows that this is equivalent with the condition that

$$\bar{u}/u_* > 13.5 \tag{4.10}$$

approximately.

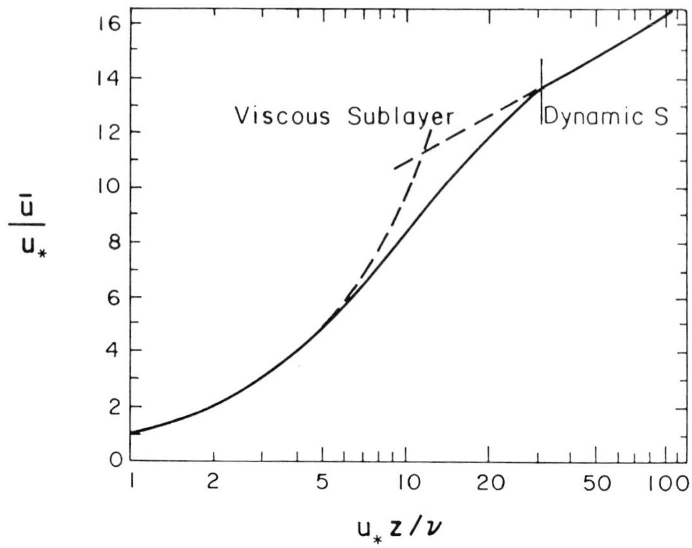

Fig. 4.1. The mean velocity profile in turbulent flow over a smooth surface as observed experimentally. The solid curve describes (4.123) for $z_+ < 5$, and (4.3) with (4.9) for $z_+ > 30$.

Mean Specific Humidity

The dimensional approach used to derive (4.1) and (4.5) can now be extended to derive expressions for the specific humidity profile and those of other quantities such as temperature, CO_2, etc. In the dynamic sublayer these are passive admixtures of the air and they do not affect the dynamics of the flow.

A decrease in specific humidity with elevation suggests that there is an upward water vapor flux. Accordingly, the rate of decrease of water vapor concentration with elevation in a fluid of density ρ is related to the flux of water vapor at the surface, $E = \rho(\overline{w'q'})$, and to the dynamics of the flow. The dynamics of the flow is governed by $d\bar{u}/dz$, τ_0 and $(z - d_0)$; since these three variables are inter-related by (4.5) only two of them, say τ_0 and $(z - d_0)$, are required to describe the flow. Five variables with four basic dimensions admit one dimensionless ratio. It can be written as

$$\frac{E}{u_*(z - d_0)\rho(d\bar{q}/dz)} = -k_v \tag{4.11}$$

which, just like (4.1) and (4.5), has also been found to be approximately invariant. In (4.11) k_v is the von Kármán constant for water vapor, which can also be written as

$$k_v = a_v k. \tag{4.12}$$

This ratio a_v of the von Kármán constants is, in fact, also the ratio of the eddy diffusivity for water vapor and the eddy viscosity under neutral conditions of the semi-empirical or K-theory of turbulence; it is thus the inverse of the turbulent Schmidt (cf. Prandtl) number under neutral conditions. Reynolds's analogy requires that a_v equal unity. Experiments have shown that a_v is probably equal to or slightly

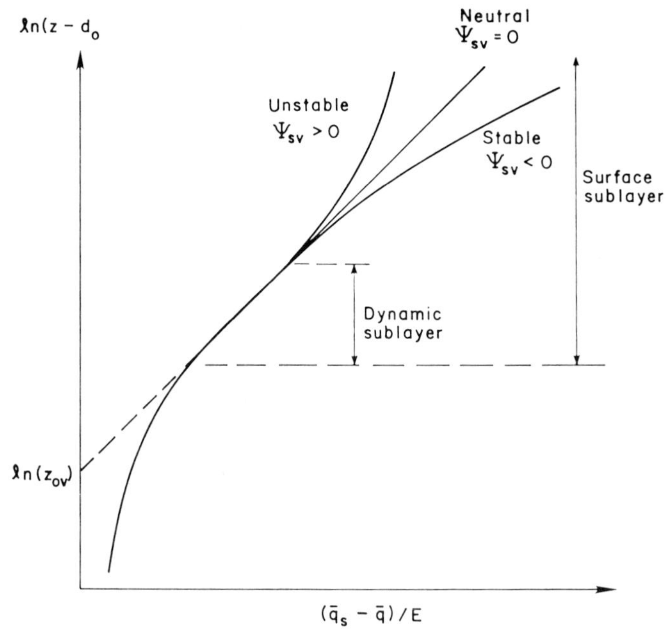

Fig. 4.2. Schematic illustration of the normalized mean humidity profile in the dynamic and in the surface sublayer (see (4.33′)). Analogous curves describe the wind speed and temperature profiles (see (4.34′) and (4.35′)).

larger than unity. For example, Pruitt *et al.* (1973) found that $a_v = 1.13$ (with $k = 0.42$). Dyer's (1974) review, however, suggests that $a_v = 1.0$ is still a reasonable assumption for practical purposes.

Integration of (4.11) between two arbitrary levels z_1 and z_2 within the dynamic sublayer yields

$$\bar{q}_1 - \bar{q}_2 = \frac{E}{a_v k u_* \rho} \ln\left(\frac{z_2 - d_0}{z_1 - d_0}\right). \tag{4.13}$$

If $\bar{q}_s$ is the value of $\bar{q}$ at the surface, the profile can also be written as

$$\bar{q}_s - \bar{q} = \frac{E}{a_v k u_* \rho} \ln\left(\frac{z - d_0}{z_{0v}}\right) \quad \text{for } z \gg z_{0v} \tag{4.14}$$

where z_{0v} is the water vapor roughness length; z_{0v}, which is merely an integration constant, can be visualized as the level above d_0 where $\bar{q}$ would assume its surface value if the logarithmic profile were extrapolated downward outside its actual range of validity (see Figure 4.2); it is the zero intercept of $(\bar{q}_s - \bar{q})$ data measured in the dynamic sublayer, and plotted versus $(z - d_0)$ on semi-log paper. The direct determination of z_{0v} by means of (4.14) has not been possible until now because there are very few profile data available on $\bar{q}$, or for that matter on any other scalar as well, which would be suitable for this purpose. Most of the presently available information on z_{0v} has been obtained indirectly from bulk transfer equations for scalar admixtures. This matter is dealt with in detail in Section 5.2 on the basis of data analyses presented in Section 4.4.

Although z_{0v} is related to z_{0m}, it is in fact quite different from it mainly because the transfer of momentum and that of a passive scalar admixture are not similar in the interfacial sublayer. This is due to the fact that in the immediate vicinity of the surface mass and heat transfer are controlled by molecular diffusion, whereas momentum transfer involves pressure forces as well. In the past it has sometimes been assumed that $z_{0v} = z_{0m}$, but this assumption can result in considerable error.

In the derivation of (4.11) the distance from the surface is taken as $(z - d_0)$ rather than z; this is done to account for the fact that it may not be easy to define the coordinate origin, especially when the surface is covered with densely placed roughness elements. However, when z is much larger than the heights of the roughness elements the exact value of d_0 is immaterial and it may be omitted from the formulation.

The parameter d_0 appearing in (4.11), (4.13) and (4.14) is evidently the same as that appearing in (4.5). Still, in the light of the different nature of z_{0m} and z_{0v}, it is not unreasonable to also suppose that the displacement height for water vapor should be different from d_0 of the wind profile. This might particularly be the case when the distribution of sources or sinks for evaporation from the roughness elements is markedly different or dissimilar from that of the momentum sinks. Unfortunately, until now this matter has not been investigated in detail. It should be noted though, that in the present similarity argument leading to (4.11), the term $(z - d_0)$ enters the formulation, because it is one of the variables governing the dynamics of the flow, and thus the momentum transfer as given by (4.5). It should also be noted that the logarithmic profile is valid only when the height z is considerably larger than the characteristic height of the roughness elements. But then, as mentioned, the exact value of d_0 is less important. The logarithmic profile equations are, in fact, relatively insensitive to the accuracy of d_0. Hence, as a practical working hypothesis, supported by the similarity argument, except perhaps for tall vegetation, the displacement height for water vapor or any other scalar admixture may probably be assumed to be the same as d_0 for momentum transfer.

In the early derivations it was customary to obtain the logarithmic humidity profile directly from the wind profile by means of Reynolds's analogy applied with an eddy diffusivity. Sverdrup (1937) was among the first to derive an equation similar to (4.11) but in terms of vapor pressure and with $k_v = k$ and $d_0 = -z_0$; moreover, in his integrated form (cf., (4.14)) he included a molecular diffusion layer near the surface with an assumed thickness proportional to (ν/u_*). He also reported that various experimental data sets taken over the ocean by Wüst (1937) and also those by R. B. Montgomery (cf., 1940) and F.L. Black all suggested that the humidity or vapor pressure is a linear function of the logarithm of height. Later, more exacting experimental confirmations of the logarithmic humidity profile were made by Pasquill (1949a) and Rider (1954). The logarithmic equation is the basis for the profile method of calculating evaporation under neutral conditions. Elimination of u_* between (4.13) or (4.14) and (4.2), (4.3) or (4.6), yields the well-known profile equations, as proposed variously by Thornthwaite and Holzman (1939), Sverdrup (1946), Pasquill (1949b) and Rider (1957).

Mean Potential Temperature

An argument similar to that for the specific humidity produces

$$\frac{H}{u_*(z - d_0)\rho c_p(\mathrm{d}\bar{\theta}/\mathrm{d}z)} = -a_h k \tag{4.15}$$

where H is the flux of sensible heat at the surface into the air and where a_h is the analog of a_v for sensible heat which has usually been found to be equal to or slightly larger than unity. Some experiments (e.g., Businger *et al.*, 1971) have yielded a_h as large as 1.35 with $k = 0.35$, and it appears (e.g., Yaglom, 1977) that the matter is far from settled. Nevertheless, for many practical applications, it is probably not unreasonable to assume that $a_h = 1.0$ with $k = 0.4$. Integration of (4.15) produces

$$\bar{\theta}_1 - \bar{\theta}_2 = \frac{H}{a_h k u_* \rho c_p} \ln\left(\frac{z_2 - d_0}{z_1 - d_0}\right) \tag{4.16}$$

or

$$\bar{\theta}_s - \bar{\theta} = \frac{H}{a_h k u_* \rho c_p} \ln\left(\frac{z - d_0}{z_{0h}}\right) \quad \text{for } z \gg z_{0h} \tag{4.17}$$

where z_{0h} is the roughness length for sensible heat. The comments made in connection with z_{0v} apply also here (see also Chapter 5).

It should be added that in view of the smallness of Γ_d, for most practical purposes in the dynamic sublayer the potential temperature $\bar{\theta}$ appearing in (4.15)–(4.17) may be replaced by the temperature $\bar{T}$(cf., (3.36)).

b. The Power Law Approximation

The logarithmic profile functions are transcendental, so that in the solution of certain turbulent transport problems they present serious mathematical difficulties. Therefore, as a close approximation, it is sometimes convenient to describe the mean profiles by simple power functions of elevation. For the mean wind speed, one has

$$\bar{u} = a\, z^m \tag{4.18}$$

where m and $a = (\bar{u}_1/z_1^m)$ are assumed to be constants for given conditions of atmospheric turbulence and of surface roughness; $\bar{u}_1$ is the mean wind speed at a reference level z_1. Equations such as (4.18) have been in use at least as early as 1876, ever since adequate wind measurements became available (Stevenson, 1880, in Chapter 2). Even though the power law does not appear to have a theoretical justification, it has been found that with proper values of the parameters it can describe experimental wind profiles rather well.

Equation (4.18) has been applied over smooth and rough surfaces, but since in nature most surfaces are rough, only this case is considered here. Through analogy with (4.6) for rough surfaces, (4.18) has also been written in the following form

$$\bar{u} = C_p u_* \left(\frac{z - d_0}{z_0}\right)^m \tag{4.19}$$

where C_p is a parameter. An equation similar to (4.19) was implicit in the work of

Prandtl and Tollmien (1924). The parameters C_p and m can be determined by fitting (4.19) to the more accurate (4.6) over the range of elevations which are of interest (e.g., Calder, 1949). As reviewed by Brutsaert and Yeh (1970b), most turbulent transfer studies point to $m = 1/7$ as a typical value, under neutral conditions, in the lower atmosphere. This is in accordance with Blasius's (e.g., Schlichting, 1960) empirical resistance law for turbulent flow in smooth pipes. However, m is not a universal constant and it depends weakly on z, z_0 and on the turbulence intensity. Various values have been proposed for C_p. Frost (1946) reasoned that (with $d_0 = 0$)

$$C_p = m^{-1} \tag{4.20}$$

by postulating that over a rough surface the mixing length is given by $z^{1-m} z_0^m$. From comparison of numerical experiments on evaporation with advection on the basis of (4.3) and (4.18), Brutsaert and Yeh (1070a, b) concluded that when m is close to (1/7) (with $d_0 = 0$), C_p may be taken close to 5.5 to 6.0. For values of m slightly different from (1/7) it was suggested that

$$C_p = (6/7m) \tag{4.21}$$

may be useful as a working equation over water surfaces.

Attempts have been made to relate power law parameters to atmospheric stability. However, since the power law is probably not a universal profile function, the applicability of any such relationships is limited. Deacon (1949) started from the hypothesis that $d\bar{u}/dz$, rather than $\bar{u}$, is a power function of z, but the resulting form for the mean profiles is not as convenient for mathematical analysis of diffusion problems as (4.18).

4.2. THE SURFACE SUBLAYER

a. The Mean Profiles

The lower part of the boundary layer above the roughness obstacles, where the flow is relatively unaffected by the viscosity and the structure of the individual obstacles, and by the Coriolis force, is commonly referred to as the surface sublayer (Figure 3.1). Thus, besides the factors governing the turbulent transfer in the dynamic sublayer, and used in (4.5), (4.11) and (4.15), also the stability of the atmosphere, that is, the effect of the buoyancy resulting from the effective vertical density gradient, must be considered. In similarity analysis, the objective is to formulate this buoyancy effect as a dimensionless variable.

There are several possible ways of doing this. Equation (3.34) shows that the acceleration per unit vertical displacement of a particle of air in an atmosphere at rest is given by the effective density stratification, which is to a close approximation

$$-\frac{g}{T_a}\left(\frac{\partial \theta}{\partial z} + 0.61\, T_a \frac{\partial q}{\partial z}\right) \tag{4.22}$$

where T_a is a mean reference temperature of the air near the surface. On the other hand, (3.64) and (3.81) show that the mean rate of contribution, by the fluctuating body – or buoyancy – forces to the turbulent kinetic energy in a horizontally homogeneous surface sublayer, is to a close approximation

$$\frac{g}{\rho T_a}\left[\left(\frac{H}{c_p}\right) + 0.61\ T_a E\right]. \tag{4.23}$$

Apparently, the density stratification has an effect on the turbulence only when there is a turbulent heat flux $\overline{w'\theta'}$ and/or vapor flux $\overline{w'q'}$. Equations (4.11) and (4.15) indicate that for given conditions of atmospheric stability this is the case in the surface sublayer, since the effect of buoyancy constitutes the only difference between the dynamic and the surface sublayer. These equations also show that for these same conditions (4.23) is proportional to (4.22). Hence, both (4.22) and (4.23) can serve equally well to construct the desired dimensionless variable that describes the effect of the atmospheric stability on turbulent transport.

The dimensionless ratios in (4.5), (4.11) and (4.15) have the variables $(z - d_0)$ and u_* in common. Accordingly, one can hypothesize that in stratified turbulent flow any dimensionless characteristic of the turbulence depend only on the following: the height above the virtual surface level, $z - d_0$; the shear stress at the surface, τ_0; the density ρ; and the production rate of turbulent energy resulting from the work of the buoyancy forces shown in (4.23). These four quantities, which can be expressed in terms of three basic dimensions, viz. time, length and air mass, can be combined into one dimensionless variable. This variable, which was proposed by Monin and Obukhov (1954) [for $d_0 = 0$], is

$$\zeta = \frac{z - d_0}{L} \tag{4.24}$$

where L, Obukhov's (1946) (Businger and Yaglom, 1971) stability length is defined traditionally by

$$L = \frac{-u_*^3\rho}{kg\left[\left(\frac{H}{T_a c_p}\right) + 0.61\ E\right]} \tag{4.25}$$

in which von Kármán's constant and the minus sign were originally introduced for convenience. Thus, L is positive for stable, negative for unstable, and infinitely large for neutral conditions. Originally $z = -L$ was interpreted as 'the height of the sublayer of dynamic turbulence'; in the turbulent kinetic energy equation (3.81), $z = -L$ could be considered to be the height at which the shear production is approximately equal to the buoyant energy production term, if the mean profiles were, in fact, logarithmic in the whole surface layer. Since this is not the case $-L$ is somewhat larger than this height. In the original formulation of Obukhov's length the turbulent flux term did not contain the effect of water vapor; but this is easily remedied by the inclusion of the term $0.61\ T_a E$.

With this hypothesis, the dimensionless characteristic of the turbulence (4.11), describing the flux-profile relationship of water vapor, is in the surface sublayer,

$$-\frac{ku_*(z - d_0)\rho}{E}\frac{\mathrm{d}\bar{q}}{\mathrm{d}z} = \phi_{sv}(\zeta) \tag{4.26}$$

in which ϕ_{sv} is supposedly a universal function of ζ. Similarly, for the mean wind gradient of (4.5) and the mean potential temperature gradient of (4.15), one has the analogous functions $\phi_{sm}(\zeta)$ and $\phi_{sh}(\zeta)$ as follows,

$$\frac{k(z - d_0)}{u_*} \frac{d\bar{u}}{dz} = \phi_{sm}(\zeta), \tag{4.27}$$

$$-\frac{ku_*(z - d_0)\rho c_p}{H} \frac{d\bar{\theta}}{dz} = \phi_{sh}(\zeta). \tag{4.28}$$

Clearly, in the dynamic sublayer, or under neutral conditions, when $|\zeta| \ll 1$ (but $z - d_0 \gg z_0$) these functions become $\phi_{sv} = a_v^{-1}$, $\phi_{sm} = 1$ and $\phi_{sh} = a_h^{-1}$. Equations (4.26)–(4.28) lead to the following profiles in the surface sublayer.

$$\bar{q}_1 - \bar{q}_2 = \frac{E}{ku_*\rho} [\Phi_{sv}(\zeta_2) - \Phi_{sv}(\zeta_1)], \tag{4.29}$$

$$\bar{u}_2 - \bar{u}_1 = \frac{u_*}{k} [\Phi_{sm}(\zeta_2) - \Phi_{sm}(\zeta_1)], \tag{4.30}$$

$$\bar{\theta}_1 - \bar{\theta}_2 = \frac{H}{ku_*\rho c_p} [\Phi_{sh}(\zeta_2) - \Phi_{sh}(\zeta_1)]. \tag{4.31}$$

The function Φ_{sv} is defined by

$$\Phi_{sv}(\zeta) = \int \phi_{sv}(x)\, dx/x \tag{4.32}$$

and Φ_{sm} and Φ_{sh} by analogous integrals.

The flux-profile relationships (4.29)–(4.31) can also be written as more obvious extensions of the logarithmic profiles to non-neutral conditions.

$$\bar{q}_1 - \bar{q}_2 = \frac{E}{a_v k u_* \rho} \left[\ln\left(\frac{\zeta_2}{\zeta_1}\right) - \Psi_{sv}(\zeta_2) + \Psi_{sv}(\zeta_1) \right], \tag{4.33}$$

$$\bar{u}_2 - \bar{u}_1 = \frac{u_*}{k} \left[\ln\left(\frac{\zeta_2}{\zeta_1}\right) - \Psi_{sm}(\zeta_2) + \Psi_{sm}(\zeta_1) \right], \tag{4.34}$$

$$\bar{\theta}_1 - \bar{\theta}_2 = \frac{H}{a_h k u_* \rho c_p} \left[\ln\left(\frac{\zeta_2}{\zeta_1}\right) - \Psi_{sh}(\zeta_2) + \Psi_{sh}(\zeta_1) \right], \tag{4.35}$$

where, similar to the suggestion by Panofsky (1963), the Ψ_s functions are defined as

$$\Psi_{sv}(\zeta) = \int_{(z_{0v}/L)}^{\zeta} [1 - a_v \phi_{sv}(x)]\, dx/x, \tag{4.36}$$

$$\Psi_{sm}(\zeta) = \int_{(z_{0m}/L)}^{\zeta} [1 - \phi_{sm}(x)]\, dx/x, \tag{4.37}$$

$$\Psi_{sh}(\zeta) = \int_{(z_{0h}/L)}^{\zeta} [1 - a_h \phi_{sh}(x)]\, dx/x. \tag{4.38}$$

The values of ζ_2 and ζ_1 in all the above equations must be large as compared to (z_0/L); over a very rough surface, such as a tall forest canopy, these relations cannot be expected to be valid if this condition is not satisfied.

Again, (cf., (4.14)] when the surface values, i.e., $\bar{q} = \bar{q}_s$, $\bar{u} = 0$, and $\bar{\theta} = \bar{\theta}_s$ are used, the profiles should be written as

$$\bar{q}_s - \bar{q} = \frac{E}{a_v k u_* \rho} \left[\ln\left(\frac{z - d_0}{z_{0v}}\right) - \Psi_{sv}(\zeta) \right], \tag{4.33'}$$

$$\bar{u} = \frac{u_*}{k} \left[\ln\left(\frac{z - d_0}{z_{0m}}\right) - \Psi_{sm}(\zeta) \right], \tag{4.34'}$$

$$\bar{\theta}_s - \bar{\theta} = \frac{H}{a_h k u_* \rho c_p}\left[\ln\left(\frac{z - d_0}{z_{0h}}\right) - \Psi_{sh}(\zeta)\right] \tag{4.35'}$$

where z_{0v}, z_{0m} and z_{0h} are the roughness lengths for water vapor, momentum and sensible heat, which are treated in detail in Chapter 5. Equation (4.33′) is shown schematically in Figure 4.2.

It should be noted that in the literature the ϕ_s functions of (4.26)–(4.28) have also been taken as functions of the Richardson number instead of ζ. The Richardson (1920) number is an alternative stability parameter, which may be written as

$$\text{Ri} = \frac{g}{T_a}\left[\frac{(\mathrm{d}\bar{\theta}/\mathrm{d}z) + 0.61\ T_a(\mathrm{d}\bar{q}/\mathrm{d}z)}{(\mathrm{d}\bar{u}/\mathrm{d}z)^2}\right] \tag{4.39}$$

if the effect of water vapor is included. The form of this parameter can be obtained by taking the ratio of the buoyancy production term (4.23) and the mechanical production term in the turbulent energy Equation (3.81), and by assuming Reynolds's analogy so that the ϕ_s's in (4.26)–(4.28) are assumed to be equal. This is not quite the case. When the production terms are left in their original form, the parameter is referred to as the flux Richardson number.

$$R_f = \frac{g}{T_a}\frac{\overline{w'(\theta' + 0.61\ T_a q')}}{(\overline{u'w'})\ \ (\mathrm{d}\bar{u}/\mathrm{d}z)} \tag{4.40}$$

The variables used in ζ, Ri and R_f are all related, so that theoretically any one of the three can be used to characterize the effect of the stability of the atmosphere on the turbulence. Ri has the advantage that it contains only gradients which can be determined experimentally; however, it varies with elevation. In addition, R_f has the disadvantage that it contains both turbulent convariances and the gradient of the mean wind. ζ is directly proportional to the elevation, since L which contains only surface fluxes is height independent. Most of the recent work dealing with the stability of the surface layer has tended to be in terms of L.

b. Some Flux-Profile Functions

The nature of the 'universal' functions ϕ_{sv}, ϕ_{sm} and ϕ_{sh} has been the subject of much theoretical and experimental research. Numerous functions have been proposed both in terms of ζ and of Ri; some were derived by semi-empirical arguments, such as interpolation between the known case for neutral conditions and that for extremely unstable conditions or free convection, already studied by Prandtl (1932), Obukhov (1946), and Priestley (1954); others were based entirely on experimental data. Reviews of these formulations have been given by Monin and Yaglom (1971) and Yaglom (1977) among others.

Unfortunately, however, the number of studies devoted specifically to the water vapor function ϕ_{sv} is relatively small compared to those on ϕ_{sm} and ϕ_{sh}. At present, it is generally assumed that

$$\phi_{sv}(\zeta) = \phi_{sh}(\zeta). \tag{4.41}$$

This is supported by the data used by Crawford (1965) and Dyer (1967). (See Figures

4.3 and 4.4.) Also, the ϕ_{sv} functions which Pruitt *et al.* (1973) fitted to their data are quite similar to the ϕ_{sh} fuctions proposed by others (cf., Yaglom, 1977). Thus, for most practical purposes (4.41) should be adequate. Nevertheless, at this point it should be noted that theoretical (Warhaft, 1976) and experimental (Verma *et al.*, 1978) evidence has been emerging, which especially for stable conditions, casts some doubt on (4.41). However, the issue is still unclear (Brost, 1979; Hicks and Everett, 1979) and more work will be needed to fully resolve it.

In addition to the uncertainty regarding (4.41), there appears to be a wide variety in the available functions ϕ_{sm} and ϕ_{sh}; still, calculations of the resulting fluxes from observed profiles are often relatively insensitive to variations in their functional form. Therefore, no attempt is made here to present a complete review. Only some typical functions are given which are consistent with available data, and which are suitable for flux computations.

Unstable Conditions

For $\zeta \leq 0$ the ϕ_s functions can be well described by an empirical formulation, suggested independently by Businger (1966) and Dyer (Businger *et al.*, 1971). Its general form is

$$\phi_{sv} = a_v^{-1}\,(1 - \beta_{sv}\zeta)^{-1/2}, \tag{4.42}$$

$$\phi_{sm} = (1 - \beta_{sm}\zeta)^{-1/4}, \tag{4.43}$$

$$\phi_{sh} = a_h^{-1}\,(1 - \beta_{sh}\zeta)^{-1/2}. \tag{4.44}$$

Several variations have been presented. At first, Dyer (1967) found that $\phi_{sv} = \phi_{sh} = (1 - 15\zeta)^{-0.55}$ with $k = 0.40$; subsequently Dyer and Hicks (1970) and Hicks (1976b) with $k = 0.41$ and Paulson (1970), Miyake *et al.* (1970), and Paulson *et al.* (1972) with $k = 0.40$ found that $\beta_{sv} = \beta_{sm} = \beta_{sh} = 16$ and $a_v = a_h = 1$ generally yield good agreement with profile and turbulent flux data. On the other hand, Businger *et al.* (1971) concluded from their measurements that $\beta_{sm} = 15$, $a_h = 1.35$ and $\beta_{sh} = 9$, but this result was accompanied by a proposed alteration of von Kármán's constant to $k = 0.35$; Högström's (1974) experiments confirmed these results. Smedman and Högström (1973) found that their humidity data could be described well by (4.42) with $k = 0.4$, $a_v = 1.0$ and $\beta_{sv} = 9$ up to an instability of $\zeta = -7.62$. Although the formulation of Businger *et al.* (1971) appears to be an anomaly, the matter is still the subject of considerable uncertainty (e.g., Yaglom, 1977) and better experimental data will be required to resolve it. Until these become available, $k = 0.4$ with

$$\phi_{sv} = \phi_{sh} = \phi_{sm}^2 = (1 - 16\zeta)^{-1/2} \quad \text{for } \zeta < 0 \tag{4.45}$$

should be adequate for practical purposes. Equation (4.45) can be compared with experimental data in Figures 4.3, 4.4, and 4.5.

Equations (4.36)–(4.38) can readily be integrated with (4.42)–(4.44) to yield the following profile functions for (4.33)–(4.35).

$$\Psi_{sv}(\zeta) = 2\ln\left[\frac{(1 + x^2)}{(1 + x_{0v}^2)}\right], \tag{4.46}$$

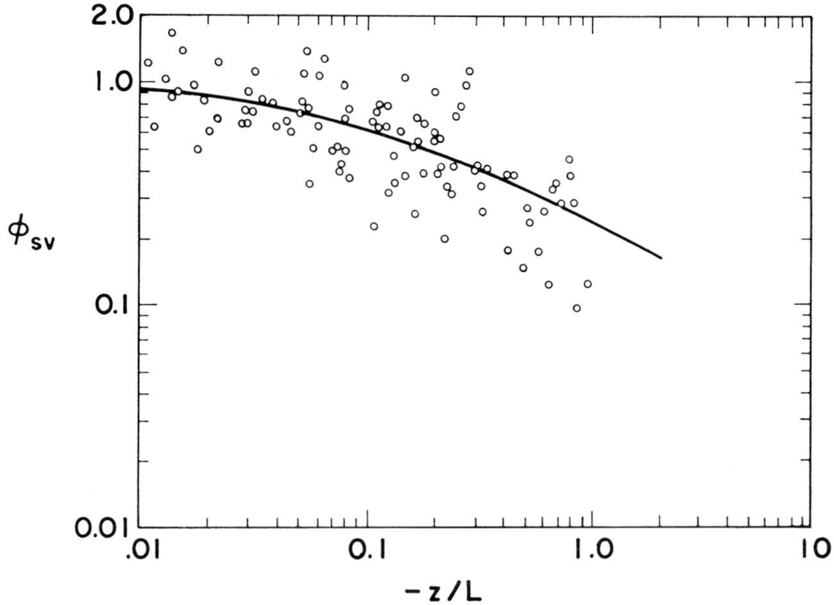

Fig. 4.3. Values of the similarity function for water vapor $\phi_{sv}=\phi_{sv}(\zeta)$ as calculated with $k = 0.4$ by Dyer (1967) from experimental data taken at Kerang in Australia. The curve represents (4.45).

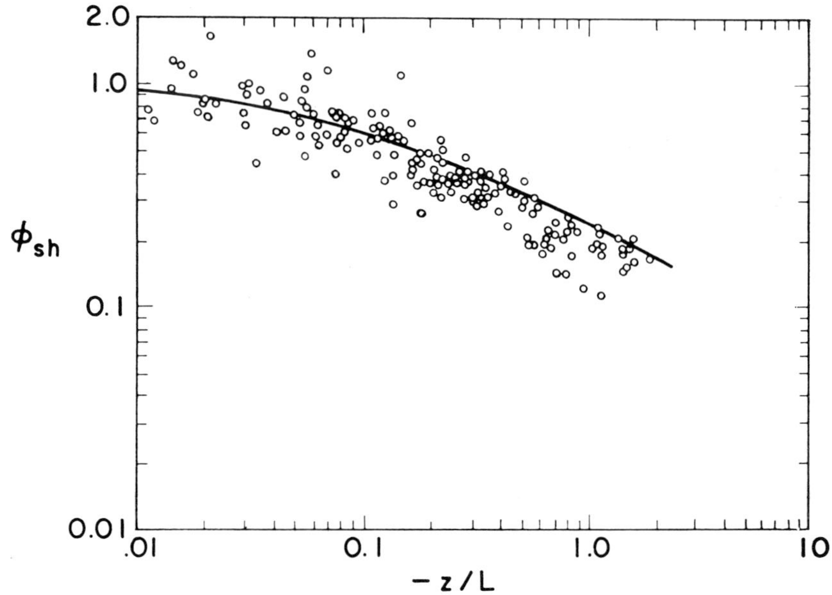

Fig. 4.4. Values of the similarity function for sensible heat $\phi_{sh}=\phi_{sh}(\zeta)$ as calculated with $k = 0.4$ by Dyer (1967) from experimental data taken at Kerang and at Hay in Australia. The curve represents (4.45).

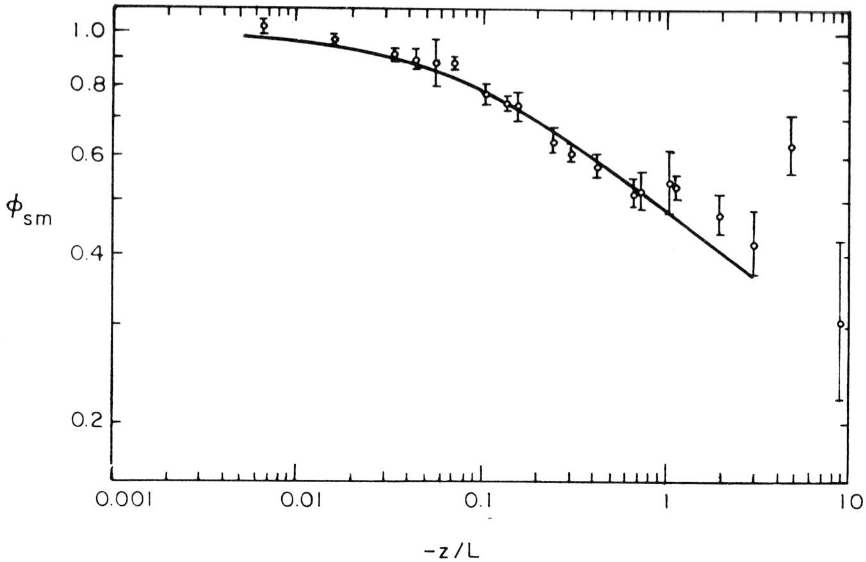

Fig. 4.5. Geometric mean values of the similarity function for momentum $\phi_{sm}=\phi_{sm}(\zeta)$ from experimental data at Hay and at Gurley in Australia. The curve represents (4.45) (adapted from Dyer and Hicks, 1970).

$$\Psi_{sm}(\zeta) = \ln\left[\frac{(1+x)^2(1+x^2)}{(1+x_{0m})^2(1+x_{0m}^2)}\right] - 2\arctan(x) + 2\arctan(x_{0m}), \quad (4.47)$$

$$\Psi_{sh}(\zeta) = 2\ln\left[\frac{(1+x^2)}{(1+x_{0h}^2)}\right], \quad (4.48)$$

where $x = (1 - \beta\zeta)^{1/4}$ in which β represents β_{sv}, β_{sm}, β_{sh}, in (4.46), (4.47) and (4.48), respectively, and where $x_{0v} = (1 - \beta_{sv}\zeta_{0v})^{1/4}$ with $\zeta_{0v} = (z_{0v}/L)$, and analogous definitions for x_{0m} and x_{0h}. Since the roughness lengths are normally much smaller than L, the lower limit in (4.36)–(4.38) is often (e.g., Paulson, 1970) taken as zero. The integral profile functions are then

$$\Psi_{sv}(\zeta) = 2\ln\left[\frac{(1+x^2)}{2}\right], \quad (4.49)$$

$$\Psi_{sm}(\zeta) = 2\ln\left[\frac{(1+x)}{2}\right] + \ln\left[\frac{(1+x^2)}{2}\right] - 2\arctan(x) + \frac{\pi}{2}, \quad (4.50)$$

$$\Psi_{sh}(\zeta) = 2\ln\left[\frac{(1+x^2)}{2}\right]. \quad (4.51)$$

If (4.45) is used, one has $x = (1 - 16\zeta)^{1/4}$.

Stable Conditions

One of the earliest forms for the ϕ_s functions, intended for near-neutral conditions, i.e., small $|\zeta|$, was obtained by Monin and Obukhov (1954) simply by a series expansion and retention of the first term only, viz.

$$\phi_{sv} = a_v^{-1}(1 + \beta_{sv}\zeta), \tag{4.52}$$

$$\phi_{sm} = (1 + \beta_{sm}\zeta), \tag{4.53}$$

$$\phi_{sh} = a_h^{-1}(1 + \beta_{sh}\zeta), \tag{4.54}$$

where β_{sv}, β_{sm} and β_{sh} are empirical constants. Clearly, with (4.26)–(4.28) these functions yield profile functions consisting of a logarithmic and a linear term. Accordingly, the functions of (4.33)–(4.35) are

$$\Psi_{sv}(\zeta) = -\beta_{sv}(\zeta - \zeta_{0v}), \tag{4.55}$$

$$\Psi_{sm}(\zeta) = -\beta_{sm}(\zeta - \zeta_{0m}), \tag{4.56}$$

$$\Psi_{sh}(\zeta) = -\beta_{sh}(\zeta - \zeta_{0h}) \tag{4.57}$$

in which the ζ_{0v}, ζ_{0m} and ζ_{0h} terms are often negligibly small for practical calculations.

At first it was hoped that the log-linear profile might be valid over a wide ζ range, but it was soon discovered (e.g., Taylor, 1960), in the case of (4.53) and (4.54), that they are applicable only for $(-\zeta) < 0.003$ under unstable conditions. Presently, there is some consensus that the log-linear profiles can be used to fit the data for moderately stable conditions when $\zeta < 1$. McVehil (1964) found, experimentally, that (4.53) and (4.54) are valid for $0 < \zeta < 1.8$ and that $\phi_{sm} = \phi_{sh}$ with $\beta_{sm} = \beta_{sh} = 7$ and $a_h = 1$. Webb (1970) concluded that the log-linear profiles are valid over the range $0 < \zeta < 1$ with $\phi_{sv} = \phi_{sm} = \phi_{sh}$ and $\beta_{sv} = \beta_{sm} = \beta_{sh} = 5.2$ for $k = 0.41$. Businger *et al.* (1971) found, over the same range of ζ, that (4.53) and (4.54) are valid; but with $k = 0.35$ the parameters were derived to be $\beta_{sm} = 4.7$, $\beta_{sh} = 6.35$ and $a_h = 1.35$.

While the log-linear function appears to be suitable to describe moderately-stable conditions when $\zeta < 1$, there is still no agreement on the values of its parameters and on the form of the ϕ_s functions for very stable conditions when $\zeta > 1$. Webb (1970) and Kondo *et al.* (1978) found that for $\zeta > 1$, ϕ_{sm} becomes a constant of the order of 6, whereas Hicks (1976b) inferred a value of about 8 around $\zeta = 10$ (see Figure 4.6). Theoretically (e.g., Monin and Yaglom, 1971, p. 437), under very stable conditions the ϕ_s functions should be proportional to ζ. A data analysis presented by Hicks (1976b) supported this and suggested that for $\zeta > 10$, $(d\bar{u}/dz)$ is approximately equal to $0.8\ u_*/kL$. Somewhat more disturbing is that, while the study of Businger *et al.* (1971) indicates that $\phi_{sh} < \phi_{sm}$, those of Hicks (1976b) and Kondo *et al.* (1978) suggest that $\phi_{sh} > \phi_{sm}$. All this illustrates the fact that even less is known about the flux-profile relationships under stable conditions than under unstable conditions. Often, however, under stable conditions the turbulent fluxes tend to be small, so that for routine flux calculations from observed profiles the exact formulation of the ϕ_s functions is not crucial. Hence, for the time being for practical flux computations, it is probably sufficiently accurate to assume $k = 0.4$ with

$$\phi_{sv} = \phi_{sm} = \phi_{sh}\begin{cases} = 1 + 5\zeta & \text{for } 0 < \zeta < 1 \\ = 6 & \text{for } \zeta > 1. \end{cases} \tag{4.58}$$

Equation (4.58) can be compared with experimental data in Figure 4.6.

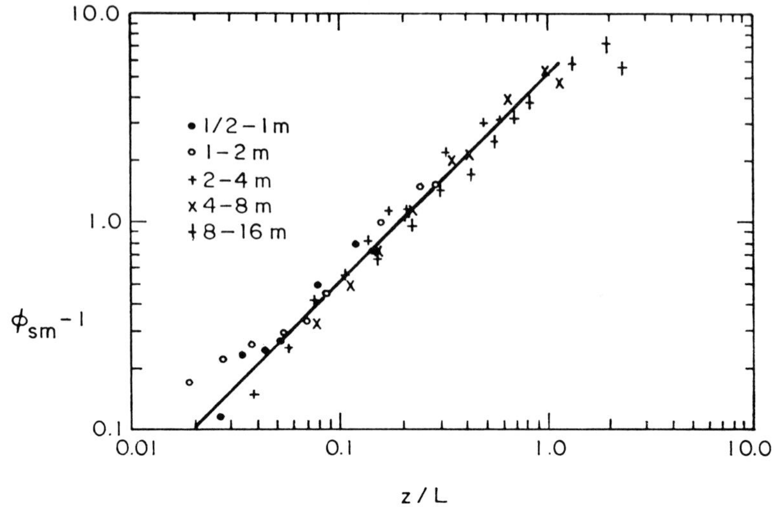

Fig. 4.6. The dependence of (ϕ_{sm} − 1) on ζ under stable conditions, as calculated from experimental data at Hay in New South Wales. The straight line represents the first of (4.58) (adapted from Hicks, 1976).

4.3. BULK PARAMETERIZATION OF THE WHOLE ABL

a. Similarity for the Mean Profiles in the Outer Sublayer

The surface sublayer occupies only the lowest 10 percent or so of the atmospheric boundary layer; higher up additional variables must be considered which govern the turbulent transport, and the similarity arguments of the surface sublayer are no longer adequate. One way to approach the problem above the surface sublayer is to simply make a formal extension of the Monin–Obukhov similarity model, and to hypothesize that the mean profiles can be expressed by equations similar to (4.26) (4.28), namely

$$-\frac{ku_*(z-d_0)\rho}{E}\frac{\mathrm{d}\bar{q}}{\mathrm{d}z}=\phi_{bv},\tag{4.59}$$

$$\frac{k(z-d_0)}{u_*}\frac{\mathrm{d}\bar{u}}{\mathrm{d}z}=\phi_{bmx},\tag{4.60}$$

$$\frac{k(z-d_0)}{u_*}\frac{\mathrm{d}\bar{v}}{\mathrm{d}z}=\phi_{bmy},\tag{4.61}$$

$$-\frac{ku_*(z-d_0)\rho c_p}{H}\frac{\mathrm{d}\bar{\theta}}{\mathrm{d}z}=\phi_{bh}.\tag{4.62}$$

However, here the universal ϕ functions depend not only on ζ but on a number of additional dimensionless variables composed of variables whose effect is not neglible at higher elevations. Several of the most important variables are discussed in what follows.

As is the case for wind-tunnel experiments over a flat plate, the thickness of the boundary layer, δ, is an important parameter. But for the atmosphere it is not always easy to define δ unambiguously and several length scales are available that provide a measure of this thickness.

Rotational (or Ekman-) Height Scale

One thickness scale results from the analysis of a steady, horizontally homogeneous, barotropic, and neutral planetary boundary layer, whose motion is described by (3.69) and (3.70) with (3.73). Inspection immediately suggests that, if u_* is used to scale the horizontal velocity terms, $(u_*/|f|)$ is the appropriate scaling length for the vertical coordinate z. Accordingly, one can define

$$\delta_r = K_r u_*/|f| \tag{4.63}$$

in which the subscript r denotes 'rotational', since the length involves the Coriolis parameter, and K_r is taken as a constant under neutral conditions. $K_r = 0.195$ was already obtained by Rossby (1932). In the similarity study of Kazanski and Monin (1960), it was assumed that $K_r = k$, but it has since generally been found that K_r should be smaller, namely somewhere between 0.15 and 0.30. As will be seen below, the exact determination of K_r is not very easy. The possibility that under non-neutral conditions K_r be a function of $u_*/(L|f|)$ has been discounted (e.g., Zilitinkevich and Deardorff, 1974).

Observed (or Inversion) Height Scale

In the earlier similarity models for a non-neutral atmosphere δ_r was considered as the only boundary-layer thickness scale. However, this scale has some drawbacks. In the real atmosphere the thickness of the atmosphere is often greatly affected by the history of the airmass and by other factors which may be unrelated to the internal boundary-layer dynamics. Among these complicated factors there is the diurnal heating, non-stationary motion, large-scale advection, and non-uniformity of temperature and humidity. Moreover, near the equator, where f approaches zero, the use of δ_r becomes quite meaningless. The inadequacy of δ_r as the only length scale was pointed out by Deardorff (1972); his numerical study showed that even under slightly unstable atmospheric conditions near the surface, δ should not be determined by (4.63), but rather by direct observation. This means that, whenever possible, δ should literally be taken as the level, up to which dynamic and thermodynamic effects resulting from the earth's surface are clearly manifested in the profiles of mean wind, temperature and specific humidity. Since this is usually related to the height of the inversion, this second measure of the boundary-layer thickness is denoted by δ_i. Examples of temperature profiles from which δ_i can be determined are shown in Figure 4.7. In practice (e.g., Garratt and Francey, 1978), under unstable conditions δ_i can be taken as the top of the 'mixed' layer, that is the depth of convective mixing above the surface. In many situations this mixed layer is capped by an inversion, so that δ_i may then be taken conveniently as the height of the main inversion base. However, this is not generally the case, and especially over land in the morning during the breakdown period of a nocturnal inversion the thickness of the surface convection layer is evidently a better scale. Under stable conditions δ_i can be taken as the thick-

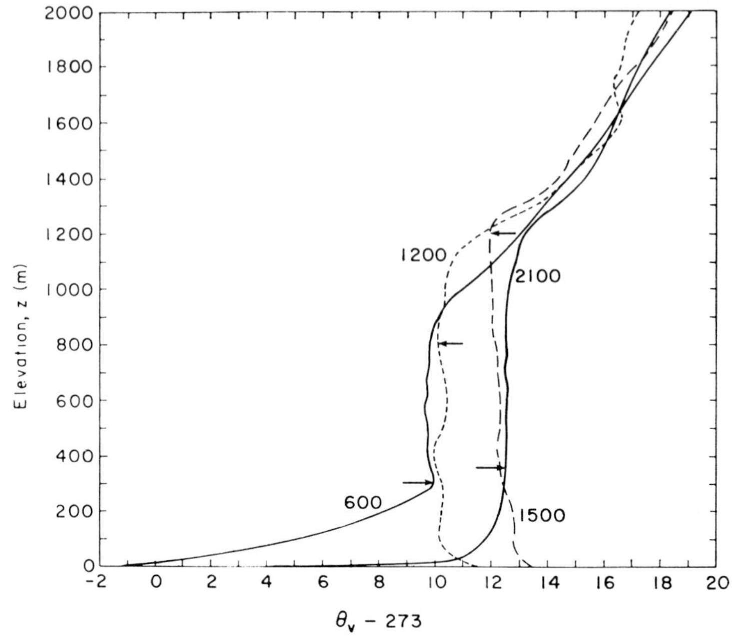

Fig. 4.7. Example of profiles of the virtual potential temperature in and above the atmospheric boundary layer. The estimated values of the height δ_i of the boundary layer are indicated by arrows; the numerical values on the profiles indicate times of day. The observations were made by Clarke *et al.* (1971) on July 16, 1967 (Day 33) at Hay, N.S.W., Australia.

ness of the inversion layer at the surface, indicated by the height to which significant cooling has extended in the $\bar{\theta}$ profile, or by the height of the lowest maximum (nocturnal jet) in the $\bar{u}$ profile. It should be stressed that the exact determination of δ_i for a real atmosphere is not always easy. Under unstable atmospheric conditions the thickness of the planetary boundary layer often changes quite drastically and rapidly; this intermittency is probably a manifestation of convective plumes, causing entrainment into the overlying inversion, and of the breaking of wave-like structures.

In the past few years the consensus has been growing that, except under near-neutral conditions, δ_i is the more important length scale. The main reason for this is that the similarity functions in which δ_i is used to represent the thickness of the boundary layer exhibit less scatter with most experimental data than those in which δ_r is used. The δ_i scale also has the advantage that it allows the direct determination of any external variable aloft at the proper altitude, and that it contains neither the usually unknown quantity u_* nor the Coriolis parameter f, which becomes meaningless near the equator.

Baroclinicity

Another characteristic of a real atmosphere is, that the pressure gradient may change with elevation. It appears that this vertical gradient of the horizontal pressure gradient, that is the geostrophic wind shear or the baroclinicity, affects the turbulence structure, as revealed by the mean wind profile (e.g., Clarke and Hess, 1974). The baroclinicity is related to the horizontal temperature gradients by the thermal wind equations (e.g., Haltiner and Martin, 1957), which may be written approximately as

$$\frac{\partial u_g}{\partial z} = -\frac{g}{fT}\frac{\partial T}{\partial y}, \tag{4.64}$$

$$\frac{\partial v_g}{\partial z} = \frac{g}{fT}\frac{\partial T}{\partial x}, \tag{4.65}$$

where T is the absolute temperature. These equations show that baroclinicity is evidence of thermal inhomogeneity in the horizontal, or *vice versa*. The baroclinicity is not likely to be constant with z, but this change is often quite small, so that to a first approximation it may probably be ignored.

There are several ways of organizing $z - d_0$, L and the above four variables into five dimensionless parameters. When the Coriolis effect is felt to be predominant, under nearly steady near-neutral conditions, one might choose $\eta_r = (z - d_0)/\delta_r$, $\mu_r = \delta_r/L$, $\nu_0 = \delta_i/\delta_r$, $\beta_{xr} = f^{-1}\, \partial u_g/\partial z$ and $\beta_{yr} = f^{-1}\, \partial v_g/\partial z$. The parameters η_r and μ_r have been introduced by Kazanski and Monin (1960), ν_0 by Wyngaard *et al.* (1974) and β_{xr}, β_{yr} by Yordanov and Wippermann (1972). (Instead of β_{xr} and β_{yr}, Arya and Wyngaard (1975) suggested the use of $M_x = f\,\delta_i\beta_{xr}/u_*$ and $M_y = f\,\delta_i\beta_{yr}/u_*$.) With these variables the similarity function of (4.59) is

$$\phi_{bv} = \phi_{bv}(\eta_r, \mu_r, \nu_0, \beta_{xr}, \beta_{yr}) \tag{4.66}$$

and the same for ϕ_{bmx}, ϕ_{bmy} and ϕ_{bh}.

A second combination, which is more suitable when the observed height δ_i is the predominant ABL thickness parameter, might consist of the following, $\eta_i = (z - d_0)/\delta_i$, $\mu_i = \delta_i/L$, $\nu_0 = \delta_i/\delta_r$, $\beta_{xi} = -(g/T)(\delta_i/K_r u_*)^2\,(\partial T/\partial y) = \beta_{xr}\nu_0^2$ and $\beta_{yi} = \beta_{yr}\nu_0^2$. The parameters η_i and μ_i were introduced by Zilitinkevich and Deardorff (1974); β_{xi} and β_{yi}, which were proposed by Brutsaert and Mawdsley (1976) can be obtained from the horizontal temperature gradient and they do not contain the Coriolis parameter. With these parameters the similarity function of (4.59) is

$$\phi_{bv} = \phi_{bv}(\eta_i, \mu_i, \nu_0, \beta_{xi}, \beta_{yi}). \tag{4.67}$$

Because (4.66) and (4.67) involve the same primitive variables, the two formulations are physically equivalent. Besides these two, there are, of course, other possible dimensionless combinations of the arguments of ϕ_{bv}. Moreover, it is conceivable that other variables should be considered; however, this will require further advances in the understanding of the structure of the atmospheric boundary layer.

In the outer reaches of the boundary layer, well above the surface layer, it is convenient to use reference values of $\bar{q}$, $\bar{u}$, $\bar{v}$ and $\bar{\theta}$ at the outer edge of the boundary layer, rather than at some other level. Accordingly, integration of (4.59) through (4.62) yields the following defect laws for the mean profiles

$$\bar{q}_\delta - \bar{q} = \frac{E}{ku_*\rho}\,\Phi_{bv}, \tag{4.68}$$

$$\bar{u} - \bar{u}_\delta = \frac{u_*}{k}\,\Phi_{bmx}, \tag{4.69}$$

$$\bar{v} - \bar{v}_\delta = \frac{u_*}{k} \Phi_{bmy}, \tag{4.70}$$

$$\bar{\theta}_\delta - \bar{\theta} = \frac{H}{k u_* \rho c_p} \Phi_{bh}, \tag{4.71}$$

where the Φ functions depend on the parameters of (4.66) or on those of (4.67) depending on which may be preferable. These functions should reduce to zero at the top of the boundary layer where $\eta = 1$ (since $d_0 \ll \delta$), $\bar{q} = \bar{q}_\delta$, $\bar{u} = \bar{u}_\delta$, $\bar{v} = \bar{v}_\delta$ and $\bar{\theta} = \bar{\theta}_\delta$. The defect lawcon cept was proposed by von Kármán (1930) for the velocity profile in neutral channel flow; in the present context it is also directly suggested by the geostrophic departure appearing in (3.69) and (3.70) with (3.73), if $\bar{u}_\delta = u_g$ and $\bar{v}_\delta = v_g$.

Unlike the Φ_s functions of the surface sublayer, the Φ_b functions of (4.68) through (4.71) depend on a large number of parameters; this makes it quite difficult to sort out the effect of each of them in the analysis of experimental data. Also, in the real atmosphere, unlike in the surface sublayer, effects due to unsteadiness, diurnal changes, horizontal advection, etc. in the outer sublayer are almost impossible to eliminate. Consequently, at present the general formulation of the flux-profile functions for the outer reaches of the ABL is still in the developmental stage. Great strides have been made in the past decade, and a large amount of information is now becoming available as a result of experimental studies and theoretical analyses by means of simple slab models, K theory, and higher order closure methods. However, a review of these efforts is beyond the scope of the present work.

b. Bulk Transfer Equations for the ABL

One of the main practical results of the development of similarity theory for the outer sublayer until now has been the formulation of bulk transfer coefficients and heat and mass transfer coefficients for the whole ABL. Actually, drag coefficients have been used by engineers for a long time in connection with channel flow and flat-plate turbulent boundary-layer problems. Lettau (1959), who considered the analogy between friction in the atmospheric boundary layer and in pipe flow, was probably the first to introduce the drag coefficient concept for the atmosphere on the basis of similarity with experimental data. The subsequent interest in bulk transfer coefficients has stemmed from the need for simple methods to parameterize the surface fluxes u_*, H and E in numerical models in terms of so-called external variables; external variables are those observed at the top of the boundary layer and near the surface. This interest has continued as more experimental data have become available to check various similarity hypotheses.

The bulk transfer coefficients for two-dimensional channel- and flat plate-type turbulent boundary layers are obtainable quite naturally by applying the principle of asymptotic matching of the profiles of the outer sublayer and of the surface sublayer. This method has been applied formally to a neutral, steady, horizontally uniform ABL by Csanady (1967) and by Blackadar and Tennekes (1968), but it was already implicit in the earlier work of Kazanski and Monin (1961). A more straightforward technique to derive the bulk transfer coefficients consists of joining the profile of the outer sublayer with the logarithmic profile of the surface sublayer. In this joining or 'patching' technique the profile for the surface sublayer is assumed *a priori*, whereas in asymptotic matching no such commitment is required; in fact, the logarithmic profile is a result of the analysis. Albeit simple joining is perhaps less rigorous, in the literature all extensions of the bulk transfer relations for a neutral boundary layer, to include more realistic conditions due to buoyancy or baroclinicity, have been based on this method. In what follows, a joining method is given that was used by Brutsaert and Mawdsley (1976) for the water vapor transfer coeffcient.

Derivation of Their General Form

In his derivation of the logarithmic velocity profile in a neutral turbulent boundary layer, Millikan (1938) assumed the existence of a finite region of overlap in which both the formulation for the wall layer and the velocity defect law are valid. Assuming Millikan's hypothesis also to be valid for the mean profile of any admixture in a stratified boundary layer, one can simply join (4.33)–(4.35) with (4.68) through (4.71) to eliminate $\bar{q}$, $\bar{u}$, $\bar{v}$, and $\bar{\theta}$ in the overlap region, and to obtain a relationship between the surface fluxes E, u_* and H and the remaining external parameters; the latter are the values of the humidity, wind speed and temperature at the lower and upper boundaries of the ABL. For the specific humidity profile one proceeds as follows.

If (4.33) and (4.68) both describe the humidity profile in the region of overlap (See Figure 3.1), (4.68) must have the same z-dependency as (4.33). Thus, within the region of overlap, Φ_{bv} must have the following form (the r and i subscripts can be omitted for the time being, since the argument is identical for both sets of dimensionless variables)

$$\Phi_{bv}(\eta, \mu, \nu_0, \beta_x, \beta_y) = a_v^{-1} [\ln(\eta) - \Psi_{sv}(\eta\mu) + D] \tag{4.72}$$

where D is a similarity function of all remaining variables not involving z,

$$D = D(\mu, \nu_0, \beta_x, \beta_y) \tag{4.73}$$

unspecified at this point and to be determined experimentally or otherwise. Equating $\bar{q}_2$ in (4.33) with $\bar{q}$ in (4.68) with (4.72), one immediately obtains the bulk transfer equation for water vapor

$$\bar{q}_1 - \bar{q}_\delta = \frac{E}{a_v k u_* \rho}\left[\ln\left(\frac{\delta}{z_1 - d_0}\right) + \Psi_{sv}(\zeta_1) - D\right] \tag{4.74}$$

or, if $\bar{q}_1$ is measured at the surface, $z = d_0$, simply

$$\bar{q}_s - \bar{q}_\delta = \frac{E}{a_v k u_* \rho}\left[\ln\left(\frac{\delta}{z_{0v}}\right) - D\right]. \tag{4.75}$$

Equation (4.74) is appropriate when the surface data $\bar{q}_1$ refer to shelter level, whereas (4.75) is used when the actual wet surface specific humidity $\bar{q}_s$ is available. An equation similar to (4.75) was already given by Zilitinkevich (1969), but his derivation implied certain unnecessary restrictions of its applicability.

For a given baroclinicity and for known values of $\bar{q}$ at $z = \delta$ and near the surface, (4.74) or (4.75) can only be applied to calculate the evaporation rate E, provided u_*

and H (to determine D from μ and ν_0) are also known. Hence, unless u_* and H are available from some other measurements, (4.74) or (4.75) cannot be applied without the analogous formulations for the wind speed and the temperature. By using the same joining technique one obtains the following for the x- and y- components for the mean wind speed at $z = \delta$.

$$\bar{u}_\delta = \frac{u_*}{k}\left[\ln\left(\frac{\delta}{z_0}\right) - B\right], \tag{4.76}$$

$$\bar{v}_\delta = -\frac{u_*}{k}A \tag{4.77}$$

and either one of the following bulk transfer equations for sensible heat

$$\bar{\theta}_1 - \bar{\theta}_\delta = \frac{H}{a_h k u_* \rho c_p}\left[\ln\left(\frac{\delta}{z_1}\right) + \Psi_{sh}(\zeta_1) - C\right], \tag{4.78}$$

$$\bar{\theta}_s - \bar{\theta}_\delta = \frac{H}{a_h k u_* \rho c_p}\left[\ln\left(\frac{\delta}{z_{0h}}\right) - C\right]. \tag{4.79}$$

The drag coefficient can be determined from (4.76) and (4.77) as $u_*^2/(\bar{u}_\delta^2 + \bar{v}_\delta^2)$. The symbols A, B and C are chosen for historical reasons as an extension of the rotational similarity used by Kazanski and Monin (1961) and Zilitinkevich *et al.* (1967); they denote similarity functions analogous to D in (4.73) and they are unspecified at this point. Note that in some publications the meaning of A and B is reversed from that used here.

Available Similarity Functions A, B, C and D

These similarity functions depend on at least four parameters as shown in (4.73), and probably on some others as well, that have not been considered until now. Unfortunately, the experimental determination of the fluxes, the mean profiles and all the other relevant variables in a real atmosphere is a difficult task subject to many errors. In addition to experiments, it is possible to determine the similarity functions by numerical modeling of the ABL. Although such models permit sensitivity analyses resulting from changes of one variable at a time, a good deal of uncertainty remains, especially for stable conditions. As a result, most of the available information is restricted to μ_r- and μ_i-dependency. Although the β_x- and β_y-dependency has been studied for A, B (Clarke and Hess, 1974; Arya and Wyngaard, 1975; Kondo, 1977) and C (Garratt and Francey, 1978), the results obtained are difficult to interpret for practical application. The ν_0-dependency of A, B and C has been considered in numerical models (e.g., Arya, 1977) but experimental data (e.g., Garratt and Francey, 1978) suggest that it is not a very important parameter. Therefore, at present, it appears practical to express A, B, C and D only as functions of one parameter, $\mu = \delta/L$, reflecting the stability of the atmosphere.

In applications it is necessary to decide whether δ_r, δ_i or possibly even some other assumed height is the most appropriate scale to represent the thickness of the ABL. The corresponding stability parameters are

$$\mu_r = \delta_r/L \tag{4.80}$$

and

$$\mu_i = \delta_i / L. \tag{4.81}$$

Let the similarity functions of the former be denoted by $A_r = A_r(\mu_r)$, $B_r = B_r(\mu_r)$, $C_r = C_r(\mu_r)$, $D_r = D_r(\mu_r)$, and those of the latter by $A_i = A_i(\mu_i)$, $B_i = B_i(\mu_i)$, $C_i = C_i(\mu_i)$, $D_i = D_i(\mu_i)$. As mentioned, in recent years a consensus has been emerging that δ_i-based similarity is preferable over that based on δ_r. Except for the study of Clarke and Hess (1973), most papers dealing with this issue lead to the conclusion that on the unstable side δ_i-based scaling produces least scatter in the calculated A, B, C and D. On the stable side, similarity modeling appears to perform so poorly, and the surface fluxes are often so small that it probably does not matter much whether δ_i or δ_r is utilized. Thus, herein major attention is paid to the δ_i-based similarity model.

(*i*) *Rotational height similarity*. Because of its historic priority, and because under neutral and stable conditions it performs equally well, if sometimes not better, a brief review of the rotational height scaling is still in order. The bulk transfer equations for the rotational similarity scheme are obtained from (4.74) though (4.79) by replacing δ in the logarithmic terms by $(u_*/|f|)$. For example, (4.74) becomes

$$\bar{q}_1 - \bar{q}_\delta = \frac{E}{a_v k u_* \rho}\left[\ln\left(\frac{u_*}{(z_1 - d_0)|f|}\right) + \Psi_{sv}(\zeta_1) - D_r(\mu_r)\right] \tag{4.82}$$

and analogous equations for (4.75) though (4.79). But, as it is possible to interpret the definitions of $\bar{q}_\delta$, $\bar{u}_\delta$, $\bar{v}_\delta$ and $\bar{\theta}_\delta$ in various ways, in the literature there is some lack of consistency in the application of this similarity scheme. Most authors have taken u_g and v_g as practical estimates of $\bar{u}_\delta$ and $\bar{v}_\delta$. Clarke and Hess (1974) also considered observed winds at $z = (0.15\, u_*/|f|)$, but they found that in either procedure the errors were equally large. Zilitinkevich (1969) suggested that $\bar{\theta}_\delta$ and $\bar{q}_\delta$ should be observed at a level proportional to $(u_*/|f|)$, but that for practical purposes it is probably sufficiently accurate to observe them at some constant level such as 1 km or 850 mb. Clarke (1970) took $\bar{\theta}_\delta$ as the $\bar{\theta}$ at the level where the $\bar{u}$ profile displays its maximum; Arya (1975) assumed that under unstable conditions $\bar{\theta}_\delta$ is to be measured where the $\bar{\theta}$ profile has its maximum, and that for stable conditions it occurs at $z = 0.25\, u_*/|f|$, which he considered to be the upper bound on the height of a stable boundary layer. Brutsaert and Chan (1978) after comparing several alternative procedures, decided on $\bar{q}$ and $\bar{\theta}$ measurements interpolated at $(0.15\, u_*/|f|)$, although all the other alternatives gave about the same degree of scatter.

The calculated values of $A_r(\mu_r)$, $B_r(\mu_r)$, $C_r(\mu_r)$ and $D_r(\mu_r)$ that have been published to date on the basis of experiments exhibit considerable scatter. Such data have been presented by Zilitinkevich (1969), Clarke (1970, 1972), Deacon (1973), Clarke and Hess (1974), Arya (1975), Yamada (1976) and Brutsaert and Chan (1978). The reader is referred to these papers for further details on these functions.

Suffice it to observe that the functions A_r, B_r, C_r and D_r have generally been found to be very similar to A_i, B_i, C_i, D_i provided δ_r is defined consistently as (4.63) in (4.74) through (4.79). This is due to the fact that δ_r is often of the same order as δ_i (e.g., Yamada, 1976) even though with strong diurnal changes in δ_i and u^*, $\nu_0 = (\delta_i/\delta_r)$

may conceivably vary over more than a decade at a given location. Hence, for practical purposes, the rotational similarity functions, for which no expressions are given here, may be deduced from those for A_i, B_i, C_i and D_i which are presented in the next section. It should be noted however, that A_r, B_r, C_r and D_r are the equivalent of A, $(B - \ln K_r)$, $(C - \ln K_r)$ and $(D - \ln K_r)$ in (4.74) through (4.79); these $\ln K_r$ terms arise from the fact that, traditionally, since Kazanski and Monin (1961), in the rotational formulations δ in the logarithmic terms of (4.74) through (4.79) is replaced by $u_*/|f|$ instead of by δ_r (see (4.82)). Thus, the relationships between the two sets of similarity functions is to a first approximation

$$\begin{aligned} A_r(\mu_r) &= A_i(\mu_r), \quad B_r(\mu_r) = B_i(\mu_r) - \ln K_r \\ C_r(\mu_r) &= C_i(\mu_r) - \ln K_r, \qquad D_r(\mu_r) = D_i(\mu_r) - \ln K_r \end{aligned} \tag{4.83}$$

provided, again, μ_r is defined with δ_r as given in (4.63), and K_r is taken to be of the order of 0.15 to 0.3 rather than unity.

(*ii*) *Observed height similarity.* This scheme was introduced by Zilitinkevich and Deardorff (1974). Its formulation is obtained by replacing δ in (4.74) through (4.79) simply by δ_i, the observed height scale. For example, (4.74) becomes

$$\bar{q}_1 - \bar{q}_\delta = \frac{E}{a_v k u_* \rho}\left[\ln\left(\frac{\delta_i}{z_1 - d_0}\right) + \Psi_{sv}(\zeta_1) - D_i(\mu_i)\right] \tag{4.84}$$

and analogous expressions are obtained for (4.75) through (4.79). Also, for this scheme, in the literature there have been several different interpretations of $\bar{q}_\delta$, $\bar{u}_\delta$, $\bar{v}_\delta$ and $\bar{\theta}_\delta$. For unstable conditions Melgarejo and Deardorff (1974) determined δ_i as the level up to which $\bar{q}$ and $\bar{\theta}$ are relatively constant, that is, the level just below a rapid decrease in $\bar{q}$ and a rapid increase in $\bar{\theta}$ with elevation. For stable conditions they made two estimates of the boundary-layer thickness: one from the temperature profile as the level up to which significant cooling has propagated from the surface; and one determined from the lowest maximum in the wind profile. Both estimates of δ_i, under stable conditions, were found to give the same large amount of scatter. Arya (1975) assumed that under unstable conditions $\bar{\theta}_\delta$ can be taken as the minimum $\bar{\theta}$ observed in the profile, but he assumed $\bar{u}_\delta = u_g$ and $\bar{v}_\delta = v_g$; thus, in a sense, he used mixed similarity criteria. Brutsaert and Chan (1978) compared different alternatives for unstable conditions; they concluded that with radiosonde measurements made at discrete intervals, a practical procedure, consisting of taking δ_i as the height of the lowest value of the potential temperature profile and $\bar{q}_\delta$ and $\bar{\theta}_\delta$ as the readings at this level, was satisfactory. Yamada (1976) took δ_i as the top of the mixed layer during daytime and as the height of the surface inversion during night time. However, for $\bar{u}_\delta$, $\bar{v}_\delta$ and $\bar{\theta}_\delta$ he took the vertical average of the geostrophic wind and of the virtual potential temperature; this idea was based on the suggestion of Arya and Wyngaard (1975) that the averaged geostrophic wind scale should remain nearly independent of geostrophic shear, that is baroclinicity. These wind scales can be obtained by integrating (3.69) and (3.70) with (3.73), viz.

$$\langle v_g \rangle = \langle \bar{v} \rangle - \frac{u_*^2}{f\delta_i}, \tag{4.85}$$

$$\langle u_g \rangle = \langle \bar{u} \rangle \tag{4.86}$$

where

$$\langle F \rangle = \int_{z_0}^{\delta_i} F \, dz/\delta_i. \tag{4.87}$$

Similarly, Yamada (1976) took the temperature scale $\langle \bar{\theta}_v \rangle$ instead of $\bar{\theta}_{v\delta}$. He found that the scatter in the similarity functions thus calculated, was considerably smaller than in previous calculations. He also calculated A_i, B_i and C_i by using $\bar{u}_\delta$, $\bar{v}_\delta$ and $\bar{\theta}_\delta$ at $z = \delta_i$, and showed that the curves through the data were similar to those based on the average scales; however, the degree of scatter was noticeably increased.

Besides a likely reduction of the dependence of the similarity functions on baroclinicity, there are other reasons for preferring the use of the averaged variables over their value at $z = \delta_i$. The layer-averaged values are not as susceptible to sampling errors and, as pointed out by Arya (1977), they are more appropriate when the similarity parameterization is to represent large-scale areal or regional averages. Although they are very similar to A_i, B_i, C_i, D_i, let the similarity functions based on layer-averaged values by denoted by A_{im}, B_{im}, C_{im}, D_{im}. For example, for water vapor the formulation (4.74) can be written as

$$\bar{q}_1 - \langle \bar{q} \rangle = \frac{E}{a_v k u_* \rho}\left[\ln\left(\frac{\delta_i}{z_1 - d_0}\right) + \Psi_{sv}(\zeta_1) - D_{im}(\mu_i)\right] \tag{4.88}$$

and analogous expressions are obtained for (4.75) through (4.79).

All similarity functions calculated from experimental data, which have appeared in the literature are in fair agreement, although the data points display an inordinate amount of scatter. Typical results are shown in Figures 4.8 through 4.12.

Several empirical and theoretical functions have been published. Several authors have proposed the following to represent their results, under unstable conditions,

$$B_i = a \ln(-\mu_i) + b, \tag{4.89}$$

$$C_i = c \ln(-\mu_i) + d \tag{4.90}$$

where a, b, c and d are constants. For example, Clarke and Hess (1973) found by least squares, with $k = 0.4$ and $a_h = 1$, that $a = c = 1$, $b = -0.71$ and $d = 1.82$; $\bar{u}_\delta$ was taken at $0.15 u_*/|f|$ and $\bar{\theta}_\delta$ at $0.25\ u_*/|f|$. Wyngaard *et al.* (1974) postulated on the evidence of their numerical model, that under unstable conditions and in the absence of geostrophic shear, an essentially shear-free convective layer exists above the surface sublayer; as a result they derived, with $k = 0.35$ and $a_h = 1.35$, that for $\mu_i < 5$, $A_i = 0$, (4.89) and (4.90) with $b = 0$ and $a = c = d = 1$. They found that B_i gave a good fit with experimental observations, but that C_i is too small. Arya (1975) concluded that (4.90) of Wyngaard *et al.* (1974) for C_i but with $a_h = 1$, represented his data quite well, whereas (4.89) for B_i seemed to overestimate slightly. Garratt and Francey (1978) used the same functional form to fit their calculated points of C_{im} on the basis of several experimental data sets. By least squares linear regression with (4.90) and $k = 0.41$, $a_h = 1$, they obtained $c = 0.46$ and $d = 4.88$, for $\mu_i < 1$.

Mawdsley and Brutsaert (1977) reasoned that, since the temperature is nearly constant in the outer convection layer, it may probably be assumed that the Monin–Obukhov surface layer formulation for the profiles is a good approximation in the

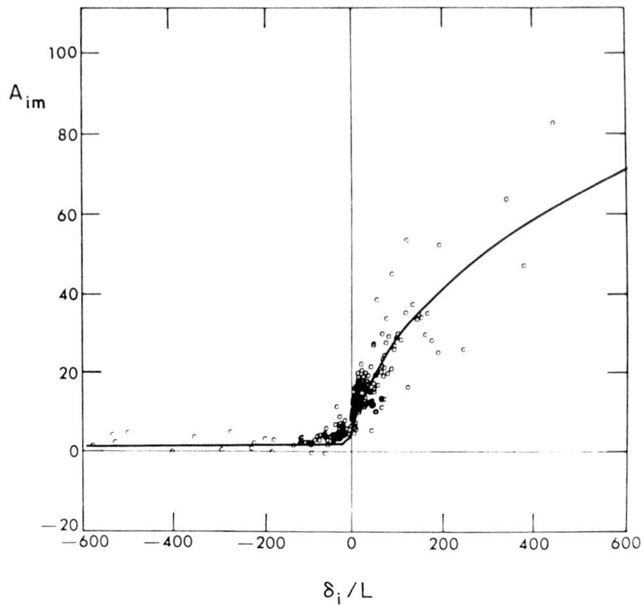

Fig. 4.8. Similarity function A_{im} versus $\mu_i = \delta_i/L$ on the basis of experimental data taken at Hay, New South Wales by Clarke *et al.* (1971). Vertically averaged geostrophic wind of (4.85) is used as wind scale $\bar{v}_\delta$. The solid line represents (4.94) (from Yamada, 1976).

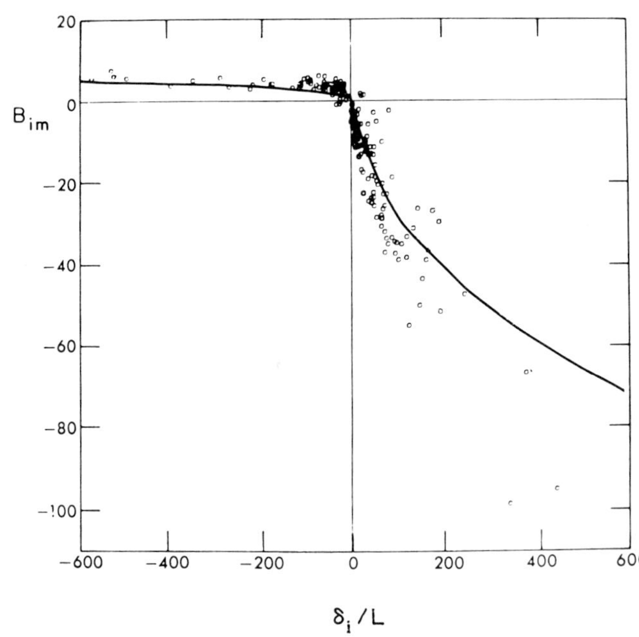

Fig. 4.9. Similarity function B_{im} versus μ_i. Vertically averaged geostrophic wind of (4.86) is used as wind scale $\bar{u}_\delta$. The solid line represents (4.95) (from Yamada, 1976).

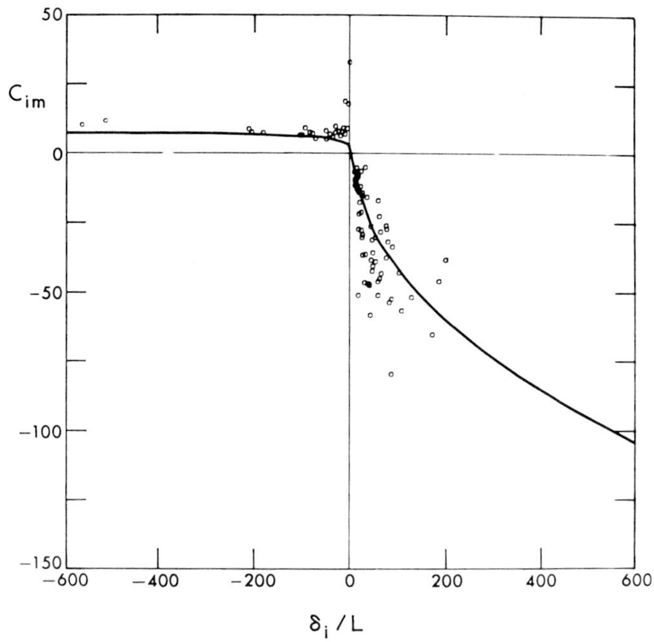

Fig. 4.10. Similarity function C_{im} versus μ_i. Vertically averaged potential temperature is used as temperature scale $\bar{\theta}_\delta$. The solid line represents (4.96) (from Yamada, 1976).

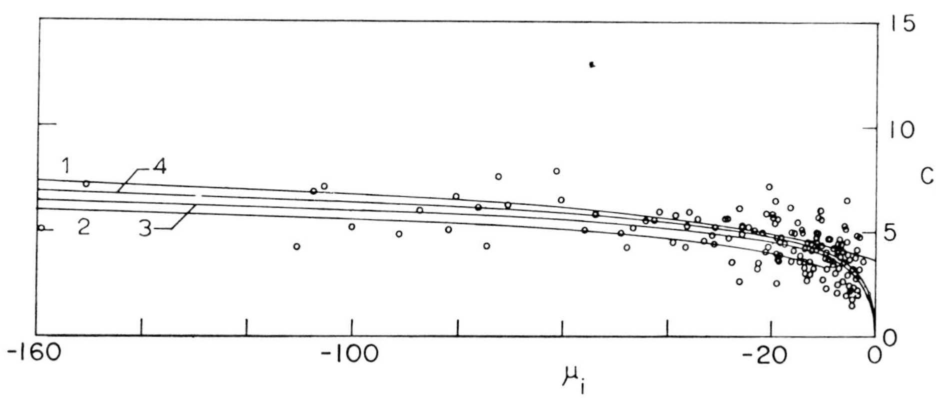

Fig. 4.11. Similarity function C_i versus μ_i, calculated from experimental data taken over the East China Sea [AMTEX]. Curve 1 represents (4.96) of Yamada (1976); curve 2 is (4.90) with the constants of Wyngaard *et al.* (1974); curve 3 is C_i in (4.92) of Mawdsley and Brutsaert (1977); curve 4 represents (4.97) (from Brutsaert and Chan, 1978).

outer sublayer as well. This assumption leads to

$$\begin{aligned} A_i &= 0 && \text{for } \mu_i \ll 0 \\ B_i &= \Psi_{sm}(\mu_i) && \text{for } \mu_i \leq 0 \\ C_i &= \Psi_{sh}(\mu_i) && \text{for } \mu_i \leq 0 \end{aligned} \tag{4.91}$$

which were applied with (4.45), (4.50) and (4.51). Thus, they proposed with $k = 0.4$ and $a_h = 1$,

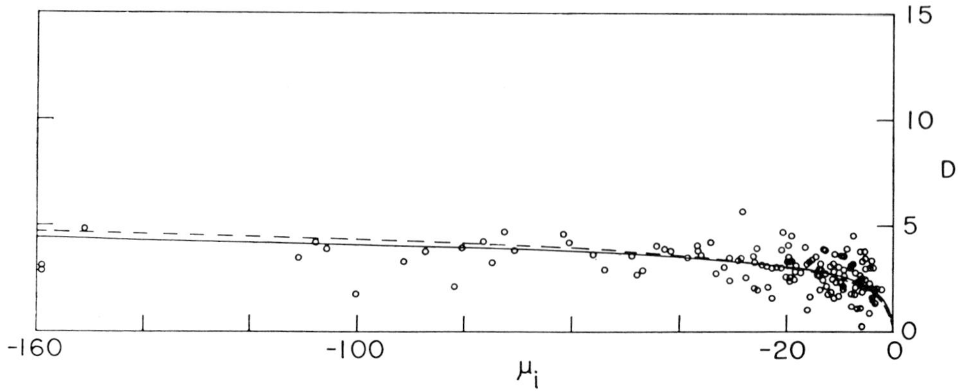

Fig. 4.12. Values of the similarity function $D_i(\mu_i)$ calculated from experimental data taken over the ocean. The solid curve represents D_i in (4.97) and the dashed line represents (4.98) (from Brutsaert and Chan, 1978).

$$\begin{aligned} A_i &= 0 \qquad \text{for } \mu_i < -147 \\ B_i &= 2 \ln\left[\frac{(1+x)}{2}\right] + \ln\left[\frac{(1+x^2)}{2}\right] - 2 \arctan (x) + \frac{\pi}{2} \\ &\qquad \text{for } \mu_i \le 0 \\ C_i &= 2 \ln\left[\frac{(1+x^2)}{2}\right] \qquad \text{for } \mu_i \le 0 \end{aligned} \tag{4.92}$$

where $x = (1-16\,\mu_i)^{1/4}$. Clearly, for $\mu_i \ll 0$, (4.92) produces (4.90) with $c = 1$ and $d = 1.39$. The upper limit of validity of A_i in (4.91) was decided upon by inspection of experimental data from different sources. For the remainder of the μ_i range the following empirical functions were fitted to the data

$$\begin{aligned} A_i &= 5 - \ln(1 - \mu_i) \qquad \text{for } -147 \le \mu_i \le 0 \\ &= 5 + 2.2 \ln(1 + \mu_i) \qquad \mu_i > 0 \\ B_i &= -2.2 \ln(1 + \mu_i) \qquad \mu_i > 0 \\ C_i &= -7.6 \ln(1 + \mu_i) \qquad \mu_i > 0. \end{aligned} \tag{4.93}$$

Yamada (1976) presented a set of empirical equations covering the whole μ_i range (with $k = 0.35$ and $a_h = 1.35$)

$$A_{im} = \begin{cases} 2.85\,(\mu_i - 12.47)^{1/2} & \mu_i > 35 \\ 3.02 + 0.3\mu_i & 0 \le \mu_i \le 35 \\ 3.02\,(1 - 3.29\mu_i)^{-1/3} & \mu_i \le 0, \end{cases} \tag{4.94}$$

$$B_{im} = \begin{cases} -2.94\,(\mu_i - 19.94)^{1/2} & \mu_i > 35 \\ 1.855 - 0.38\mu_i & 0 \le \mu_i \le 35 \\ 10 - 8.145\,(1 - 0.008376\mu_i)^{-1/3} & \mu_i \le 0, \end{cases} \tag{4.95}$$

$$C_{im} = \begin{cases} -4.32\,(\mu_i - 11.21)^{1/2} & \mu_i > 18 \\ 3.665 - 0.829\mu_i & 0 \le \mu_i \le 18 \\ 12 - 8.335\,(1 - 0.03106\mu_i)^{-1/3} & \mu_i \le 0. \end{cases} \tag{4.96}$$

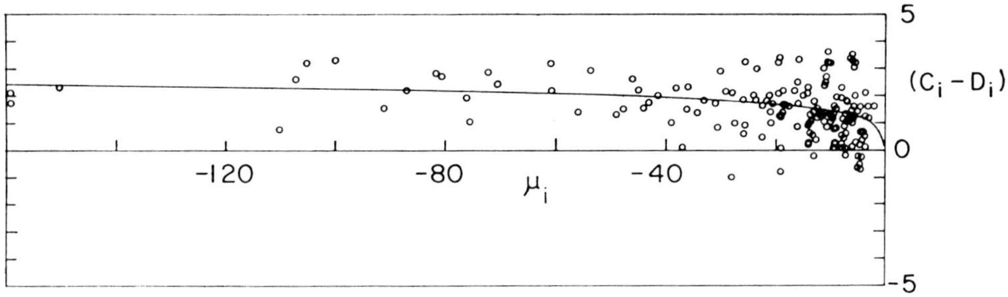

Fig. 4.13. Values of $(C_i - D_i)$ calculated from data taken over the ocean. The curve represents $(C_i - D_i)$ obtained from (4.97) (from Brutsaert and Chan, 1978).

These equations can be compared with experimental data in Figures 4.8, 4.9 and 4.10.

Very little work has been done to date on the water vapor function D_i. Brutsaert and Chan (1978) analyzed the data of the Amtex experiments and concluded that the following give a good description of these data (see Figures 4.11 and 4.12).

$$\left.\begin{aligned} C_i &= 1.06\ \Psi_{sh}(\mu_i) \\ D_i &= 0.685\ \Psi_{sh}(\mu_i) \end{aligned}\right\} \text{ for } \mu_i \leq 0 \tag{4.97}$$

where Ψ_{sh}, defined in (4.38), was taken as $2 \ln [(1 + x^2)/2]$ like in (4.92). An almost equally good representation of the data can be obtained with the function proposed for B_i in (4.91), viz.

$$D_i = \Psi_{sm}(\mu_i) \quad \text{for } \mu_i \leq 0 \tag{4.98}$$

where Ψ_{sm} is given by (4.50). Equations (4.97) show that D_i is smaller than C_i, or

$$D_i = 0.646\ C_i \quad \text{for } \mu_i \leq 0. \tag{4.99}$$

Figure 4.13 shows the difference $(C_i - D_i)$ for the data points of Figures 4.11 and 4.12, together with the curve $0.375\ \Psi_{sh}(\mu_i)$ obtained from (4.97). Equation (4.98) suggests that the boundary-layer profile of the mean specific humidity is more similar to that of the mean wind profile component in the direction of u_*, than to that of the mean potential temperature. It has been impossible until now to derive a D_i function for stable conditions, due to the lack of suitable data.

As can be seen from this brief review, there is still no consensus in the literature concerning the optimal functional form to describe A_i, B_i, C_i and D_i. The large scatter in the determination of these functions is now generally attributed to experimental errors in the available measurements. Thus, with time it should be possible to improve on this. Nonetheless, the various functions that have been proposed are not drastically different. This is illustrated in Figure 4.11 which shows curves of $C_i(\mu_i)$ calculated by means of (4.90), (4.92), (4.96) and (4.97). The difference between these curves is within the scatter, which is typical for most other studies as well (e.g., Figure 4.10).

4.4. THE INTERFACIAL SUBLAYERS

a. Similarity for the Mean Profiles

An interfacial sublayer may be defined as the sublayer of the turbulent ABL immediately adjacent to the surface but below the dynamic sublayer. In this sublayer, the universal logarithmic profiles are not valid, and there are as many different types of flow as there are types of surfaces. Close to the surface, some, if not all, of the following features must be considered: (i) The flow is not fully turbulent; thus, while the turbulence is not damped out completely, the flow may be affected by the viscosity and the transport of scalar admixtures may depend on their molecular diffusivities. (ii) Except for smooth surfaces, the nature and the placement of the roughness elements of the surface profoundly affect the flow pattern; the flow takes place between and, in the case of vegetation, even through the obstacles. (iii) On the whole, Reynolds's analogy is less likely to be valid. This is due to the fact that momentum transport involves not only viscous shear but also local pressure gradients related to form drag on the roughness obstacles, whereas transport of inert or passive admixtures such as water vapor at the surface can only take place by molecular diffusion. Reynolds's analogy may also not be valid as a result of the difference in the distribution of sinks or sources of momentum, sensible heat and water vapor on the surface.

Within the framework of similarity modeling, the mean profiles can be described formally by means of expressions which are analogous to those used in the other sublayers of the ABL. This can be done either in gradient form (like (4.26), (4.59) etc.) or directly (like (4.29), (4.68) etc.) as follows

$$\bar{q}_s - \bar{q} = \frac{E}{ku_*\rho}\,\Phi_{0v} \tag{4.100}$$

$$\bar{u} = \frac{u_*}{k}\,\Phi_{0m} \tag{4.101}$$

$$\bar{\theta}_s - \bar{\theta} = \frac{H}{ku_*\rho c_p}\,\Phi_{0h} \tag{4.102}$$

in which k is retained to facilitate the analogy with the other sublayers. Theoretically, the profile functions Φ_{0v}, Φ_{0m}, and Φ_{0h} are universal functions, but they depend on a large number of variables.

In view of the features mentioned above, these variables can briefly be listed as the distance from the lowest reference level, z; some measure of the thickness of the interfacial sublayer h; the turbulence dynamics as manifested by the surface shear stress, u_*; the viscosity of the air, ν; the molecular diffusivity of the water vapor κ_v; the thermal diffusivity, κ_h; in the case of rough surfaces, variables describing the size, shape, arrangement, density and rigidity of the roughness obstacles; and in the case of vegetation some additional variables describing the size, shape, distribution, and density of the foliage elements and branches. These variables can be organized as dimensionless parameters, so that formally one can infer

$$\Phi_{0v} = \Phi_{0v}\left(\frac{z}{h}, h_+, Sc, \gamma_{b1}, \gamma_{b2}, \ldots, \gamma_{sv1}, \gamma_{sv2}, \ldots\right), \tag{4.103}$$

$$\Phi_{0m} = \Phi_{0m}\left(\frac{z}{h}, h_+, \gamma_{b1}, \gamma_{b2}, \ldots \gamma_{sm1}, \gamma_{sm2}, \ldots\right), \tag{4.104}$$

$$\Phi_{0h} = \Phi_{0h}\left(\frac{z}{h}, h_+, \mathrm{Pr}, \gamma_{b1}, \ldots \gamma_{sh1}, \gamma_{sh2}, \ldots\right), \tag{4.105}$$

where $h_+ = (u_* h/\nu)$ is a Reynolds number; $\mathrm{Sc} = \nu/\kappa_v$ the Schmidt number; Pr the Prandtl number; $\gamma_{b1}, \gamma_{b2}, \ldots$ dimensionless parameters describing the bulk geometry of the roughness elements; $\gamma_{sv1}, \gamma_{sv2}, \ldots$ dimensionless parameters describing the small-scale structure and geometry of the roughness obstacles as regards water vapor transport; and similarily $\gamma_{sm1}, \gamma_{sh1}, \ldots$ for momentum and heat transport, respectively.

Clearly, the number of dimensionless parameters required for a unified treatment of all types of surfaces is much too large to be practical. Moreover, present knowledge of the Φ_0 functions is limited to their dependence on only a few of the parameters.

Although most natural surfaces are intermediate or transitional cases, for a more detailed analysis, it is convenient to distinguish three representative types of surfaces. These are, as shown in Figure 3.1, the smooth surface, the surface with bluff roughness elements and the surface with permeable and fairly densely placed roughness obstacles. These cases are treated below in Sections 4.4c, d, and e, respectively.

b. Interfacial Bulk Transfer Equations for Scalar Admixtures

The mean profiles of scalars such as $\bar{q}$ and $\bar{\theta}$ in interfacial sublayers have not been studied very much. Rather, most experimental studies have concentrated on the determination of specific fluxes, E and H, in terms of measurements of $\bar{q}$ and $\bar{\theta}$ at the wall, $z = 0$, and at an elevation z, which is well within the fully turbulent boundary layer. To deduce the specific features of the transport in the interfacial sublayer, the analysis of such measurements usually involves the decomposition of the bulk transfer equation into a part pertaining to the interfacial and a part pertaining to the dynamic or surface sublayer. Just like in Section 4.3b, this requires the assumption of a region of overlap or, at least, a boundary of contact between the interfacial and the dynamic sublayer, at $z = h$, where the respective profile equations can be 'joined' or 'patched'. One of the first to apply this joining technique was Sverdrup (1937), who assumed that the interfacial sublayer over the ocean is in fact laminar. As will be seen below, this is an oversimplification. In general, the concept of a composite bulk transfer equation can be developed as follows.

Specifying $z = h$ where $\bar{q} = \bar{q}_h$ and $\bar{\theta} = \bar{\theta}_h$, in (4.100) and (4.102), one can write

$$\bar{q}_s - \bar{q}_h = \frac{E}{k u_* \rho} \Phi_{0v}\left(\frac{z}{h} = 1, \ldots\right), \tag{4.106}$$

$$\bar{\theta}_s - \bar{\theta}_h = \frac{H}{k u_* \rho c_p} \Phi_{0h}\left(\frac{z}{h} = 1, \ldots\right) \tag{4.107}$$

in which the remaining variables between the brackets are the same as given in (4.103) through (4.105). Note that, in past papers on the subject, $k\,\Phi_{0v}^{-1}(z/h = 1, \ldots)$ and $k\,\Phi_{0h}^{-1}(z/h = 1, \ldots)$ have sometimes been called the interfacial mass transfer coefficient or interfacial Dalton number,

$$\mathrm{Da}_0 = \frac{E}{\rho u_*(\bar{q}_s - \bar{q}_h)} \tag{4.108}$$

and the interfacial heat transfer coefficient or interfacial Stanton number

$$\mathrm{St}_0 = \frac{H}{\rho u_* c_p(\bar{\theta}_s - \bar{\theta}_h)}. \tag{4.109}$$

In addition, it will prove convenient to introduce an interfacial drag coefficient $k\Phi_{0m}^{-1}(z/h = 1, \ldots)$ as follows

$$\mathrm{Cd}_0 = \frac{u_*^2}{\bar{u}_h^2}. \tag{4.110}$$

where $\bar{u}_h$ is the mean velocity at $z = h$. At $z = h$, also the dynamic sublayer formulations should apply. For the specific humidity this is from (4.13)

$$\bar{q}_h - \bar{q}_r = \frac{E}{a_v k u_* \rho} \ln\left(\frac{z_r - d_0}{h - d_0}\right) \tag{4.111}$$

where $\bar{q}_r$ is the specific humidity at z_r which is any level within the dynamic sublayer.

The unknown and untractable quantity $\bar{q}_h$ can be eliminated simply by assuming continuity in $\bar{q}$ at $z = h$ between the two adjacent sublayers. Thus, from (4.106) and (4.111) one obtains

$$\bar{q}_s - \bar{q}_r = \frac{E}{k u_* \rho}\left[\Phi_{0v}\left(\frac{z}{h} = 1, \ldots\right) + a_v^{-1} \ln\left(\frac{z_r - d_0}{h - d_0}\right)\right] \tag{4.112}$$

or in view of (4.6), (4.108) and (4.110)

$$\bar{q}_s - \bar{q}_r = \frac{E}{u_* \rho}\left[\mathrm{Da}_0^{-1} - a_v^{-1}\,\mathrm{Cd}_0^{-1/2} + (a_v k)^{-1} \ln\left(\frac{z_r - d_0}{z_{0m}}\right)\right] \tag{4.113}$$

The bulk transfer equation is often formulated conveniently in terms of a mass transfer coefficient, or Dalton number, Ce_r. This is defined by

$$E = \mathrm{Ce}_r \rho \bar{u}_r(\bar{q}_s - \bar{q}_r) \tag{4.114}$$

where the subscript r refers to the reference level z_r where $\bar{u}$ and $\bar{q}$ are measured. In a similar way the drag coefficient Cd_r is defined by

$$\mathrm{Cd}_r = \frac{u_*^2}{\bar{u}_r^2}. \tag{4.115}$$

The bulk transfer Equation (4.113) yields thus the water vapor transfer coefficient

$$\mathrm{Ce}_r = \frac{\mathrm{Cd}_r^{1/2}}{(\mathrm{Da}_0^{-1} - a_v^{-1}\mathrm{Cd}_0^{-1/2} + a_v^{-1}\mathrm{Cd}_r^{-1/2})}. \tag{4.116}$$

To facilitate comparison with other work it should be noted that several authors (e.g., Owen and Thomson, 1963; Chamberlain, 1966) have written this result in terms of a quantity $B = [a_v(\mathrm{Da}_0^{-1} - a_v^{-1}\mathrm{Cd}_0^{-1/2})]^{-1}$ so that

$$\mathrm{Ce}_r = \frac{a_v\,\mathrm{Cd}_r^{1/2}}{B^{-1} + \mathrm{Cd}_r^{-1/2}}. \tag{4.116'}$$

In a similar way, from (4.107), (4.109), (4.110) and (4.16) and defining a heat transfer coefficient (or Stanton number) Ch_r, as follows,

$$H = \mathrm{Ch}_r \rho \bar{u}_r c_p (\bar{\theta}_s - \bar{\theta}_r) \tag{4.117}$$

one obtains

$$\mathrm{Ch}_r = \frac{\mathrm{Cd}_r^{1/2}}{(\mathrm{St}_0^{-1} - a_h^{-1}\mathrm{Cd}_0^{-1/2} + a_h^{-1}\mathrm{Cd}_r^{-1/2})}\,. \tag{4.118}$$

The importance of (4.116) and (4.118) is that they allow the study of the transport characteristics of scalars in the interfacial sublayer; equations like them have been used widely in the analysis of experimental data to evaluate $(\mathrm{Da}_0^{-1} - a_v^{-1}\mathrm{Cd}_0^{-1/2})$ $[=(a_v B)^{-1}]$ or $(\mathrm{St}_0^{-1} - a_h^{-1}\,\mathrm{Cd}_0^{-1/2})$. The determination of these interfacial transfer coefficients is treated in Sections 4.4c, d and e, for three different types of surfaces.

Equations (4.116) and (4.118) were derived for $z = z_r$ in the dynamic sublayer or in a neutral surface sublayer. It is straightforward to include the effect of atmospheric stability by using (4.33) instead of (4.13). Since usually $(h - d_0) \ll |L|$ so that $\Psi_{sm}[(h - d_0)/L] \approx \Psi_{sv}[(h - d_0)/L] \approx 0$, one obtains with $z = z_r$, as a reference level in the non-neutral surface sublayer

$$\mathrm{Ce}_r = \frac{\mathrm{Cd}_r^{1/2}}{(\mathrm{Da}_0^{-1} - \mathrm{a}_v^{-1}\mathrm{Cd}_0^{-1/2}) + (a_v k)^{-1}\left[\ln\left(\dfrac{z_r - d_0}{z_{0m}}\right) - \Psi_{sv}(\zeta_r)\right]}\,. \tag{4.119}$$

and an analogous expression for Ch_r. Note that, especially in the case of tall vegetation, it may happen that $(h - d_0)$ is not much smaller than $|L|$, and that the functions Φ_{0v}, Φ_{0m}, Φ_{0h} should depend also on an atmospheric stability parameter. However, practically nothing is known about this at present, and in what follows the interfacial transfer functions are assumed to be insensitive to atmospheric stability.

As mentioned, Sverdrup (1937) was probably the first to propose a transport model consisting of two sublayers, namely one near the wall with molecular diffusion, and one above the molecular layer with purely turbulent transfer; the layer nearest the water surface was assumed to be steady and uniform, but its thickness had to remain a parameter. A conceptually similar model was proposed by Kitaygorodskii and Volkov (1965). Sheppard (1958) did not make use of a separate molecular diffusion sublayer, but he assumed that transport occurs through molecular and turbulent diffusion side by side from $z = 0$ on up throughout the dynamic sublayer, with a diffusivity $(ku_* z + \kappa_v)$; this yielded a result similar to (4.116) with $a_v = 1$ and, instead of $(\mathrm{Da}_0^{-1} - a_v^{-1}\mathrm{Cd}_0^{-1/2})$ a term $[\ln(ku_* z_0/\kappa_v)]/k$ approximately. In the light of later studies, which are covered in the following three sections (4.4c, d and e) Sheppard's proposal is not very realistic.

c. Smooth Surfaces: The Viscous Sublayer

Momentum Transfer

A hydrodynamically smooth surface can be defined by the criterion

$$z_{0+} < 0.13 \tag{4.120}$$

approximately, where $z_{0+} \equiv (u_* z_0/\nu)$ is the roughness Reynolds number. In nature, most surfaces do not satisfy this criterion, but often open water at low wind speeds, snow and plane and regular ice surfaces are smooth. Over a smooth surface the interfacial sublayer is commonly referred to as the viscous sublayer. The upper boundary of this sublayer, implied by Figure 4.1, is given by

$$h = 30\,\nu/u_*. \tag{4.121}$$

Substitution in (4.3) with (4.9) produces, in accordance with (4.10), $(\bar{u}_h/u_*) = 13.5$, that is

$$\mathrm{Cd}_0^{-1/2} = 13.5 \tag{4.122}$$

approximately.

For $z_+(\equiv u_* z/\nu) < 5$ to 7, that is the lower part of the viscous sublayer, the mean velocity profile is linear,

$$\bar{u} = u_* z_+ \tag{4.123}$$

and there is a transitional layer higher up, over the range $5 \leq z_+ \leq 30$, between the linear and the logarithmic profile. In the literature (e.g., Monin and Yaglom, 1971; p. 282) interpolation formulae have been proposed to cover the whole range from linear to logarithmic, as shown in Figure 4.1; however, for practical purposes, it is often sufficiently accurate to simply extrapolate both (4.3) with (4.9) and (4.123) into the transitional zone; as indicated by the dashed lines in Figure 4.1, these equations are then applied above and below their intersection point at $z_+ = 11$, respectively. Accordingly, one would obtain $(\bar{u}_h/u_*) = 11$ which is close to (4.122).

Interfacial Transfer Coefficients for Scalars

Among the dimensionless variables listed in (4.103), it is clear that only Sc can have any effect on Da_0; indeed, over a smooth surface h_+ is constant as shown in (4.121) and the geometry of the surface roughness plays no role. Similarly, St_0 depends only on Pr. In the literature, there is good agreement on the appropriate form of the combined term $(\mathrm{Da}_0^{-1} - a_v^{-1}\,\mathrm{Cd}_0^{-1/2})$ needed in (4.116) or its heat transfer analog for (4.118). A number of expressions proposed in the past is listed in Table 4.1. Even though presumably these expressions were developed independently and for different

TABLE 4.1
Some expressions for the interfacial transfer coefficients for smooth surfaces, when $z_{0+}(= u_* z_0/\nu) < 0.13$, approximately

Reference	$(\mathrm{Da}_0^{-1} - a_v^{-1}\mathrm{Cd}_0^{-1/2})\ [= (a_v B)^{-1}]$
Friend and Metzner (1958)	$11.8\,(\mathrm{Sc} - 1)\,\mathrm{Sc}^{-1/3}$
Petukhov and Kirillov (1958)	$12.7\,\mathrm{Sc}^{2/3} - 12.7$
Petukhov *et al.* (1961)	
Kader and Yaglom (1972)	$12.5\,\mathrm{Sc}^{2/3} - 10.24$
Yaglom and Kader (1974)	
Kondo (1975)	$11.6\,\mathrm{Sc}^{2/3} - 12.05$
Brutsaert (1975a)	$13.6\,\mathrm{Sc}^{2/3} - 13.5$

Note: For heat transfer Da_0, Sc, a_v are replaced by St_0, Pr, a_h, respectively.

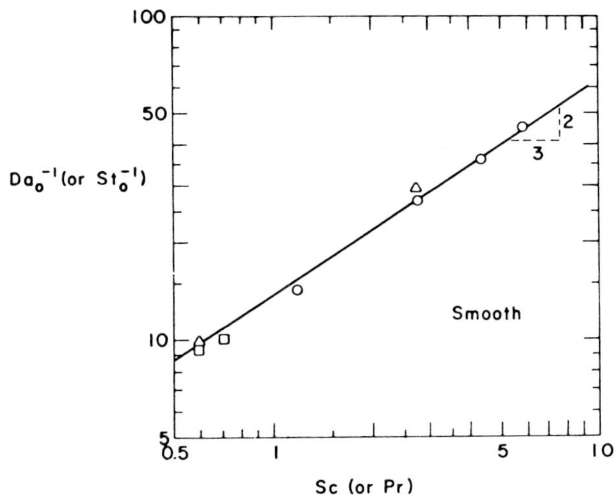

Fig. 4.14. Values of Da_0(or St_0) for a smooth surface obtained by means of (4.116) and (4.122) with $a_v = 1$, plotted versus Schmidt (or Prandtl) number. Open circles represent the median of the data of Dipprey and Sabersky (1963); triangles, Chamberlain (1968); squares, Mangarella *et al.* (1971) for heated water wind waves (from Brutsaert, 1975a).

conceptual models, they exhibit a striking similarity. No doubt, this is due to the availablity of good-quality experimental data for smooth surfaces; it also shows that the interfacial transfer coefficent for a smooth surface is relatively insensitive to its derivation.

As an illustration, consider the expression obtained in Brutsaert (1975a). The underlying conceptual model involves molecular diffusion into random-lived eddies, whose length- and time- scales are assumed to be given by Kolmogorov's theory for microscale turbulence. By assuming a linear mean velocity profile at the wall given by (4.123), the solution of the molecular diffusion equation for transport into a microscale eddy in contact with the wall yields

$$E = C_S \rho \kappa_v^{2/3}(\bar{q}_s - \bar{q}_h)\, u_* \nu^{-2/3} \tag{4.124}$$

where C_S is an empirical constant. This gives an interfacial Dalton number [cf., (4.108)]

$$\mathrm{Da}_0 = C_S\, \mathrm{Sc}^{-2/3}. \tag{4.125}$$

Equations (4.125) and (4.122) with $a_v = 1$, yield satisfactory agreement with experimental data as shown in Figure 4.14. The straight line with slope (2/3) in Figure 4.14 indicates that $C_S^{-1} = 13.6$, approximately, so that

$$\mathrm{Da}_0^{-1} - a_v^{-1}\mathrm{Cd}_0^{-1/2} = 13.6\,\mathrm{Sc}^{2/3} - 13.5. \tag{4.126}$$

In an interesting application of this result, Merlivat and Coantic (1975) and Merlivat (1978) studied the importance of the molecular diffusivity on the rate of evaporation into the turbulent atmosphere. They found that (4.116) with (4.126) yields satisfactory results in the description of the fractionation of the stable isotopes of water $H_2^{16}O$, $H_2^{18}O$ and HDO, during evaporation from a smooth water surface when $z_{0+} < 1$. These three isotopes, with different molecular weights, have different molecular diffusivities and thus Schmidt numbers, so that their evaporation rates are also different.

d. Surfaces with Bluff Roughness Elements

In the present context, roughness elements can be called 'bluff', when they are impermeable obstacles, with a height which is not large as compared to their aspect width normal to the mean flow. Examples of this type of rough surface, are a plowed field, rigid vegetation with very large leaves (e.g., cabbage and beet plants), irregular ice surfaces and developing water waves.

Momentum Transfer

A surface can be considered hydrodynamically rough when the following criterion is satisfied

$$z_{0+} > 2 \tag{4.127}$$

approximately.

The flow between and around bluff roughness elements consists of different types of wake- and cavity-flow, varying considerably from one place to another with convective accelerations and decelerations. Furthermore, the flow pattern is bound to be quite specific for any given type of roughness configuration. Consequently, it is not practically feasible to formulate a general similarity relationship (4.101) for the mean velocity profile between the roughness obstacles $\bar{u}$, in terms of a reasonable number of geometric parameters $\gamma_{b1}, \gamma_{b2} \ldots$ as indicated in (4.104). Except for the simplest of surfaces (e.g., rectangular rib roughnesses) until now the effect of the geometry of the roughnesses and their mean size h_0, has commonly been combined and lumped into the roughness parameter z_0. This is equivalent with the assumption that instead of (4.104), one has

$$\Phi_{0m} = \Phi_{0m}\left(\frac{z}{h_0}, z_{0+}\right) \tag{4.128}$$

to formulate the wind profile in (4.101).

Accordingly, one would expect that Cd_0 in (4.110) be at least a function of the roughness Reynolds number z_{0+} [defined in (4.7)]. However, especially at high Reynolds numbers, the momentum transfer at a bluff-rough surface is primarily a result of form drag involving local pressure gradients and less a result of viscous shear. Thus, even the z_{0+}-dependency of Cd_0 could conceivably be weak. This matter has not been studied in the past and apparently nothing is known about this dependency. Nevertheless, experimental velocity profiles in the neighborhood of rough walls (e.g., Paeschke, 1937; Liu *et al.*, 1966, Figure 4.11) suggest that, while there is much variation, the logarithmic profile usually is not valid below a lower level where $(\bar{u}/u_*)$ is of the order of 5, approximately. (The observed variation runs between approximately 4 and 8.) Hence, if this is taken as a crude but practical estimate of $(\bar{u}_h/u_*)$ at the upper boundary of the interfacial sublayer, (4.110) yields

$$\mathrm{Cd}_0^{-1/2} \simeq 5 \tag{4.129}$$

approximately.

Interfacial Transfer Coefficients for Scalars

If the effects of the thickness of the interfacial sublayer and of the geometrical parameters are all lumped into the roughness length z_0, it is clear, in view of (4.103), that Da_0 must be a function of z_{0+} and Sc. Similarly, St_0 is a function of z_{0+} and Pr. The different expressions for $(\mathrm{Da}_0^{-1} - a_v^{-1}\mathrm{Cd}_0^{-1/2})$ that have been proposed in the literature, are not in as good mutual agreement as those for smooth surfaces. This is due to the fact that rough surfaces have been studied less, and also to the fact that the z_{0+}-dependency must certainly be affected by the geometrical configuration and the nature of the roughness elements. Several empirical and theoretical expressions are presented in Table 4.2.

TABLE 4.2
Review of some published expressions for the interfacial transfer coefficients for bluff-rough surfaces, i.e., for $z_{0+} > 2$, approximately

Reference	$\mathrm{Da}_0^{-1} - a_v^{-1}\,\mathrm{Cd}_0^{-1/2}[=(a_vB)^{-1}]$
Dipprey and Sabersky (1963) $(1.2 \leq \mathrm{Pr} \leq 6)$	$10.25\, z_{0+}^{0.20}\, \mathrm{Sc}^{0.44} - 8.48$
Owen and Thomson (1963) $(0.7 \leq (\mathrm{Sc, Pr}) \leq 6)$	$2.40\, z_{0+}^{0.45}\, \mathrm{Sc}^{0.8}$
Sheriff and Gumley (1966) $\mathrm{Pr} = 0.7$	$7.78 z_{0+}^{0.199} - 4.65$
Dawson and Trass (1972) $(300 \leq \mathrm{Sc} \leq 4600)$	$12.87\, z_{0+}^{0.25}\, \mathrm{Sc}^{0.58} - 8$
Yaglom and Kader (1974) $(0.7 \leq (\mathrm{Sc, Pr}) \leq 9)$	$0.55\, h_{0+}^{1/2}\, (\mathrm{Sc}^{2/3} - 0.2) + 9.5 - 2.12 \ln (h_0/z_0)$
Brutsaert (1975a) $(0.6 \leq (\mathrm{Sc, Pr}) \leq 6)$	$7.3\, z_{0+}^{1/4}\, \mathrm{Sc}^{1/2} - 5$

Note: For heat transfer Da_0, Sc, a_v are replaced by St_0, Pr, a_h, respectively.

A few comments on these formulations are in order. The empirical expression of Sheriff and Gumley (1966) is rather similar to the theoretical result of Brutsaert (1975a), but it is only applicable to heat transfer in air. The empirical expression of Dawson and Trass (1972) is less suited for application in the atmosphere, because the investigated Schmidt numbers were much too high. Dipprey and Sabersky (1963) were probably the first to present a thorough analysis of heat transfer data in a fashion suitable for the application of (4.116), (4.118) and (4.119). The constant 8.48 in their expression originates from $(\bar{u}_h/u_*) = (\ln 30)/k$, and it is obtained from (4.3) with the assumption that the upper boundary of the interfacial sublayer lies at $h = 30\, z_0$, which they inferred from Nikuradse's relationship $h_{0s} = 30\, z_0$ for the sand grain roughness; however, as is shown in Chapter 5, the height of roughness obstacles is closer to 7 z_0 or 8 z_0, which with (4.3) would yield (4.129). On the other hand, Owen and Thomson (1963) stated that $(\bar{u}_h/u_*)$ is so small that it is indistinguishable from zero. It is clear that both positions, namely $(\bar{u}_h/u_*) = 8.48$ and $(\bar{u}_h/u_*) = 0$ are too extreme, since h is likely to be smaller than 30 z_0, and since it is definitely larger than zero. At any rate, if in all the expressions of Table 4.2, $\mathrm{Cd}_0^{-1/2}$ were taken to be the same, the powers of z_{0+} and Sc would also more nearly be the same.

Both theoretical expressions shown in Table 4.2 were derived by means of conceptual models for the flow near the wall, with similarity considerations. The theoretical result of Yaglom and Kader (1974) was derived on the basis of an eddy diffusivity model for the interfacial sublayer, as an extension of an earlier model for smooth surfaces. This model is based on the observation that, very close to a smooth wall, the power series expansion in z for the mean profile of $\bar{\theta}$ or $\bar{q}$ consists of a linear term followed by a fourth-order term. Thus it was assumed that, within the viscous sublayer, transfer takes place with an effective diffusivity which is equal to the molecular diffusivity very close to the wall, and proportional to z^3 farther away. This led to an interfacial Dalton (cf., Stanton) number

$$\mathrm{Da}_0 = (h_0\, u_*/\nu)^{-1/2}(b_1'\mathrm{Sc}^{2/3} - b_2')^{-1} \tag{4.130}$$

in which b_1' and b_2' are empirical constants and h_0 is the mean height of the roughness obstacles; the 2/3 power of Sc is evidently a carry-over from the smooth-surface model. The quantity corresponding to $a_v^{-1}\,\mathrm{Cd}_0^{-1/2}$ was put equal to $[(a_v k)^{-1} \ln (h_0/z_0) - C_y]$, where C_y is another empirical constant. The values of these three constants are given in Table 4.2.

The theoretical model of Brutsaert (1975a) is based on the notion that transfer at the surface takes place by molecular diffusion into internal Kolmogorov-scale eddies; these eddies are renewed intermittently after random times of contact, during which they are assumed to be stagnant between the roughness elements. The renewal process was assumed to be analogous to the randomly-repeated cycle of 'bursting' and 'ejection' of fluid, followed by an 'inrush' of fresh fluid, which has been observed in various flow visualization studies (e.g., Kim *et al.*, 1971; Corino and Brodkey, 1969; Grass, 1971). The solution of this molecular diffusion problem yields the following evaporation rate

$$E = C_R \rho \kappa_v^{1/2}\, u_*^{3/4}(\bar{q}_s - \bar{q}_h)\,(\nu z_0)^{-1/4} \tag{4.131}$$

where C_R is the only empirical constant. By virtue of (4.108), this result can be written as

$$\mathrm{Da}_0 = C_R z_{0+}^{-1/4}\,\mathrm{Sc}^{-1/2}. \tag{4.132}$$

The combination of (4.132) with (4.129) and $a_v = 1$ is in good agreement with experimental data for a wide variety of rough surfaces. This can be seen in Figures 4.15, 4.16 and 4.17, where the Da_0^{-1} values obtained by means of (4.116) with (4.129) and $a_v = 1$ are plotted versus z_{0+}. These data consist of measurements of heat transfer, of transport of radioactive vapor of Thorium-B and of water to and from surfaces with various types of roughness elements. The medians of all these Da_0^{-1} data, reduced to $z_{0+} = 10$, are plotted versus Sc in Figure 4.18; the slope of the line is (1/2), as

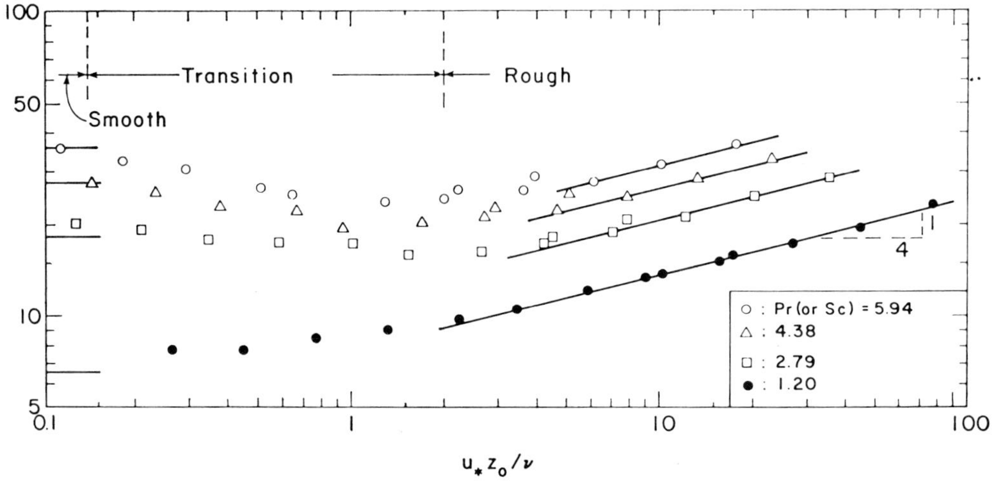

Fig. 4.15. Data of Dipprey and Sabersky (1962) processed as $[(Cd_r^{1/2}/Ce_r) - (Cd_r)^{-1/2} + 5]$, versus roughness Reynolds number z_{0+}. The straight lines in the smooth (slope zero) and the rough (slope 1/4) domains fit (4.126) and (4.133), respectively. The ordinate corresponds to Da_0^{-1} in the rough domain but to $(Da_0^{-1} - 8.5)$ in the smooth domain (from Brutsaert, 1975a).

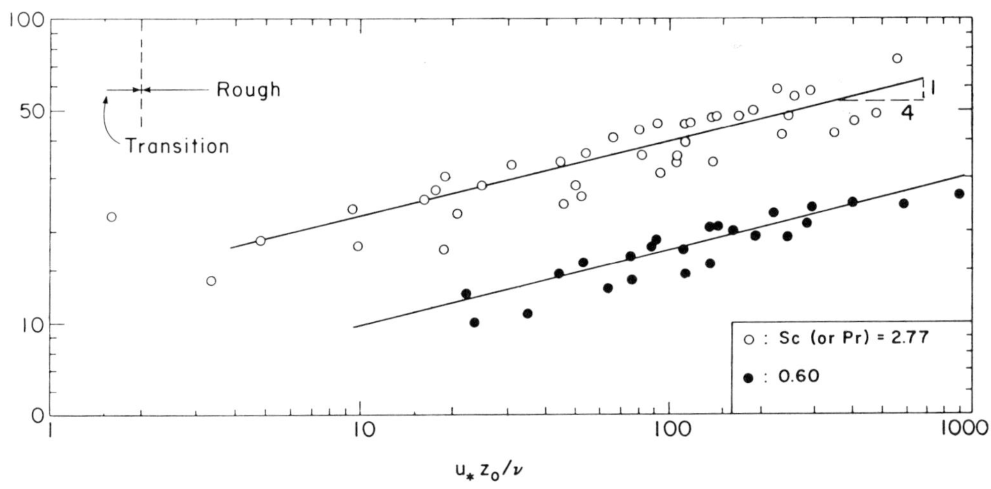

Fig. 4.16. Data of Chamberlain (1968) processed as Da_0^{-1} by means of (4.116) and (4.129) with $a_v = 1$, versus roughness Reynolds number z_{0+}. The straight lines, representing the median of the data, have a slope of (1/4), in agreement with (4.132) (from Brutsaert, 1975a).

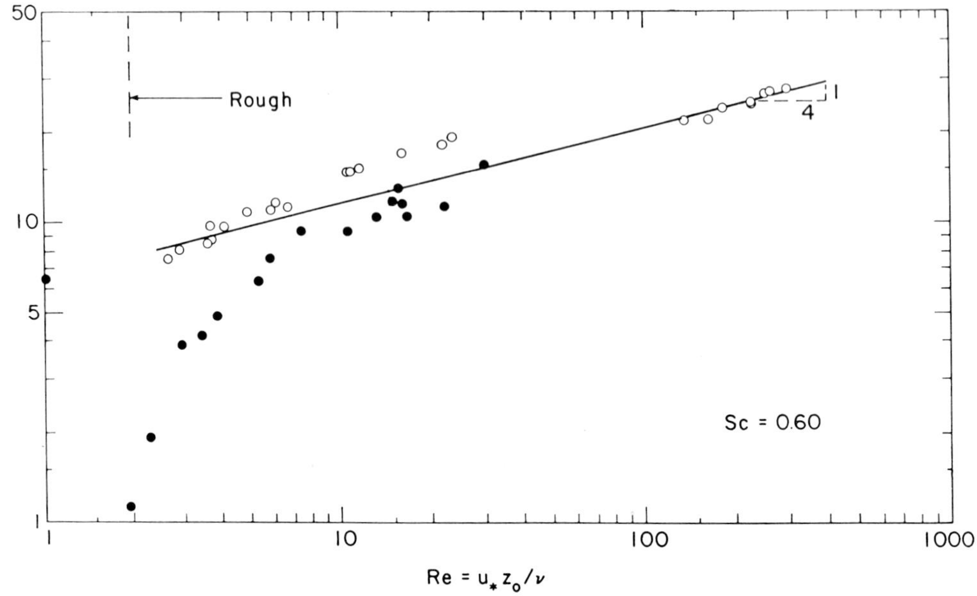

Fig. 4.17. Evaporation data of Mangarella *et al.* (1971), obtained for a water surface with wind waves; the data are processed as Da_0^{-1} by means of (4.116) and (4.129) with $a_v = 1$ and plotted versus roughness Reynolds number z_{0+}. The open circles refer to heated water, and the solid circles to isothermal conditions (from Brutsaert, 1975a).

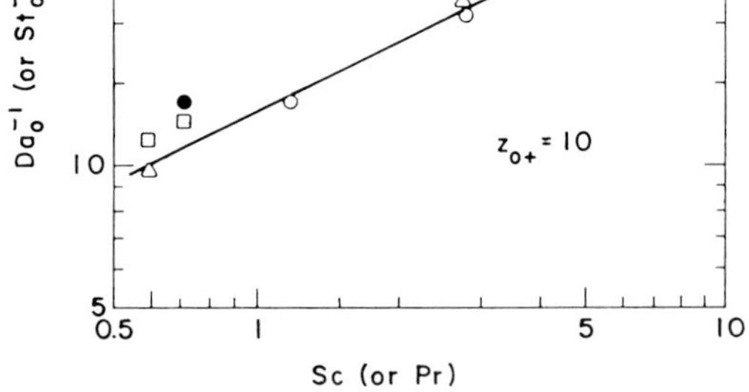

Fig. 4.18. Values of Da_0^{-1} for rough surfaces, which are reduced to $z_{0+} = 10$, and obtained by means of (4.116) and (4.129) with $a_v = 1$; they are plotted versus Schmidt number. Open circles represent the median of the data of Dipprey and Sabersky (1962); triangles, Chamberlain (1968); squares, Mangarella *et al.* (1971) for wind waves; solid circle, Nunner (1956) (from Brutsaert, 1975a).

required by (4.132). The value of the empirical constant resulting from Figure 4.18 is $C_R^{-1} = 7.3$, approximately. Substitution of this value in (4.132) yields thus with (4.129) and $a_v = 1$, the result for bluff-rough surfaces

$$\mathrm{D}a_0^{-1} - a_v^{-1}\,\mathrm{Cd}_0^{-1/2} = 7.3\,z_{0+}^{1/4}\,\mathrm{Sc}^{1/2} - 5. \tag{4.133}$$

The experimental data used in the determination of C_R cover the range $0.6 \leq (\mathrm{Sc}, \mathrm{Pr}) \leq 6$, which is the range of interest to meteorology.

The result of (4.133) has been tested by Merlivat (1978) by means of an isotopic method. It yielded good agreement with experimental evaporation rates of stable isotopes of water, namely $H_2^{16}O$, HDO, and $H_2^{18}O$, from a wavy water surface for $z_{0+} \geq 1$.

It must be pointed out that the available experimental data only cover Reynolds numbers, z_{0+}, smaller than 1,000. Thus, (4.133) and the other expressions in Table 4.2 as well, may not be valid for very rough surfaces, for example, to describe heat transfer when z_0 is of the order of a meter or more. In addition, the water surface data were taken under conditions of developing wind waves. The expressions may thus not be valid over water surfaces with swell, that is, with slowly decaying waves; therefore, more research is needed to study these problems.

The expressions of Table 4.1 are restricted to $z_{0+} < 0.13$ approximately, and those of Table 4.2 to $z_{0+} > 2$ approximately. There is no theory available to model the transitional regime from smooth to rough flow. For practical purposes, it will probably suffice to apply a suitable interpolation. Such interpolations have been used by Yaglom and Kader (1974), Kondo (1975) and Brutsaert and Chan (1978) in various ways. Simpler still is Merlivat's (1978) criterion; she observed that in the case of water waves $z_{0+} = 1$ can be used as the upper limit of validity of (4.126) and the lower limit for (4.133).

e. Surfaces with Permeable Roughnesses: The Canopy Sublayer

Major parts of the earth's surface are covered with vegetation. Most plants are not typically bluff obstacles. Rather, vegetational surfaces consist of roughness elements which are characteristically of a more permeable and fibrous nature, and which are often placed quite densely.

Momentum Transfer

There are many different types of vegetation so that it is quite difficult to make broad generalizations in terms of the formulation of (4.101) with (4.104). It is generally accepted that in the leafy part of a dense, uniform, tall vegetational canopy, the profiles of the mean wind speed $\bar{u}(z)$ and the mean horizontal shear stress $\tau(z)$ are decaying functions with depth below the top of the vegetation (see Figure 4.19 through 4.22). This was already observed around 1925 by Geiger (1961; p. 327). Several profile functions have been proposed in the literature. The usual way to derive such profile functions, is to start with the assumption that the branches and leaves are diffusely distributed, so that they act as a continuum-like sink for momentum. The horizontal equation of motion in the absence of a pressure gradient is then (cf., (3.69)]

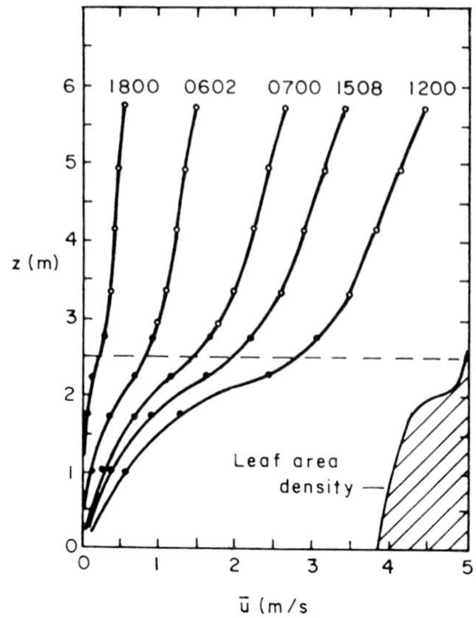

Fig. 4.19. Some representative wind speed profiles for 10 min. periods, with the beginning time as indicated, for a 2.5m high maize stand, taken near Ithaca, N.Y. (adapted from Wright and Brown, 1967).

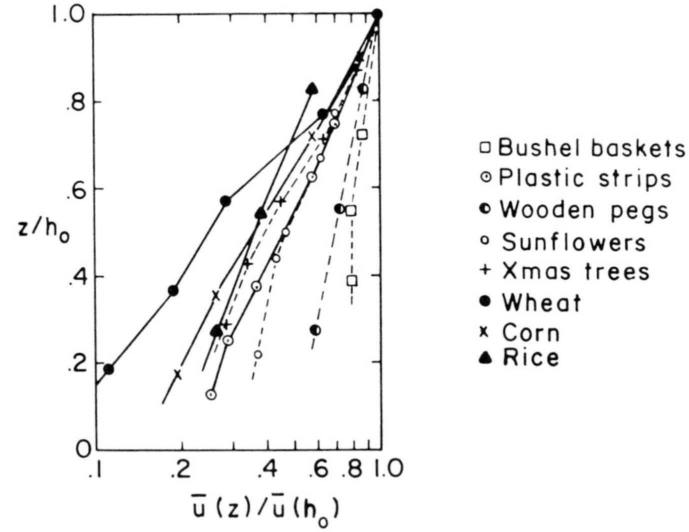

Fig. 4.20. Some examples of observed wind profiles within simple canopies (adapted from Cionco, 1978).

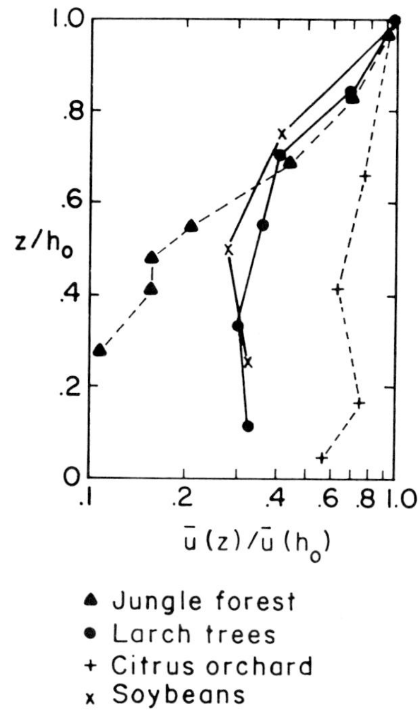

Fig. 4.21. Some examples of observed wind profiles within complex canopies (adapted from Cionco, 1978).

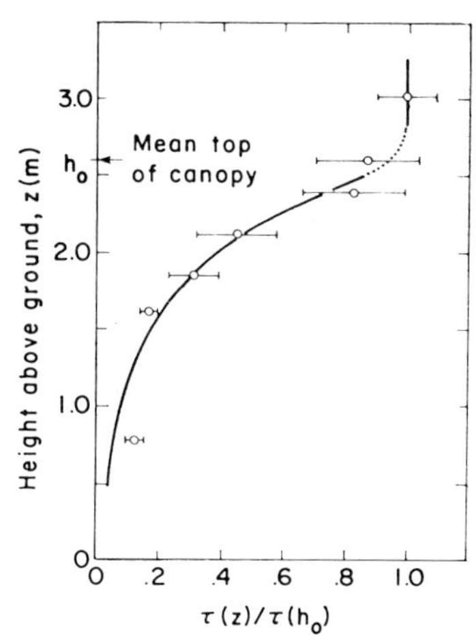

Fig. 4.22. The measured variation of the eddy momentum flux $-\overline{(u'w')}$ with height within a maize (corn) canopy in Illinois. The curve is the exponential profile (4.142) with $a_d = 2$ (adapted from Hicks and Sheih, 1977).

$$-\frac{d\tau}{dz} + D_f = 0 \tag{4.134}$$

where $\tau(z)$ is the horizontal shear stress in the air and D_f the momentum sink term, that is the drag experienced by the foliage per unit volume of air. In practically all derivations it is further assumed, first, that the shear stress is proportional to the velocity gradient through the definition of the eddy viscosity

$$K_m = \frac{(\tau/\rho)}{(d\bar{u}/dz)} \tag{4.135}$$

and second, that D_f is proportional to $\bar{u}^2$ as follows

$$D_f = A_f C d \rho \bar{u}^2/2 \tag{4.136}$$

where $A_f = A_f(z)$ is the surface area (both sides) of leaves per unit volume of air, and Cd_f is a foliage drag coefficient. A_f is related to the leaf area index, i.e., the area (one side) of foliage per unit area of ground surface by the integral LAI $= \int_0^{h_0} A_f \, dz/2$. For a given A_f and z in a given type of vegetation Cd_f is probably a function of Reynolds number $Re_L = (\bar{u}L_f/\nu)$, where L_f is a characteristic size scale of the leaves as follows $Cd_f = a \, Re_L^b$ where a and b are approximately constants. For single objects of various shapes and angles of incidence, values of a and b can be estimated (e.g., Schlichting, 1960). The matter is complicated in actual canopies, however, due to the distribution of angles of attack of the leaves, the variety in their shapes and their mutual interference, depending on the foliage density of $A_f(z)$. Thom (1971) concluded that Cd_f is proportional to $\bar{u}^{-1/2}$ for an artificial crop consisting of cylinders and for beans; also Inoue and Uchijima (1979) arrived at the same conclusion for rice and maize, but the scatter was quite large. Seginer *et al.* (1976) found that $b = 0$ for a model canopy of slender rods. It should be noted that, in a continuum-like treatment consisting of a parameterization by means of mean wind profile equations, the details of the flow and the exact value of Cd_f are of less concern.

Equation (4.134) with (4.135) and (4.136) was probably first applied around 1960 by Ordway *et al.* (1963) and also by Tan and Ling (1963), who solved it numerically. Depending on the assumed form for K_m and $A_f Cd_f$, various wind velocity profile functions can be obtained. With the assumption of a constant mixing length l_c in the canopy, so that $K_m = l_c^2 \, |d\bar{u}/dz|$, and $A_f Cd_f =$ const, one obtains the exponential profile, introduced independently by Inoue (1963) and Cionco (1965),

$$\bar{u} = \bar{u}(h_0) \exp\left[-a_w\left(1 - \frac{z}{h_0}\right)\right] \tag{4.137}$$

where $\bar{u}(h_0)$ is the mean velocity at $z = h_0$, and a_w is an extinction or absorption parameter. With the assumption $(K_m/\bar{u}) = [K_m(h_0)/\bar{u}(h_0)]$ in the canopy and $A_f Cd_f =$ const, Cowan (1968) obtained a hyperbolic sine profile

$$\bar{u} = \bar{u}(h_0) \left[\frac{\sinh(a'_w z)}{\sinh(a'_w h_0)}\right]^{1/2} \tag{4.138}$$

where a'_w is a parameter; this result satisfies the condition that $\bar{u} = 0$ at ground level. Finally, if both K_m and $A_f Cd_f$ are taken as constants, one obtains (Landsberg and James, 1971; Thom, 1971)

$$\bar{u} = \bar{u}(h_0)\left[1 + a''_w\left(1 - \frac{z}{h_0}\right)\right]^{-2} \tag{4.139}$$

where a''_w is another parameter.

The above three profile functions were derived on the basis of markedly different assumptions concerning K_m or $A_f Cd_f$ which are not easy to verify. Nevertheless, all three functions give a mean velocity which is decaying with depth and the differences among them, when fitted to data, are usually well within the scatter observed in field experiments. A more serious matter is that, in some canopies, with the leafy part near the top and rather empty trunk space near the ground surface, there have been observations of a secondary wind speed maximum (e.g., Lemon *et al.*, 1970; Oliver, 1975); monotonically decaying profiles such as (4.137) through (4.139) are inadequate to describe this. Some attempts have been made to deal with this problem (e.g., Kondo and Akashi, 1976; Shaw, 1977) but difficulties remain. However, in problems of turbulent transport between vegetation and atmosphere, this may be of less importance; the air flow above the vegetation is mainly interacting with layers near the top of the canopy, and less with layers near the soil surface.

At present, the exponential profile is most widely used and its applicability is best understood. Thus, some elaboration on it is of interest. A simpler derivation of the exponential profile, which does not invoke any formal mixing length hypothesis can be given as follows. The starting point is (4.134), and it is assumed that the momentum sink term is proportional to the intensity of turbulence, that is the turbulent kinetic energy $\bar{e}_t$ [cf., (3.64)]; the latter may, in turn, be assumed to be proportional to the covariance of the turbulent velocities, $-\overline{u'w'} = u_{*c}^2$. This then gives

$$D_f = 2\gamma_d \rho u_{*c}^2 \tag{4.140}$$

where γ_d is a parameter, reflecting the removal efficiency of momentum by the foliage; it may depend on u_{*c} and z, but for a uniform canopy and high Reynolds number flows, as a first approximation, it can probably be taken as a constant. It is convenient to scale γ_d with the mean height of the roughness obstacles,

$$\gamma_d = a_d/h_0. \tag{4.141}$$

Integration of (4.134) with (4.140) and (4.141) yields

$$\frac{\tau(z)}{\tau(h_0)} = \frac{u_{*c}^2(z)}{u_{*c}^2(h_0)} = \exp(-2\,a_d\xi) \tag{4.142}$$

where

$$\xi = (h_0 - z)/h_0 \tag{4.143}$$

is a normalized depth coordinate pointing into the canopy. Very few measurements of $\tau(z)$ have been published; the $\overline{(u'w')}$ data of Seginer *et al.* (1976) and Finnigan and Mulhearn (1978) in artificial canopies, support the exponential form of (4.142). As shown in Figure 4.22, Hicks and Sheih (1977) found that (4.142) with $a_d = 2$ gave a good representation of their covariance measurements in a dense immature corn (maize) canopy as far down as $\xi = 0.7$. If also (4.136) is assumed to be valid, comparison with (4.140) and (4.142) immediately yields for a uniformly dense canopy

$$\bar{u} = \bar{u}(h_0) \exp(-a_w \xi). \tag{4.137'}$$

This derivation for A_f = const suggests that $a_w = a_d(1 + b/2)^{-1}$; thus, although almost nothing is known experimentally about the relationship between a_w and a_d, in fully turbulent flow, their values cannot be very different.

In Figures 4.20 and 4.21, observed profiles of the mean velocity are shown for different types of natural and model canopies. It can be seen that in simple canopies, (4.137′) may be valid down to as low as 0.1 h_0; in the more complex canopies of Figure 4.21, it is valid only in approximately the top half of the canopy.

The extinction parameter a_w has been determined for different types of canopies. For example, Cionco (1972) has suggested that a_w increases as both canopy density and element flexibility increase. In his review of available experimental data, he concluded that a_w varies between 0.4 and 0.8, approximately, for sparse rigid elements (citrus orchard, wooden pegs, bushel baskets); between 1 and 2, approximately, for moderately dense semirigid elements (corn, rice, larch, christmas trees, sunflower and plastic strips); between 2 and 4 for dense flexible elements (wheat, oats, im-

TABLE 4.3
Illustration of the change of the extinction coefficient a_w of the mean velocity profile during the growing season of maize (after Inoue and Uchijima, 1979)

Height, h_0 (cm)	a_w	Mean foliage area density A_f (cm^2/cm^3)
50	1.6	0.022
140	2.0	0.036
225	2.6	0.038
277	3.0	0.030

mature maize). Table 4.3, after Inoue and Uchijima (1979), gives an illustration of the change in a_w with plant growth. In a more conceptual framework, Kondo (1971, 1972) has related a_w to the zero-displacement height d_0, as follows

$$a_w = \frac{A_k h_0}{(h_0 - d_0)} \tag{4.144}$$

where, from experiments, A_k was found to be a constant slightly smaller than or equal to unity. On this basis, since d_0 is of the order of $(2h_0/3)$, it has been suggested (e.g., Brutsaert, 1975c) that $a_w = 3$ can serve as a representative value for a dense vegetation, when no other information is available.

The flux-profile relationship expressed as the eddy viscosity defined in (4.135) is

$$K_m = K_m(h_0) \exp(-a_m \xi) \tag{4.145}$$

where $K_m(h_0)$ is the value of K_m at $z = h_0$, and $a_m = (2a_d - a_w)$ is an extinction parameter, whose value is probably close to those of a_d and a_w (see Figure 4.23).

In past work, the values of $\bar{u}(h_0)$, $u_{*c}(h_0)$, and $K_m(h_0)$ required in (4.137), (4.142) and (4.145), respectively, have usually been taken as those obtained by simply joining these equations with the corresponding formulations of the dynamic sublayer at

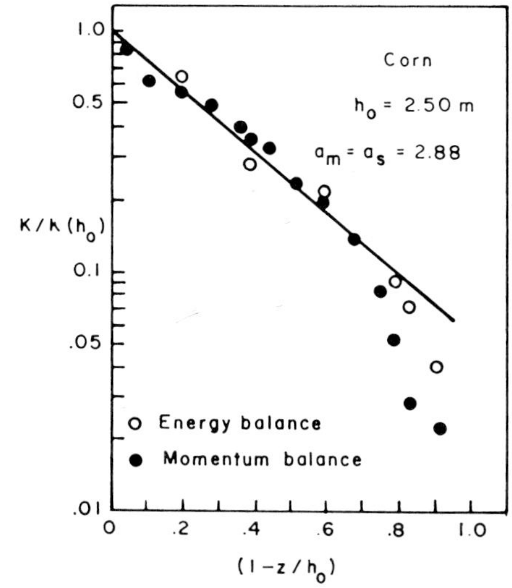

Fig. 4.23. Normalized eddy viscosities (solid circles) and eddy diffusivity (open circles) profiles in a maize stand near Ithaca, N.Y. The straight line represents (4.145) and (4.150) (adapted from Wright and Brown, 1967).

$z = h_0$. Thus it is, in fact, assumed that the mean height of the plants h_0 coincides with the lower boundary of the surface sublayer $z = h$. The assumption that the transition from the canopy sublayer to the dynamic sublayer is sudden at $z = h_0$. is probably analogous to directly joining (4.123) with (4.3) for a smooth surface, as shown by the dashed lines of Figure 4.1. Although it is an oversimplication, this assumption may provide a useful first approximation. Accordingly, the reference value $u_{*c}(h_0)$ in (4.124) is u_*; the reference value $K_m(h_0)$ in (4.145) is, by virtue of (4.135), with the logarithmic profile

$$K_m(h_0) = ku_*(h_0 - d_0). \tag{4.146}$$

Similarly, one obtains for the mean velocity reference $\bar{u}(h_0)$ by substituting $z = h_0$ in (4.6). As will be seen in Chapter 5 z_0 and d_0 are of the order of $(h_0/8)$ and $(2\ h_0/3)$, respectively; hence, if $z = h_0$ is taken as the thickness of the canopy sublayer, so that $\bar{u}_h = \bar{u}(h_0)$, the canopy layer drag coefficient Cd_0 defined in (4.110) is of the order of

$$\mathrm{Cd}_0^{-1/2} \simeq 2.5. \tag{4.147}$$

Interfacial Transfer Coefficients for Scalars

Equations (4.106) and (4.107) with (4.103) and (4.105) indicate on what types of variables the transfer coefficients Da_0 and St_0 can be expected to depend. However, for an actual vegetation, it is very difficult to be more specific beyond this formal generalization. Compared to smooth and bluff-rough surfaces, very little experimental and theoretical work has been done, which would be relevant for a general but practical formulation of bulk transfer coefficients for surfaces with permeable roughness elements. In view of the wide diversity in vegetational covers, a rigorous similarity formulation may very well be an impossible task. A better understanding of this problem will require more research. But in the context of evaporation and heat transfer at the earth's surface, the available results have some practical and useful implications.

Experimental results have been presented by Chamberlain (1966), Stewart and Thom (1973), Garratt and Hicks (1973), Garratt (1978b) and Garratt and Francey (1978), and theoretical analyses by Cowan (1968), Thom (1972) and Brutsaert (1979a). These studies show that there is a pronounced dissimilarity between the bulk transfer properties for scalars at permeable-rough surfaces and those at bluff-rough surfaces. More specifically, they show that for heat and mass transfer at a vegetational cover, the quantity $(\mathrm{Da}_0^{-1} - a_v^{-1}\,\mathrm{Cd}^{-1/2})$, $[= (a_v B)^{-1}]$ is relatively insensitive to changes in z_0 and that it depends only mildly on u_*.

For example, Chamberlain (1966) obtained a fairly complete set of data on transport of gases to and from artificial and natural grass. In the case of Thorium-B deposition on artificial grass, the values of $[a_v(\mathrm{Da}_0^{-1} - a_v^{-1}\,\mathrm{Cd}^{-1/2})]$ $(=B^{-1})$ ranged between 6.1 and 12.8 for a range of u_* between 12.8 and 200 cm s^{-1}. In the case of evaporation from wet grass, B^{-1} ranged only between 4.3 and 7.3 for u_* ranging from 15.8 to 170 cm s^{-1}. Garratt and Hicks (1973), who collected some other available evidence, pointed out that B^{-1} of the experimental data for vegetational surfaces is not very sensitive to changes in z_{0+}, this in contrast to B^{-1} for bluff roughness elements. Garratt (1978b) concluded that $k\ B^{-1}$ for heat transfer is about 2.5 ± 0.5 above flat rough terrain of savannah scrub consisting of 25 percent tall trees of average height 8 m, some 65 percent dry grass of up to 1 m, and 10 percent burnt grass and sandy soil. Although the surface temperatures of these three types of vegetation were different, for the analysis of the heat transfer data an 'overall effective surface temperature' was derived from aircraft and weighted ground-based radiometric observations. Several of these experimental results on $k\ B^{-1}$ are summarized in Figure 4.24.

In the theoretical analysis of the problem, an approach can be followed analogous to that based on (4.134), (4.135) and (4.136) for momentum transfer. Thus, the turbulent transfer of any scalar inert admixture of the flow is assumed to be governed by

$$-\frac{\mathrm{d}F}{\mathrm{d}z} + S_f = 0 \tag{4.148}$$

where F is the vertical specific flux of the admixture in the canopy air and S_f the distributed source (or negative sink) term of the admixture emanating from the foliage surfaces. Again, as is the case for momentum in (4.135) the flux-profile relationship can be expressed in the form of an eddy diffusivity

$$K_c = \frac{-(F/\rho)}{(\mathrm{d}\bar{c}/\mathrm{d}z)} \tag{4.149}$$

where $\bar{c}$ is the concentration of the physical quantity under consideration; for sensible heat it is $\bar{c} = c_p\,\bar{\theta}$ and for water vapor $\bar{c} = \bar{q}$.

Equations (4.148) and (4.149) are the basis for the many numerical models in the literature which, starting with Philip's (1964), have been used to simulate turbulent

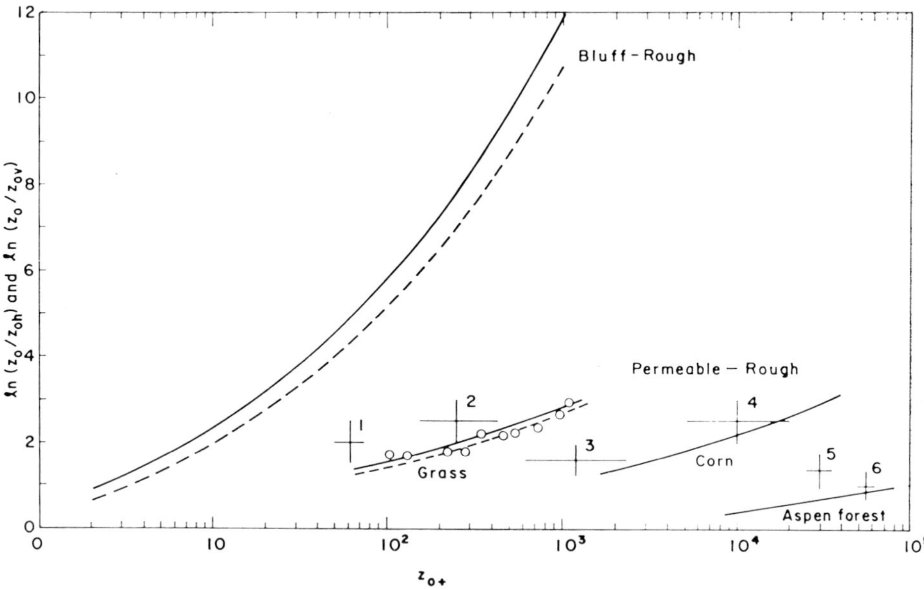

Fig. 4.24. Comparison between experimental and theoretical results on kB^{-1}, that is $[k(a_h\mathrm{St}_0^{-1}$— $\mathrm{Cd}^{-1/2})] = \ln(z_0/z_{0h})$ for heat transfer (solid curves) and $[k(a_v\mathrm{Da}_0^{-1}$— $\mathrm{Cd}_0^{-1/2})] = \ln(z_0/z_{0v})$ for evaporation from wet surfaces (dash curves), plotted versus roughness Reynolds number $z_{0+} = (u_* z_0/\nu)$. The curves for grass, maize and aspen forest are theoretical results obtained with (4.162) and (4.160) for Pr = 0.71 (Sc = 0.59), $C_L = 0.25$, $m = 0.25$ and $n = 0.36$. The numbered points represent experimental results for heat transfer and they are taken from the collections of Garratt and Hicks (1973) and Garratt and Francey (1978); 1 is short grass; 2 medium length grass; 3 bean crop; 4 savannah scrub; 5 and 6 pine forest (horizontal bars represents z_{0+} range and vertical bars the standard deviation). The data points superimposed on the theoretical curve for evaporation from wet grass were obtained by Chamberlain (1966). The curves for the bluff-rough surfaces are obtained with (4.133) [or (5.28) and (5.29)] (adapted from Brutsaert, 1979a).

transfer in canopies. In principle, numerical solutions allow considerable variability and flexibility in the assumed boundary conditions, and the assumed K_c and S_f for different types of vegetation. On the other hand, analytical solutions, whenever they are feasible, usually permit a more concise parameterization. Bulk transfer coefficients, as defined in (4.114) and (4.117), require a concentration $\bar{c}_s$ at the surfaces of the foliage elements which is constant with elevation. For this case of constant $\bar{c}_s$, different analytical solutions have been derived by Cowan (1968) (see, however, Thom (1972; p. 129)) and Brutsaert (1979a). The latter, which is more general, is summarized in what follows, as an illustration of the results that are obtainable with this type of analysis.

To solve (4.148) with (4.149), the functional form of K_c and S_f must be specified. It appears from the literature that, at least in the dense upper layers of many types of vegetational canopies, the eddy diffusivity of a scalar admixture can be described by an exponential profile, similar to $\bar{u}$, u_* and K_m (see Figure 4.23); this means that, in spite of possible fundamental objections Reynolds's analogy may serve as a working approximation, and that the flux-profile relationship for a scalar is roughly similar to that for momentum. The eddy diffusivity profile for a uniform canopy can, therefore, be described by

$$K_c = K_c(h_0) \exp(-a_s \xi) \tag{4.150}$$

where a_s is a parameter of the same order of magnitude as a_w in (4.137) or a_m in (4.145). Most determinations of $K_c(z)$ in the past, have been made by means of an energy budget method applied to different levels within the canopy. Among the first to apply this technique were Saito (1962) for wheat and Uchijima (1962) for a paddy field. Uchijima and Wright (1964) reported that Uchijima's (1962) K values for rice could be fitted by (4.150) with $a_s = 3.1$. Similar findings were reported by Brown and Covey (1966) with $a_s = 2.6$ for corn; by Denmead (1964; Brown and Covey, 1966) with $a_s = 4.25$ for pine forest; by Lemon (1965) with $a_s = 2.5$ for red clover; by Wright and Brown (1967) with $a_s = 2.88$ (and $a_s = a_w$) for maize (see Figure 4.23); by Uchijima *et al.* (1970) with a_s from 2.46 to 2.88 for corn; and by Denmead (1976b) with a_s from 2.2 to 3.3 for wheat. In a different determination of $K_c(z)$, Meroney (1970) solved the convective-turbulent diffusion equation with measurements of horizontal and vertical concentration gradients of a helium plume within a model forest canopy in a wind tunnel. Also in this case, the obtained eddy diffusivity $K_c(z)$ was quite similar to (4.150). It should be noted though, that (4.150) is probably valid only for canopies with a fairly uniform leaf area distribution $A_f(z)$; Inoue and Uchijima (1979; Figure 8) have obtained $K(z)$ profiles for a rice-canopy with a leaf-area strongly concentrated in the middle layers, which differed markedly from (4.150).

The source term S_f in (4.148) results from the flux from or to the surfaces of the individual leaves and stem segments in the canopy. It can be given by a bulk transfer equation, similarly to (4.140),

$$S_f = A_f \, \mathrm{Ct}_f \, \rho u_{*c}(\bar{c}_s - \bar{c}) \tag{4.151}$$

where Ct_f is a bulk transfer coefficient for the foliage elements. From dimensional analysis, and from known coefficients for various geometrical shapes, this coefficient probably has the form

$$\mathrm{Ct}_f = C_L \, \mathrm{Re}_{*c}^{-m} \, \mathrm{Sc}^{-n} \tag{4.152}$$

in which $\mathrm{Re}_{*c} = (u_{*c} L_f / \nu)$ is a local canopy Reynolds number, L_f a characteristic size of the foliage elements, Sc the Schmidt number; C_L, m and n are parameters which may depend on the shape, density or crowding and orientation of the leaves and on the intensity of the turbulence. In a review of likely values of these parameters and past work (Brutsaert, 1979a; Appendix B), it is shown that the powers of (4.152) probably occupy a fairly narrow range $(1/5) \leq m \leq (1/2)$ and $(1/2) \leq n \leq (2/3)$. However, the experimental evidence is still inconclusive. In (4.151) and (4.152) use is made of u_{*c} as velocity scale, just like in (4.140). It would also be possible to use the mean wind speed $\bar{u}$ as velocity scale, like in (4.136). However, in view of the similarity of (4.142) and (4.137′) it is likely that this would make little difference from a practical point of view. The local friction velocity is probably a better measure of the turbulence intensity. Still, if the velocity $\bar{u}$ were to be used in (4.151) and (4.152), the parameter

C_L could be readily adjusted by merely multiplying it by the ratio $(u_{*c}/\bar{u})^{1-m}$; in view of (4.147) this quantity is of the order of $(2.5)^{m-1}$.

When $\bar{c}_s$ is assumed constant it is convenient to normalize the concentration, viz.

$$\chi = \frac{(\bar{c}_s - \bar{c})}{(\bar{c}_s - \bar{c}_h)} \tag{4.153}$$

where $\bar{c}_h$ is the concentration in the air at $z = h_0$; thus combining (4.148) with (4.149) through (4.153) one obtains

$$\frac{d^2\chi}{d\xi^2} + C_1 \frac{d\chi}{d\xi} - C_2 e^{N\xi} = 0 \tag{4.154}$$

in which $C_1 = -a_s$,

$$C_2 = \frac{A_f C_L h_0^2}{a_c k(h_0 - d_0) \operatorname{Re}_*^m \operatorname{Sc}^n} \tag{4.155}$$

and

$$N = -a_d(1 - m) + a_s. \tag{4.156}$$

In (4.155) a_c is the analog of a_v and a_h for any scalar admixture, and Re_* is defined by

$$\mathrm{Re}_* = \frac{u_* L_f}{\nu} \tag{4.157}$$

where u_* is the friction velocity at $z = h_0$ which is assumed to be the same as in the surface sublayer above the canopy. To derive a bulk transfer coefficient the boundary conditions can be taken as

$$\begin{aligned} \chi &= 1, & \xi &= 0 \\ \chi &\to 0, & \xi &\text{ very large.} \end{aligned} \tag{4.158}$$

The second condition is obtained by assuming that near the top of the canopy the effect of the ground surface is not 'felt', so that the exact formulation of the condition at the ground is immaterial. Actually, this same assumption is already implicit in the exponential profiles (4.137), (4.142), (4.145) and (4.150). The solution of (4.154) with (4.158) allows the calculation of the specific flux at $z = h_0$, F_0, i.e., the total specific flux from the vegetational surface by means of (4.149) with (4.150). This yields

$$F_0 = -a_c k\,[(h_0 - d_0)/h_0]\,\rho u_*(\bar{c}_s - \bar{c}_h)G_0 \tag{4.159}$$

where $G_0 = (d\chi/d\xi)$ is the gradient at $\xi = 0$; for the special case when $a_s = a_d = a$ (say) this is

$$G_0 = -C_2^{1/2} K_{\lambda-1}\left(\frac{2C_2^{1/2}}{ma}\right) \Big/ K_\lambda\left(\frac{2C_2^{1/2}}{ma}\right) \tag{4.160}$$

where C_2 is defined in (4.155) and where $K_\lambda(\;\;)$ is the modified Bessel function of order $\lambda = m^{-1}$ of the second kind (e.g., Abramowitz and Stegun, 1964). The function $G_0 = G_0(C_2)$ is shown in Figure 4.25 for typical values of the parameters a and m. The specific flux F_0 corresponds to E and H in (4.108) and (4.109). Hence, the interfacial Dalton (or Stanton) number for a canopy is

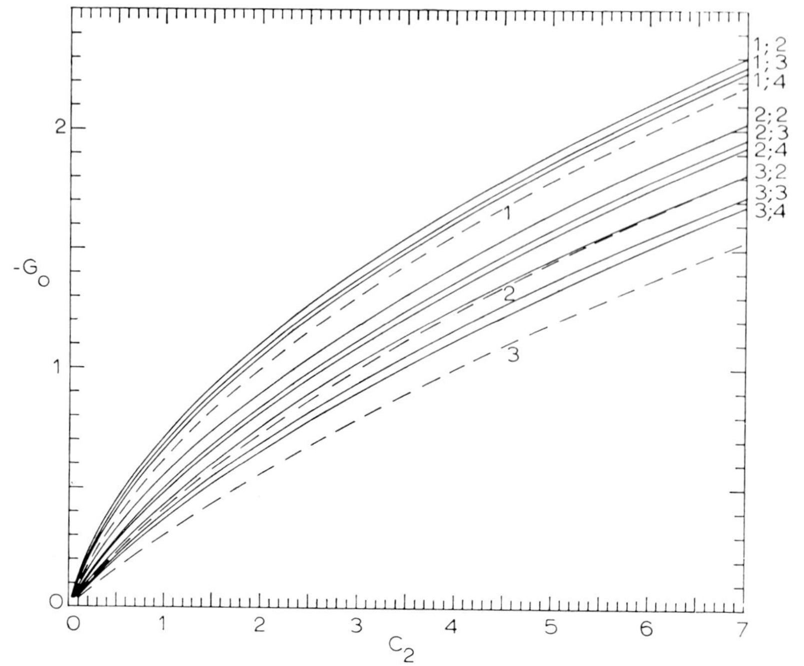

Fig. 4.25. The function $-G = -G_0(C_2)$ calculated by means of (4.160). Each solid curve is identified by 2 numbers: the first is the value of a $(=a_s = a_d)$ and the second the value of m^{-1}. The dashed curves are calculated for the simple case $N = 0$ in (4.154); the identification number is the value of a (from Brutsaert, 1979a).

$$\mathrm{Da}_0 = a_v k[(h_0 - d_0)/h_0](-G_0) \tag{4.161}$$

which can be used in (4.116) [or (4.118)], in the following form [$= (a_v B)^{-1}$],

$$\mathrm{Da}_0^{-1} - a_v^{-1}\,\mathrm{Cd}_0^{-1/2} = (a_v k)^{-1}\left\{\left[\frac{h_0}{(h_0 - d_0)(-G_0)}\right] - \ln\left[\frac{(h_0 - d_0)}{z_0}\right]\right\}. \tag{4.162}$$

When no information is available on z_0 and d_0, (4.162) can probably be approximated by $2.5\,[3/(-G_0) - 1]$.

The experimental data of Chamberlain (1966) on $(k\,B^{-1})$ for a grass surface are well suited for inverse calculations to determine the values of the parameters in (4.152). By matching the theoretical result (4.162) and (4.160) for $a = 2$ with these data (see Figure 4.26), the following were obtained $C_L = 0.25$, $m = 0.25$, and $n = 0.36$; a slightly different analysis, with the data of Thorium-B only, yielded $C_L = 0.29$, $m = 0.25$, and $n = 0.5$. Both sets of parameters produce rather similar results in direct prediction calculations. These parameter values are based on a fit of only one set of experimental data with this theoretical model. Therefore, they should be considered rather tentative and more research will be necessary. Nevertheless, as can be seen from the review of past work (Brutsaert 1979a, Appendix B) they are of the correct order of magnitude.

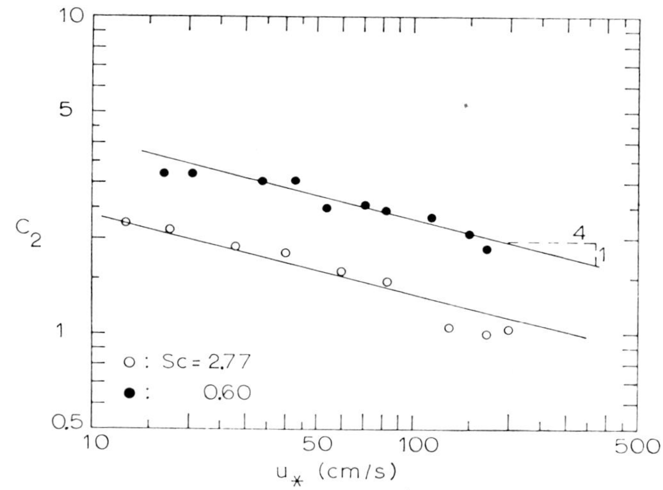

Fig. 4.26. Data on dry deposition of Thorium-B and evaporation of water from wet artificial grass obtained by Chamberlain (1966) and processed through (4.162) and (4.160) (cf., Figure 4.25) with $a = 2$ and $m = 1/4$. The parameters for the grass are $h_0 = 7.5$ cm, $A_f = 0.58$ cm^{-1}, $z_0 = 1$ cm, $d_0 = 5$ cm. The two straight lines correspond to (4.155) with $C_L = 0.25$ and $n = 0.36$ (from Brutsaert, 1979a).

As an illustration, the theoretical result for heat transfer with the parameters obtained for grass, has been applied to maize (corn) and aspen forest. These results are shown in Figure 4.24. The details of the computations and the choice of the canopy parameters a, L_f, A_f, h_0, z_0, d_0 are discussed by Brutsaert (1979a). It can be seen that the theory is in reasonable agreement with the available experimental data.

The results of the calculations shown in Figure 4.24 confirm the earlier generalized observations concerning $[ka_v(\mathrm{Da}_0^{-1} - a_v^{-1}\,\mathrm{Cd}_0^{-1/2})]$ $(\equiv k\,B^{-1})$ on the basis of the experimental evidence. They agree with the experiments (Chamberlain, 1966, 1968; Garratt and Hicks, 1973) in that $k\,B^{-1}$ as a function of z_{0+} falls in either of two categories. For surfaces with bluff roughness elements, (4.133) predicts that it rises steeply (up to $z_{0+} = 1000$); but for surfaces with densely placed permeable roughness obstacles, (4.160) predicts that it is relatively insensitive to z_{0+}. Closer inspection of the calculated examples shows that $k\,B^{-1}$ depends only slightly on u_* and it is practically independent of z_0. In fact, z_0 does not appear directly in (4.160) with (4.155). In the theoretical model $k\,B^{-1}$ can be totally independent from u_*, as some field data (e.g., Garratt, 1978b; Garratt and Francey, 1978) suggest, only if $m = 0$ in (4.152); but this matter will require more research. The theoretical solution thus shows that, in contrast to bluff-rough surfaces, for a given Sc (or Pr) the roughness Reynolds number z_{0+} is not the only nor even the main relevant parameter to classify experimental data for permeable-rough surfaces. Equation (4.155), defining C_2, contains several important variables, and it is their combination which governs the magnitude of G_0 in (4.160) and, thus, of the transfer coefficients in (4.162). The fact that all these variables (and probably some others which are not included in this simple theory) must be considered, complicates any rigorous similarity approach to this problem. As a first approximation, for practical purposes, when no other information is available, $k\,B^{-1}$

can be assumed to be of the order of 2 for scalars whose Schmidt or Prandtl number is of the order of 0.6 to 0.8. In the case of stands of large trees and forests, this quantity may be as small as 1 or even smaller.

In view of the uncertainty regarding the bulk transfer coefficients for the canopy sublayer, it would seem that, in practice, over canopies it is preferable to avoid formulations such as (4.114) or (4.117) which make use of the surface concentration, $\bar{c}_s$, $\bar{q}_s$ or $\bar{\theta}_s$. Still, there are situations where the use of the surface concentration [involving Ce_r and Ch_r (see Section 4.4b) or z_{0v} and z_{0h} (see Section 5.2a)] to formulate the surface flux is advantageous, or even unavoidable. One is the situation involving condensation, such as dew formation, or evaporation from a wet vegetation. In both cases, a knowledge of the surface temperature suffices to determine the saturation value $q_s^* = q_s^*(\bar{T}_s)$. A second, rather obvious application is the calculation of sensible heat transfer to and from a vegetational surface, when the foliage temperature is known. The temperature of the vegetational elements is rarely uniform but often a suitable average should be satisfactory. For example, the mean surface temperature, as seen by remote sensing, should be quite useful in this regard; an application of this idea to savannah scrub has been presented by Garratt (1978b) and the results are shown in Figure 4.24. A third type of situation, where surface concentration values can or should be used, is dry deposition of a pollutant, for which the vegetation acts as a perfect sink, such as Thorium-B of Chamberlain (1966); in this case the surface concentration can be taken as zero, i.e., $\bar{c}_s = 0$.

Evaporation from Dry Vegetation; Resistance Formulation

To describe evaporation, or rather transpiration from a vegetation, whose surface is not actually wetted, it is impossible to make use of bulk transfer equations such as (4.114) or related ones, because the specific humidity at the leaf surfaces q_s is unknown. An approach, widely used in agricultural micrometeorology, to overcome this difficulty, consists of replacing q_s by the saturation value $q_s^* = q_s^*(T_s)$, which is assumed to prevail in the sub-stomatal cavities of the foliage elements; in addition, a bulk stomatal resistance can be introduced to characterize the transfer between the stomatal cavities and the leaf surface. The concept of a resistance to vapor transfer at the leaf surfaces was suggested by Penman and Schofield (1951) in the context of a correction to Penman's Equation (10.15) and subsequently further developed and applied in various forms by Slatyer and McIlroy (1961), Monteith (1965, 1973), Cowan (1968), Thom (1972, 1975).

The bulk stomatal resistance r_{st} can be defined by

$$E = \rho(\bar{q}_s^* - \bar{q}_s)/r_{st}. \tag{4.163}$$

As shown in Figure 4.27, there are different ways of incorporating this concept in the surface parameterizations discussed earlier in this chapter. The transfer in the canopy air can be formulated in terms of a canopy air resistance r_{0v}, defined by

$$E = \rho(\bar{q}_s - \bar{q}_h)/r_{0v} \tag{4.164}$$

which is equivalent to (4.108), so that $r_{0v} = (u_* \mathrm{Da}_0)^{-1}$. The unknown $\bar{q}_s$ can now be eliminated by combining (4.163) with (4.164), or

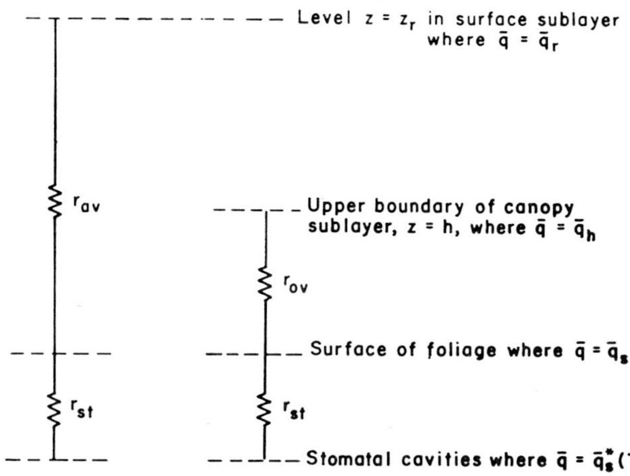

Fig. 4.27. Schematic diagram showing resistance parameters that can be used to describe transfer to and from a vegetational surface.

$$E = \rho(q_s^* - q_h)/(r_{st} + r_{0v}). \tag{4.165}$$

Hence, if one uses $\bar{q}_s^*$, which is known for a given surface temperature, instead of the unknown $\bar{q}_s$, the term $(\mathrm{Da}_0^{-1} - a_v^{-1}\,\mathrm{Cd}_0^{-1/2})[= (a_v B^{-1}]$ in (4.116), (4.119), (5.22) and other bulk transfer formulations, should be replaced by

$$(r_{st} + r_{0v} - a_v^{-1} r_{0m})\, u_*. \tag{4.166}$$

In (4.166) the canopy resistance r_{0m} for momentum is introduced for consistency of notation; it is defined by

$$r_{0m} = (u_*^2/\bar{u}_h)^{-1} \tag{4.167}$$

which is equivalent with (4.110) so that $r_{0m} = (u_*\mathrm{Cd}_0^{1/2})^{-1}$.

The transfer through the canopy air and in the air above it can also be represented by a combined resistance, r_{av} which may be called the aerodynamic resistance to water vapor; it is defined by

$$E = \rho(\bar{q}_s - \bar{q}_r)/r_{av} \tag{4.168}$$

which is equivalent to (4.114), so that, obviously, $r_{av} = (\bar{u}_r \mathrm{Ce}_r)^{-1}$. Elimination of the unknown $\bar{q}_s$ between (4.163) and (4.168) yields

$$E = \rho(\bar{q}_s^* - \bar{q}_r)/(r_{st} + r_{av}) \tag{4.169}$$

as the bulk transfer equation [cf., (4.114)] for a vegetational surface with a dry surface.

As an aside, to allow comparison with other work here, it should be noted that (4.169) is often written in terms of resistances which are quite different from r_{st} and r_{av}, viz.

$$E = \rho(\bar{q}_s^* - \bar{q}_r)/(r_c + r_a) \tag{4.170}$$

where r_c is usually called the canopy resistance and r_a the aerodynamic resistance. The latter is commonly defined by

$$r_a = (u_*^2/\bar{u}_r)^{-1} \tag{4.171}$$

which is equivalent to (4.115) and which shows that $r_a = (\bar{u}_r \mathrm{Cd}_r)^{-1}$. As illustrated, for example, in (4.119) (see also (5.10) and (5.21)), Ce_r (or r_{av}) may be quite different from Cd_r (or r_a). This means that, as also pointed out by Thom (1972), the canopy resistance r_c in (4.170) may be expected to differ from the bulk stomatal resistance r_{st} in (4.169). Of course, it is possible to define r_c by an equation similar to (4.163) in which $\bar{q}_s$ is replaced by $\bar{q} = \bar{q}(z_{0m} + d_0)$, as given by (4.33) with $z_1 - d_0 = z_{0m}$, $z_2 = z_r$. The physical meaning of all this is not easy to comprehend, so that the conceptual significance of the canopy resistance is rather obscure. In practical calculations, r_c is usually determined as the remainder or rest term after r_a is subtracted from the total resistance appearing as denominator in (4.170) or (7.169). This should present no particular difficulties. However, the unambiguous definition of r_{st} and the clear connection between r_{av} and Ce_r, make the formulation of (4.169) much preferable to that of (4.170).

The bulk stomatal resistance r_{st} and the related canopy resistance r_c have been determined in numerous experiments for various types of vegetation under different conditions. In addition, attempts have been made to relate them with such factors as the Bowen ratio, soil moisture suction in the root zone, soil moisture deficit, humidity deficit in the air and other variables, with the objective of obtaining empirical relationships (e.g., Monteith, 1965; Van Bavel, 1967; Szeicz and Long, 1969; Federer, 1977; Garratt, 1978b). Although none of the relationships obtained so far appear to be sufficiently general to be practical for predictive purposes, the bulk stomatal or canopy resistance formulation can be useful as a diagnostic index and in certain simulation models.

CHAPTER 5

The Surface Roughness Parameterization

5.1. THE MOMENTUM ROUGHNESS

The momentum roughness, z_{0m}, is an important parameter, not only for the wind profile, but it is also essential in the calculation of z_{0v} for water vapor, z_{0h} for heat and the roughness parameters for other scalars. For a smooth surface, that is when $z_{0+} < 0.13$, approximately, the momentum roughness is simply given by

$$z_{0m} = 0.135\ \nu/u_*. \tag{5.1}$$

The earth's surface, especially its land portion, is usually rough. For a rough surface, that is when $(u_* z_0/\nu) > 2$, approximately, one has

$$z_{0m} = z_0. \tag{5.2}$$

For very rough surfaces, it is also necessary to introduce the (zero-plane) displacement height d_0, as shown in (4.5) and (4.6). With the present state of knowledge, there is still no good substitute for the experimental determination of z_0 and d_0 from wind velocity profile measurements. However, in the absence of such measurements, it may sometimes be necessary to estimate these parameters from simple geometric characteristics of the surface.

a. Land Surfaces

A few values of z_0 for various common surfaces are given in Table 5.1.

Many studies have been conducted to relate z_0 with measurable characteristics of the surface. The most obvious, and also the simplest and available surface characteristic, that comes to mind for this purpose, is the mean height of the roughness obstacles, h_0. Paeschke (1937) was probably the first to consider this for crop-covered surfaces. His results showed that for rough snow, various grassy surfaces, wheat, fallow and beets, the ratio

$$(h_0/z_0) = 7.35 \tag{5.3}$$

gave a good fit with the wind profile data. This was later confirmed by Tanner and Pelton (1960) who proposed $z_0 = h_0/7.6$, approximately and by Plate (1971) who reviewed additional data. For a surface covered with transverse square bars, Moore obtained $(h_0/z_0) = 7.5$, which was confirmed by Perry and Joubert (1963). The median of Chamberlain's (1968) data on artificial bluff and wave-like elements

TABLE 5.1
Examples of roughness parameters for various surfaces

Surface description	z_0 (cm)	Reference
Mud flats, ice	0.001	Sutton (1953)
Smooth tarmac (airport runway)	0.002	Bradley (1968)
Large water surfaces ('average' conditions)	0.01–0.06	Numerous references
Grass (lawn up to 1 cm high)	0.1	Sutton (1953)
Grass (airport)	0.45	Kondo (1962)
Grass (prairie in Nebraska)	0.65	Kondo (1962)
Grass (artificial, 7.5 cm high)	1.0	Chamberlain (1966)
Grass (thick up to 10 cm high)	2.3	Sutton (1953)
Grass (thin up to 50 cm)	5	Sutton (1953)
Wheat stubble plain (18 cm, Kansas)	2.44	Businger *et al.* (1971)
Grass (with a few scattered bushes and clumps of trees; regional value for Salisbury Plain, England)	4	Deacon (1973)
1–2m high vegetation (Cape Canaveral, Florida)	20	Fichtl and McVehil (1970)
Trees (10–15m high) (Cape Canaveral, Florida)	40–70	Fichtl and McVehil (1970)
Savannah scrub (25% trees $\simeq$ 8m; 65% dry grass $\lesssim$ 1m; burnt grass and sand 10%)	40	Garratt (1978b)
Large city (Tokyo, Japan)	165	Yamamoto and Shimanuki (1964)

yielded $(h_0/z_0) = 8$, approximately, with extremes of 16.6 and 4.2, whereas his (Chamberlain, 1966) data on artificial grass gave $(h_0/z_0) = 7.5$.

In reality, the matter is not as simple as suggested by Paeschke's (5.3); invariably, more detailed analysis has shown that (h_0/z_0) is, in fact, a fairly complicated function of other surface characteristics as well. Several formulations have been derived for bluff obstacles (e.g., Lettau, 1969; Wooding *et al.*, 1973) and for permeable obstacles (e.g., Takeda, 1966; Cowan, 1968; Seginer, 1974) for z_0, and d_0 as well, involving such characteristics as height, frontal area, surface density, concentration, and other geometrical or drag parameters. Unfortunately, the accuracy of these formulations, while probably better than (5.3), is still quite low. For example, Seginer (1974) (see Figure 5.1) has shown that (z_0/h_0) as a function of $(\mathrm{Cd}_f\bar{A}_f h_0)$ (these symbols are defined behind (4.136); $\bar{A}_f$ is the vertically-averaged value) displays a maximum at about $(\mathrm{Cd}_f\bar{A}_f h_0) = 0.2$. Thus, below 0.2, (z_0/h_0) increases gradually with roughness density to reach a rounded peak about 1.5 to 2 times that given by (5.3); above 0.2, (z_0/h_0) decreases gradually. This phenomenon is not difficult to interpret. When the density of sparsely-placed obstacles increases, the drag increases and, therefore, also z_0; but when the placement of the obstacles becomes extremely dense, the flow may actually skim over their tops without entering the space below, so that the effective roughness decreases. Apparently, except for more complicated numerical canopy models (e.g., Seginer, 1974; Kondo and Akashi, 1976), none of the simple formulations referred to above can describe this peak in (z_0/h_0) with increasing roughness density. Another effect on z_0, which is not considered in these formulations, is that of

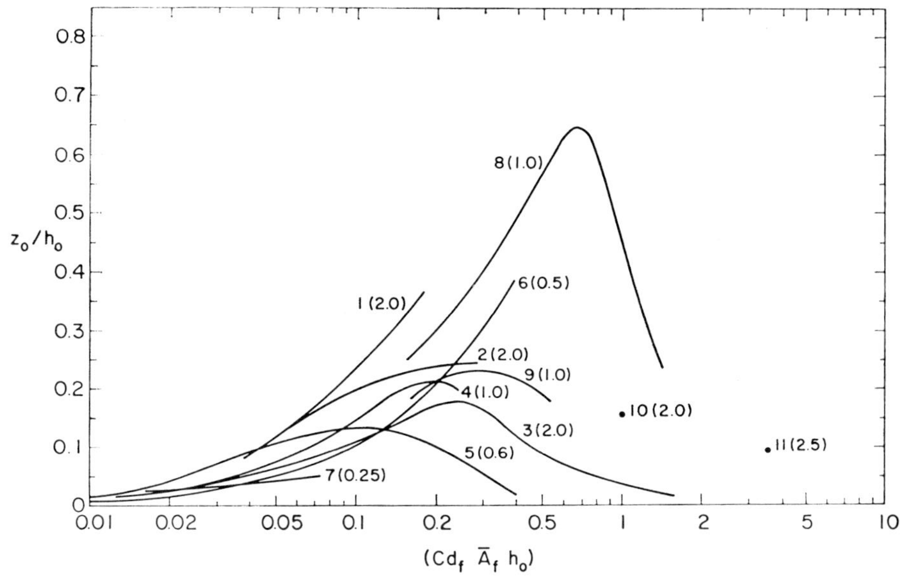

Fig. 5.1. Summary of data on the relative roughness length (z_0/h_0) as a function of ($Cd_f\bar{A}_fh_0$). The drag coefficients are rough estimates shown in parentheses. The numbers 1 and 2 denote data on 3-D baffles: 3, 2-D slats; 4, 2-D rods; 5, 3-D spheres; 6, 3-D baskets; 7, 3-D hemispheres; 8, 3-D pegs; 9, 2-D rods; 10, 3-D strips; 11, 3-D rods (adapted from Seginer, 1974).

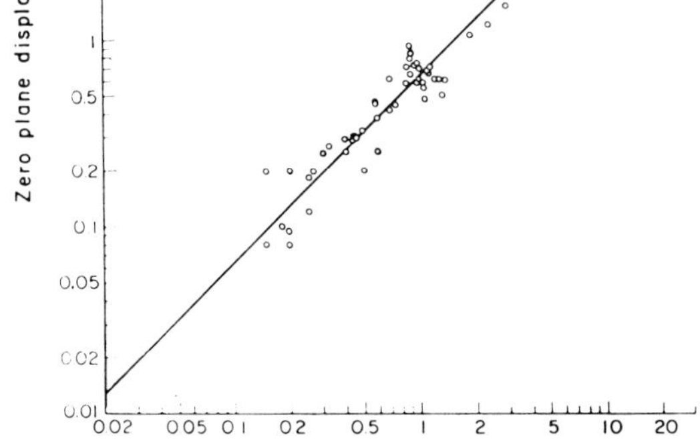

Fig. 5.2. Relationship between zero plane displacement d_0 and height h_0 of 19 different types of vegetation and bare ridged soil. The best fit straight line is indistinguishable from (5.4) (adapted from Stanhill, 1969).

the wind, which may be considerable in the case of flexible vegetation. Nevertheless, even if the available formulations for z_0 were more accurate, their application to natural surfaces would rarely be practical, because the necessary parameters are not generally known. Consequently, in the absence of wind profile data, practically the only recourse is to make use of a relationship such as Paeschke's, or one similar to it.

In comparison to (z_0/h_0), the ratio (d_0/h_0) appears to be less sensitive to the nature of the surface or to other factors (e.g., Munro and Oke, 1973). Stanhill (1969) considered data from many sources, mainly on agricultural crops (see Figure 5.2) and he derived $d_0 = 0.7\ h_0^{0.98}$ (in cm) with a correlation coefficient of 0.97; this is equivalent to $(d_0/h_0) = 0.64$ for an average $h_0 = 66$ cm. The mean of data presented by Kondo (1971) was $(d_0/h_0) = 0.68$, with extremes of 0.53 and 0.83. Hence,

$$d_0 = \frac{2}{3} h_0 \tag{5.4}$$

appears to be fairly representative for natural crop covered surfaces. Again, it is clear that the ratio (d_0/h_0) cannot really be a constant. For extremely sparsely placed roughness elements, the ground surface is the true reference and d_0 should be very close to zero. On the other hand, for very densely placed obstacles, when the flow skims over the tops, (d_0/h_0) should approach unity. Fortunately, however, the determination of d_0 is not as critical as that of z_0. Since d_0 appears in $(z - d_0)$, the profile functions of Chapter 4 are not very sensitive to its exact value, provided, of course, $z \gg z_0$. Therefore, (5.4) should be adequate as a first approximation, when no actual wind profile data are available for a more correct determination.

Interestingly enough, Paeschke's (1937) result, (5.3) has also been deduced in several ways by rather simple arguments for vegetational canopies. For example, Kondo (1971) considered the case of sparsely-placed roughness obstacles when $d_0 = 0$. Matching $\bar{u}$ and $(d\bar{u}/dz)$, obtained by the logarithmic profile (4.3) for $z > h_0$ and by Cowan's (1968) (4.138) for $z < h_0$, at $z = h_0$, he was able to show that, since $a'_w \ll 1$ for thinly placed obstacles,

$$(h_0/z_0) = e^2 \tag{5.5}$$

or 7.39. However, (Brutsaert, 1975c) Paeschke's ratio may also be derived for densely-placed obstacles, as follows. Inside a uniform but dense stand of plants the wind profile may be represented by (4.137). Matching $\bar{u}$ and $d\bar{u}/dz$, obtained by (4.3) and (4.137), at $z = h_0$, one obtains

$$\frac{h_0 - d_0}{z_0} = \exp\left[\frac{h_0}{a_w(h_0 - d_0)}\right]. \tag{5.6}$$

Thus, with (4.144) and (5.4) this yields

$$(h_0/z_0) = 3e \tag{5.7}$$

or 8.15. The similarity of (5.5) and (5.7) with Paeschke's (5.3) suggests that the ratio (h_0/z_0) is relatively insensitive to the method adopted to calculate it. It should be re-emphasized, however, that these results may only be considered as first approximations, to be used when no measured wind profiles are available for the given surface.

b. Water Surfaces

The interaction between turbulent air and a free water surface involves complicated physical phenomena. Therefore, the prediction of z_{0m} over water is still subject to some uncertainty.

It is generally accepted that, for lower wind speeds, the water surface is smooth, but it is not clear whether (5.1) is always valid. In some experiments, the surface has been observed to be 'super-smooth' with a z_{0m}-value smaller than (5.1). The issue is unresolved. Csanady (1974) has attributed this phenomenon to surface-tension effects resulting from surface films or other impurities. Kondo and Fujinawa (1972), however, showed that such discrepancies can be due to the neglect of atmospheric stability, or to the overestimation of the wind speed by means of cup anemometers and by the neglect of the surface drift of the water.

For moderately strong winds with well-developed waves, it is generally accepted that z_0 depends on the surface shear stress. Accordingly, it has been proposed by Charnock (1955) from dimensional considerations that

$$z_0 = \frac{u_*^2}{bg} \tag{5.8}$$

where b is a constant. Charnock (1958) put $b = 81$ but Hicks (1972a) obtained $b = 62.5$; Smith and Banke (1975), and also Garratt (1977), found that (5.8) with $b = 69$ provides a good description of experimental data, which is practically the same as (5.12). SethuRaman and Raynor (1975) obtained $b^{-1} = 0.016\ (\pm 0.011)$ for moderately rough conditions over the range $0.15 < z_{0+} < 4$; but for very rough conditions, when $z_{0+} > 4$ they derived $b^{-1} = 0.072\ (\pm 0.030)$ from the data. Equation (5.8) is similar to the one proposed earlier by Rossby and Montgomery (1935), but with $\bar{u}$ instead of u_*. A similar relationship with one additional parameter was proposed by Yasuda (1975)

$$z_0 = a\,u_*^b \tag{5.9}$$

where a and b are constants. With units of cm and s he obtained, from Kondo's (1975) interpretation of (5.12), the following $a = 0.01689$, $b = -1$ for $u_* \leq 6.89$ cm s^{-1} and $a = 4.86\ 10^{-4}$, $b = 1.27$ for $u_* > 6.89$ cm s^{-1}. Kondo (1977) suggested that (5.9) with $b = 1$ (i.e., $u_*/z_0 = $ const) and $a^{-1} = 1.4\ 10^3$ s^{-1} can be used as a practical approximation for $20 \leq u_* \leq 100$ cm s^{-1} or $6 \leq \bar{u}_{10} \leq 25$ m s^{-1}.

The hydrodynamical characteristics of a water surface are often described directly in terms of the drag coefficient, Cd_r, instead of by the roughness coefficient, z_0. From (4.34′), for the surface sublayer and (4.115), these two parameters are related by (over water d_0 is taken as zero)

$$Cd_r = k^2\left[\ln\left(\frac{z_r}{z_{0m}}\right) - \Psi_{sm}\left(\frac{z_r}{L}\right)\right]^{-2} \tag{5.10}$$

in which over water the height z_r, where the wind speed is measured, is commonly taken as 10 m. Clearly, when the atmosphere is near-neutral, which it is often assumed to be over extensive water surfaces, Cd_r varies much less than z_{0m}, on account of the logarithmic relationship.

Therefore, many investigators have felt that, for practical purposes, any shear stress or wind speed dependence of Cd_r is negligible, except perhaps at high winds. Nonetheless, the scatter in the available results is still considerable; the average coefficients referred to 10 m elevation, Cd_{10}, which have been obtained by various experimental methods, under neutral or near-neutral conditions, range widely from less than 1×10^{-3} to more than 2×10^{-3}. A few results are shown in Table 5.3 Upon comparing all these available results, the value

$$Cd_{10} = 1.4\ 10^{-3}, \tag{5.11}$$

which corresponds to $z_0 = 0.023$ cm, appears to be a fairly typical average. In general, lower Cd_{10} values have been obtained over shallow water (e.g., Emmanuel, 1975; Hicks *et al.*, 1974) than over the deep ocean. The method used to obtain u_* also appears to have an effect; for example, Cd_{10}, obtained by the surface slope method (Wieringa, 1974), is much larger than values obtained by profile- or eddy-correlation methods. The results of Dunckel *et al.* (1974) and Krügermeyer *et al.* (1978) illustrate that the wind-speed measurement cannot be made too closely to the surface.

A number of relationships have been inferred between Cd_{10} and wind speed or friction velocity. The following linear function has been well investigated

$$Cd_{10} = (a + b\ \bar{u}_{10})\ 10^{-3} \tag{5.12}$$

where a and b are constants. This formulation is in accord with Munk's (1955) speculation that the total drag may be the sum of a skin friction, proportional to $\bar{u}^2$, and a form drag, proportional to $\bar{u}^3$. As noted earlier, Smith and Banke (1975) have shown that (5.12) is closely equivalent with Charnock's (5.8) for z_0. Some experimentally obtained values of a and b are listed in Table 5.2 and the corresponding versions of (5.12) are shown in Figure 5.3. A similar approach is based on an assumed linear relationship between the logarithms, resulting in the following

$$Cd_{10} = a\ \bar{u}_{10}^{b}\ 10^{-3} \tag{5.13}$$

where a and b are constants. Equation (5.13) was used to describe experimental data by Wieringa (1974) with $a = 0.7$, $b = 0.3$ for $5 < \bar{u}_{10} < 15$ m s^{-1} (Curve 10 in Figure 5.3). Garratt (1977), who reviewed data from many sources, recommended (5.13) with $a = 0.51$, $b = 0.46$ for neutral conditions over the range $4 < u_{10} < 21$ m s^{-1} (Curve 11 in Figure 5.3).

One probable reason for the uncertainty surrounding the magnitude of Cd_r over water is that, besides wind speed or friction velocity, other factors must be considered as well. A major cause of disagreement among the experimental results is that many workers do not allow for atmospheric stability effects, which may be considerable. Other factors probably involve a characterization of the water waves and their stage of development, but it is likely that, at very high wind speeds, spray and rain are also important factors.

Several attempts have been made to include some of these effects. Kitaygorodskiy *et al.* (1973) felt that the roughness length z_0 itself is a legitimate characteristic of the degree of sea-wave development. Accordingly, they proposed that Cd_r be taken as a function of the roughness Reynolds number as follows

TABLE 5.2
Some values of the parameters obtained for the drag coefficient Cd_{10} over water in (5.12)

Reference	a	b (s m^{-1})	Scatter (SD)	Range $\bar{u}_{10}$ (m s^{-1})	Remarks (curve number in Figure 5.3)
Deacon and Webb (1962)	1.0	0.07		<14	Data from many sources (1)
Garratt (1977)	0.75	0.067		4 – 21	Data from many sources (reduced to neutral) (2)
Kondo (1975)	Smooth			<2	
	1.2	0.025		8 – 16	Sagami Bay off Hiratsuka (3)
	0	0.073		25 – 50	From momentum budget data and wind flume experiments (34)
Sheppard *et al.* (1972)	0.36	0.10		3 – 16	Data from Lough Neagh by profile method (5)
Smith (1974)	0.58	0.068	(± 0.24)	4 – 16	Atlantic Ocean, by eddy correlation (6)
	0.82	0.039	(± 0.20)	3 – 10	Lake Ontario (7)
Smith and Banke (1975)	0.63	0.066	(± 0.23)	3 – 21	111 combined data (8)
Wieringa (1974)	0.87	0.048		>5	Various eddy correlation methods. Data from many sources. (9)

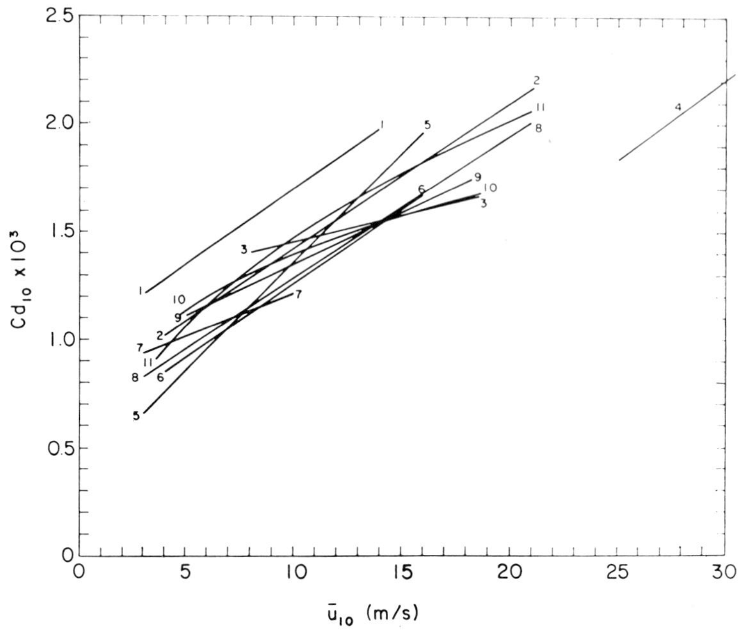

Fig. 5.3. Comparison of different forms of (5.12) and (5.13) for the drag coefficient over water as function of the mean wind speed at 10 m. The numbers indicate references given in Table 5.2 and after (5.13).

$$\mathrm{Cd}_{10} = a\, z_{0+}^{b} \tag{5.14}$$

where a and b are constants; on the basis of data from several experiments they found that $a = 1.2 \times 10^{-3}$ and $b = 0.15$ for $10^{-3} < z_{0+} < 300$. But it should be noted that any such relationship is, in fact, equivalent to (5.8) or (5.9), since, upon elimination of Cd_{10} with (5.10) for neutral conditions, it results in a relationship between z_0 and u_*. For example, Charnock's (5.8) can also be written as a drag coefficient; for $b = 69$ one obtains with (5.10), for neutral conditions

$$\mathrm{Cd}_{10} = k^2[11.9 - \ln(z_{0+}^{2/3})]^{-2}. \tag{5.8$'$}$$

Some attempts have also been made to formulate the effect of sea state by considering the coupling between the wind and the water waves. A possible similarity approach follows from Kitaygorodskiy's (1969) hypothesis that over a wavy water surface, under neutral conditions, one has for the wind profile

$$\frac{kz}{u_*}\frac{d\bar{u}}{dz} = \phi_{wm} \tag{5.15}$$

where now

$$\phi_{wm} = \phi_{wm}\left(\frac{z}{\lambda}, \frac{c}{u_*}\right), \tag{5.16}$$

in which c is the phase velocity of the dominant wave and λ the corresponding wave length. On the basis of (5.15) and with some simplifying assumptions, the following was deduced by Brutsaert (1973) by scaling the turbulent kinetic energy equation

$$\phi_{wm} = 1 + \beta\left(\frac{c}{u_*} - \alpha\right)^2 \quad \text{for} \left(\frac{c}{u_*}\right) > \alpha \tag{5.17}$$

where α and β are constants, which were found to be of the order of 29 and 0.006, respectively. Hence, integration of (5.15) with (5.17) yields a drag coefficient under neutral conditions

$$\mathrm{Cd}_r = \mathrm{Cd}_{r0}\left[1 + \beta\left(\frac{c}{u_*} - \alpha\right)^2\right]^{-2} \quad \text{for} \left(\frac{c}{u_*}\right) > \alpha \tag{5.18}$$

where Cd_{r0} denotes the drag coefficient as given by (5.10) with $\Psi_{sm} = 0$, which results from (5.18) when $(c/u_*) = \alpha$. The stated condition in (5.17) and (5.18) refers to swell, that is, the situation of a relatively weak wind over well-developed, slowly decaying waves. In a statistical analysis of experimental data over heavy swell, apparently under stable conditions, in the Atlantic, Davidson (1974) arrived at a similar result. He concluded that Cd_r is variable but that it is not a function only of wind speed, but also of c/u_*

$$\mathrm{Cd}_r = k^2\left[\ln\left(\frac{z_r}{z_0}\right) + 6.44\left(\frac{z_r}{L}\right) + 0.13\left(\frac{c}{u_*} - 26.3\right)\right]^{-2} \tag{5.19}$$

in which c is the phase speed corresponding to the wave spectrum peak. For neutral conditions, with $z_r = 10$ m and with Davidson's roughness $z_0 = 0.024$ cm, (5.19) can also be written as

$$\mathrm{Cd}_{10} = 1.4 \times 10^{-3}\left[1 + 0.0122\left(\frac{c}{u_*} - \alpha\right)\right]^{-2} \quad \text{for} \left(\frac{c}{u_*}\right) > \alpha \tag{5.20}$$

which has the same trend as (5.18). At any rate, more experimental and theoretical research will be necessary to further investigate expressions such as (5.18) and (5.20) and to arrive at a better understanding of the air flow over water waves.

5.2. THE SCALAR ROUGHNESS

The roughness length for any given passive scalar admixture is the height where the concentration assumes its surface value, when the logarithmic profile, which this scalar obeys within the dynamic sublayer, is extrapolated downward. Equations (4.14) or (4.17) can serve as definition of this concept. In the interfacial sublayer, Reynolds's analogy is not valid, so that there is no justification to use z_{0m}, derived from the mean-wind velocity profile, also for the profiles of mean specific humidity, temperature or any other scalar. The concept of a scalar roughness is useful in the concise theoretical formulation of the bulk transfer coefficients defined in (4.114) and (4.117) in terms of similarity functions for the different sublayers above the interfacial sublayer; it serves as a lower limit for the integral forms of these different similarity functions as shown, for example for water vapor, in (4.14), (4.33′) and (4.75). Consequently, it facilitates the parameterization of the transport phenomena in the interfacial sublayer.

a. Calculation from Interfacial Transfer Coefficients

Consider, for example (the argument is analogous for sensible heat or for any other scalar), the roughness length for water vapor, z_{0v}; for the nonneutral surface sublayer it is defined in (4.33′). Hence, the bulk mass transfer coefficient, defined in (4.114) can be written in terms of z_{0v} as

$$\mathrm{Ce}_r = a_v k\, \mathrm{Cd}_r^{1/2}\left[\ln\left(\frac{z_r - d_0}{z_{0v}}\right) - \Psi_{sv}\left(\frac{z_r - d_0}{L}\right)\right]^{-1} \tag{5.21}$$

where Cd_r is defined in (4.115). Combining (5.21) with (4.119) (or, more simply for the dynamic sublayer, combining (4.14) with (4.113)) one obtains for z_{0v} the following expression

$$z_{0v} = z_{0m} \exp\left[-k a_v(\mathrm{Da}_0^{-1} - a_v^{-1}\,\mathrm{Cd}_0^{-1/2})\right]. \tag{5.22}$$

This equation reduces the problem of determining the scalar roughness to that of knowing the interfacial or canopy transfer coefficients ($\mathrm{Da}_0^{-1} - a_v^{-1}\mathrm{Cd}_0^{-1/2}$). As mentioned in Chapter 4, the transfer in the interfacial sublayer is sometimes formulated in terms of the quantity B defined in (4.116′). Obviously (z_{0v}/z_{0m}) can also be expressed in terms of B as follows (cf. Chamberlain, 1966)

$$z_{0v} = z_{0m} \exp(-k\, B^{-1}). \tag{5.23}$$

The nature of the interfacial transfer coefficients has been treated in Section 4.4. However, for easy reference, the application of (5.22) is now reviewed for the three different types of surfaces.

Smooth Surfaces

For a smooth surface, when $z_{0+} < 0.13$, the 'roughness' for a scalar (or rather, the zero-intercept of $\bar{q}_s - \bar{q}_r$ versus height on a semi-log plot, as shown in Figure 4.2), can be readily calculated with any one of the expressions in Table 4.1. Taking (4.126), one obtains from (5.22) with (5.1) and $k = 0.4$

$$z_{0v} = (30\,\nu/u_*) \exp(-13.6\, k a_v\, \mathrm{Sc}^{2/3}). \tag{5.24}$$

Hence, in the lower atmosphere, when for water vapor $\mathrm{Sc} = 0.595$, this becomes,

$$z_{0v} = 0.624\,\nu/u_*. \tag{5.25}$$

Similarly, one obtains for the sensible heat roughness, with $\mathrm{Pr} = 0.71$, the following

$$z_{0h} = 0.395\,\nu/u_*. \tag{5.26}$$

Evidently, for smooth surfaces z_{0v} and z_{0h} are both somewhat larger than z_{0m} given in (5.1).

Bluff-Rough Surfaces

For surfaces with bluff roughness elements, when $z_{0+} > 2$, the interfacial transfer coefficients are treated in Section 4.4d. Substitution of (4.133) (as a typical expression) in (5.22) yields for any scalar

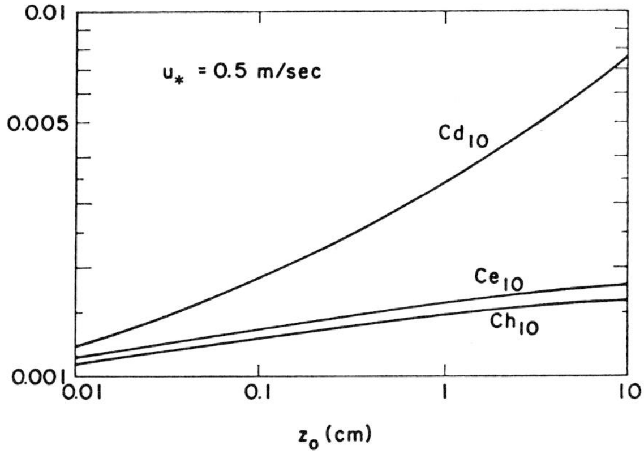

Fig. 5.4. Illustration of the dissimilarity between the drag coefficient and the bulk transfer coefficients for water vapor and sensible heat for a reference height z_r = 10 m above a bluff-rough surface with $z_{0m} = z_0$ and $d_0 = 0$ under neutral conditions ($\overline{\Psi}_{sm} = \overline{\Psi}_{sv} = \overline{\Psi}_{sh} = 0$). Cd_{10} is calculated by means of (5.10) and Ce_{10} and Ch_{10} by means of (5.21) with (5.28) and (5.29), respectively (adapted from Brutsaert, 1975a).

$$z_{0v} = 7.4\, z_0 \exp(-7.3\, k a_v z_{0+}^{1/4}\, \mathrm{Sc}^{1/2}). \tag{5.27}$$

Applied to water vapor in the lower atmosphere with Sc = 0.595, (5.27) becomes, approximately,

$$(z_{0v}/z_0) = 7.4 \exp(-2.25\, z_{0+}^{1/4}). \tag{5.28}$$

Similarly, for sensible heat, with Pr = 0.71, one has

$$(z_{0h}/z_0) = 7.4 \exp(-2.46\, z_{0+}^{1/4}). \tag{5.29}$$

Equations (5.28) and (5.29) are also shown graphically in Figure 4.24 as the dashed and solid line, respectively, for bluff surfaces. Clearly, in the rough domain over surfaces with bluff obstacles z_{0h} and z_{0v} are considerably smaller than z_0. This large difference in roughness is a manifestation of the dissimilarity between the transfer mechanisms of momentum and those of scalar admixtures right at the surface. Momentum transfer takes place not only as a result of viscous shear, but the roughness obstacles also generate an effective form drag involving local pressure gradients. The transfer of a passive scalar admixture at the wall is controlled primarily by molecular diffusion. This dissimilarity is also manifested by the difference in the bulk transfer coefficients, as illustrated in Figure 5.4.

Permeable-Rough Surfaces

The interfacial transfer coefficients needed in (5.22) for surfaces covered with permeable, slender or vegetational obstacles are treated in Section 4.4e. As an illustration, substituting the theoretical solution for a dense and uniform canopy (4.162) in (5.22) one obtains

$$\frac{z_{0v}}{z_0} = \exp\left\{-\left[\frac{h_0}{(h_0 - d_0)(-G_0)}\right] + \ln\left[\frac{(h_0 - d_0)}{z_0}\right]\right\} \tag{5.30}$$

where $G_0 = G_0(C_2)$ is given by (4.160) which is shown in Figure 4.25. C_2 is defined in (4.155) and it can be applied with $C_L = 0.25$, $m = 0.25$ and $n = 0.36$ for any scalar. For example, for sensible heat z_{0h} can be calculated by means of (5.30) with C_2 given by

$$C_2 = 1.41 \text{ LAI } [h_0/(h_0 - d_0)] \text{ Re}_*^{-0.25} \tag{5.31}$$

approximately, where $\text{LAI} = (\bar{A}_f h_0)/2$ is the leaf-area index. Calculations of the roughness of sensible heat for grass, maize and aspen forest are shown in Figure 4.24 together with experimental data for other types of vegetation.

The ratio (z_{0h}/z_0) does not display as much change for most grassy or tree-covered surfaces as it does for surfaces with bluff roughnesses. From Figure 4.24, (z_{0h}/z_0) (or (z_{0v}/z_0) for any admixture with Pr or Sc around 0.6 to 0.8) appears to be of the order of 1/7 to 1/12, but for tall trees it is probably of the order of 1/3 to 1/2, but not much larger. These values can be used as rough estimates for practical calculations, when no other information is available.

As mentioned in Section 4.4e the profile functions for the turbulent boundary layer, such as (4.14), (4.33′) and (4.75), are sometimes applied to non-wet vegetation with $\bar{q}^* = \bar{q}^*(T_s)$ instead of the actual surface specific humidity $\bar{q}_s$. For this case the roughness length for water vapor (5.22) must be applied with (4.166); this yields in the resistance notation

$$(z_{0v}/z_{0m}) = \exp\left[-k a_v u_* (r_{st} + r_{0v} - a_v^{-1}\, r_{0m})\right] \tag{5.32}$$

b. Values Over Water

Obtained from Theory

Over a smooth water surface, when $z_{0+} < 0.13$, (5.24) can probably be used. For rough water surfaces, with wind-generated waves, it has been found (e.g., Figure 4.17; see also Merlivat, 1978) that the interfacial transfer coefficients for a surface with bluff-roughness elements are applicable. Hence, in the absence of experimental data, the scalar roughness for a rough sea surface can probably be determined by means of (5.27). To apply this equation, the aerodynamic roughness z_0 of the water surface must be known. As seen in Section 5.1b, z_0 is often difficult to determine. It is possible to use various u_* of $\bar{u}_{10}$-dependent z_0 expressions such as (5.9) and (5.12). However, a first approximation for z_{0v} or z_{0h} can be obtained by simply assuming (5.11) to be valid throughout the rough regime. With this assumption, which corresponds with a constant $z_0 = 0.0228$ cm, (5.28) for water vapor yields (units are cm and s)

$$z_{0v} = 0.169 \exp\,(-1.40\; u_*^{1/4}) \tag{5.33}$$

which can be used for rough flow. For smooth flow (5.25) can be used.

In a similar way (5.29), for sensible heat yields over a rough water surface,

$$z_{0h} = 0.169 \exp\,(-1.53\; u_*^{1/4}) \tag{5.34}$$

and (5.26) can be used for a smooth surface.

For the transitional flow regime, between smooth and rough, no theoretical expressions are available. To describe this regime, which was assumed to occupy the range $2 < u_* < 20$ cm s^{-1}, Brutsaert and Chan (1978) used a simple interpolation. For example, for sensible heat, one has

$$z_{0h} = \beta z_{0h,r} + (1 - \beta)\, z_{0h,s} \tag{5.35}$$

where $z_{0h,r}$ and $z_{0h,s}$ are the values obtained with (5.34) and (5.26) respectively, and $\beta = (u_* - 2)/18$ is a weighting factor. For z_{0v} the same approach can be used with (5.35) but with $z_{0v,r}$ and $z_{0v,s}$ obtained from (5.33) and (5.25) respectively. Again, a simpler method consists of applying Merlivat's (1978) criterion $z_{0+} = 1$, as a sudden transition point from smooth to rough; this would allow the use of (5.25) and (5.26) for $u_* < 7$ cm s^{-1}, and (5.33) and (5.34) for $u_* > 7$ cm s^{-1}, without the need for any interpolation.

Some Other Estimates

Very few papers have dealt with the determination of z_{0v} and z_{0h} from experimental data over open water. Sheppard *et al.* (1972) found no substantial difference between z_{0m}, z_{0v} and z_{0h} for $z_{0+} < 10$. Since their data appear to be close to transition between smooth and rough, the differences are not very large and may have been lost in the experimental errors. As noted behind (5.26) and behind (5.29), for smooth flow z_{0v} and z_{0h} are larger than z_{0m}, whereas for rough flow, they are smaller. Hicks (1975) felt that $z_{0v} = (\kappa_v/ku_*)$ and $z_{0h} = (\kappa_h/ku_*)$, which result from Sheppard's (1958) model for interfacial transport (see final paragraph Section 4.4b) should be adequate for practical purposes. Clearly, these roughness lenghts can also be written as $z_{0v} = 4.20\,\nu u_*^{-1}$ and $z_{0h} = 3.52\,\nu u_*^{-1}$, approximately; in the light of (5.25), (5.26), (5.33) and (5.34), however, these roughness lengths probably result in an overestimation of the transfer coefficients.

As shown in (5.21) for given conditions of atmospheric stability, there is a one to one relationship between the roughness length z_{0v} (or z_{0h}) and the mass transfer coefficient Ce_r (or Ch_r). Therefore, just like for momentum (cf., (5.10)), the scalar transfer properties of a given surface can be described directly as bulk transfer coefficients Ce_r or Ch_r, without using the auxiliary concept of the roughness length. Most experimental results have been analyzed this way.

Kitaygorodskiy *et al.* (1973) have proposed the following on the basis of eddy-correlation measurements on a platform in 10 m-deep water in the Caspian Sea (cf. (5.14))

$$Ce_{10} = a\, z_{0+}^{b} \tag{5.36}$$

in which $a = 1.0 \times 10^{-3}$ and $b = 0.11$ for the range $10^{-2} < z_{0+} < 10$. In view of the unknown effects of the atmospheric stability during the experiments, and the present uncertainty in Cd_r over water, it is difficult to compare this result with those obtainable by means of theoretical equations such as, e.g., (5.21) with (5.24) and (5.27), or (4.119) with (4.126) and (4.133). Especially under light wind conditions, when $z_{0+} < 1$, the effect of atmospheric stability and of surface roughness anomalies ('super-smooth') may be considerable. Nevertheless, the ratio $(Ce_{10}/Cd_{10}) = z_{0+}^{-0.04}/1.20$, obtained from (5.14) and (5.36), exhibits a similar trend to that expected from the theoretical equa-

TABLE 5.3
Some values obtained for the bulk transfer coefficients Cd_{10}, Ce_{10} and Ch_{10}

Reference	10^3 Cd_{10}	10^3 Ce_{10}	10^3 Ch_{10}	Wind Speed Range (ms^{-1})	Site	Technique
Dunckel *et al.* (1974)	1.56	1.28	1.46	4.5 – 11	Atlantic (Bomex)	Profile
Emmanuel (1975)	1.15 (±0.2)	1.34 (±0.3)	1.1 (±0.3)	2.7 – 8	Lake Hefner (approx. 8 m deep)	Eddy correlation
	1.34 (±0.3)	1.34 (±0.4)	1.15 (±0.3)			Profile
Friehe and Schmitt (1976)		1.32	Different expressions depending on range of $\bar{u}\,\Delta T$		Data from many sources	Eddy correlation
Hicks and Dyer (1970)	1.1 (±0.1)		1.40 (±25%)	2 – 10	Bass Strait	Eddy correlation
Hicks *et al.* (1974)	Smooth		1.1	3 – 10	Coral reef (Papua, less than 2.5 m deep)	Eddy correlation
Müller-Glewe and Hinzpeter (1974)			1	<8	Baltic Sea	Eddy correlation
Pond *et al.* (1971)			1		Pacific (off San Diego)	Eddy correlation
	1.44 (±0.26)	1.18 (±0.17)		4 – 7.5	Atlantic (Bomex) and Pacific	Eddy correlation
	1.44 (±0.40)	1.20 (±0.25)			(San Diego)	Dissipation
Pond *et al.* (1974)	1.49 (±0.28)	1.36 (±0.40)		2 – 8	Arabian Sea	Profile
	1.48 (±0.21)	1.41 (±0.18)	1.47 (±0.64)	2.5 – 8	Atlantic (Bomex)	Profile
Smith (1974)	1.2	1.2 (±0.3)	1.3 (±0.5)	3 – 10	Lake Ontario	Eddy correlation
Smith and Banke (1975)	1.6		1.5 (±0.4)	High Winds 8 – 21	Sand bar off Sable Island, N.S.	Eddy correlation
Tsukamoto *et al.* (1975)	1.32	1.28	1.40	3 – 13	East China Sea (Amtex)	Eddy correlation

tions; thus, (Ce_{10}/Cd_{10}) decreases as z_{0+} increases. On the other hand, however, the ratio obtained from (5.14) and (5.36) is smaller than unity also for smooth flow when $z_{0+} < 0.13$; the theoretical equations predict a ratio larger than unity for a smooth surface and smaller than unity for a rough surface.

Some average values of Cd_{10}, Ce_{10} and Ch_{10}, obtained in various experiments during the past decade, can be compared in Table 5.3. The observed Ce_r and Ch_r values are generally smaller than the corresponding Cd_r. An observation in many studies is that, if there is any wind speed dependence of Ce_r or Ch_r, it is much smaller than that of Cd_r. This is consistent with the results shown in Figure 5.4.

It should be noted, again, that over large water surfaces the experimental determination of Ce_r and Ch_r, even more than that of Cd_r, is very demanding, and that experimental errors are always large. In Section 5.1b mention was made of the effects of the atmospheric stability and of the sea state in connection with Cd_r. However, besides these effects the determination of Ce_r and Ch_r may be even more affected by such factors as spray and the presence of surface films of contaminants. In addition, there is the difficulty of ascertaining the water surface temperature. The matter is of considerable importance and the scientific community appears to be divided. Some workers hold that the only realistic way is to parameterize by the use of normally available data, namely on the basis of submerged (or bucket water) temperatures, a few cm below the surface, as usually reported in ship observations. More recently, however, the opinion has been gaining that the true surface temperature should be employed, which is usually obtained by infrared thermometry. Clearly, these unresolved issues will require further study.

CHAPTER 6

Energy Fluxes at the Earth's Surface

Evaporation and sensible heat flux into the atmosphere require the availability of some form of energy at the earth-atmosphere interface. This question can be treated quantitatively by considering the equation for the energy budget for a layer of surface material. Depending on the nature of the surface, this layer may consist of water, or of some other substrate like soil, canopy or snow; although this layer can be taken to be infinitesimally thin, it may sometimes even comprise a lake or a vegetational canopy over its entire depth. For practical purposes, the energy budget equation, in a form somewhat more general than (1.2), is

$$R_n - L_e E - H + L_p F_p - G + A_h = \frac{\partial W}{\partial t} \tag{6.1}$$

where the energy fluxes toward the layer are taken as positive and those away from it as negative. R_n is the net radiative flux density at the upper surface of the layer, L_e the latent heat of vaporization, L_p the thermal conversion factor for fixation of carbon dioxide, F_p the specific flux of CO_2, G the specific energy flux leaving the layer at the lower boundary, A_h the energy advection into the layer expressed as specific flux, and $\partial W/\partial t$ the rate of energy storage per unit area in the layer; in the case of an ice or snow layer this last term may include the energy consumed by fusion, and L_e may have to be replaced by L_s, the latent heat of sublimation.

The exact nature of several of the terms depends on the type of layer or substrate for which the energy balance is written. Actually, for many practical purposes, several of them can be omitted so that (6.1) assumes a much simpler form. The order of magnitude and the diurnal variation of the main terms in the energy budget for different surfaces are shown in Figures 6.1 through 6.4.

The terms E and H and their parameterization are dealt with in Chapters 4 and 5. In the remainder of the present chapter a brief account is given of the physical significance of each of the other terms in (6.1) for different types of substrate. A detailed treatment of the energy budget is beyond the scope of this book. Nevertheless, some simple methods are presented which may be adequate to estimate these terms for climatological or certain engineering design purposes, or which may be useful for correlations to calculate missing data in a temporarily-interrupted record.

6.1. NET RADIATION

The net radiation can be broken down into several components, viz.

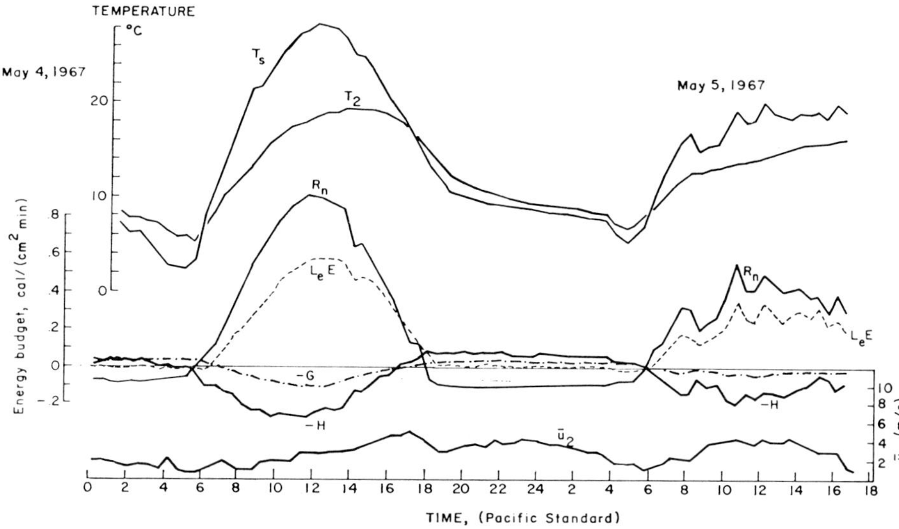

Fig. 6.1. Example of the daily cycle of the energy balance together with the surface temperature, the temperature and the mean wind speed at 2 m. for a grass covered surface at Davis, California (adapted from Pruitt *et al.*, 1968). In this case the balance equation is $R_n = L_eE + H + G$.

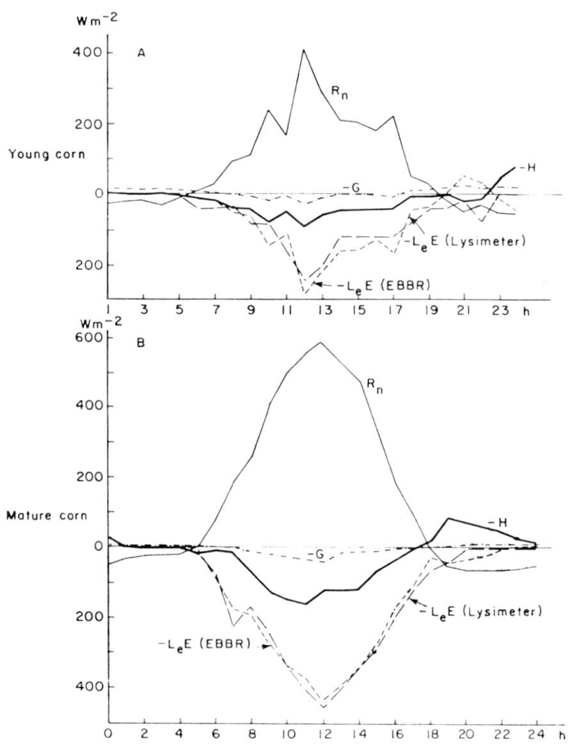

Fig. 6.2. Example of the daily cycle of the energy balance for a young (A) and a mature (B) maize canopy near Versailles, France (adapted from Perrier *et al.*, 1976).

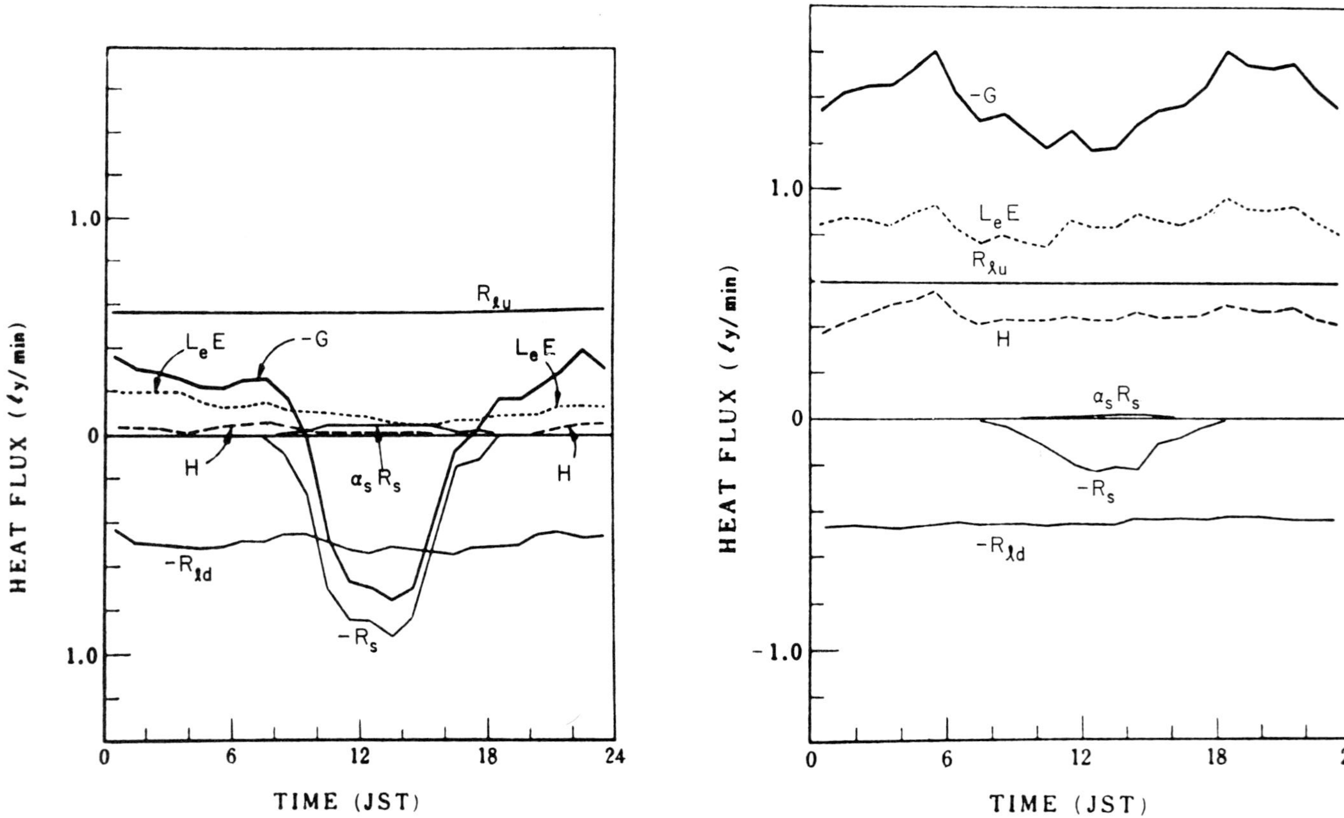

Fig. 6.3. Examples of the daily course of the energy balance in (cal min^{-1} cm^{-2}) at the surface of a deep water body. The data were obtained over the East China Sea during the Air Mass Transformation Experiment (AMTEX) on February 15 (a) and 25 (b) 1974; the time is Japanese Standard Time (adapted from Yasuda, 1975).

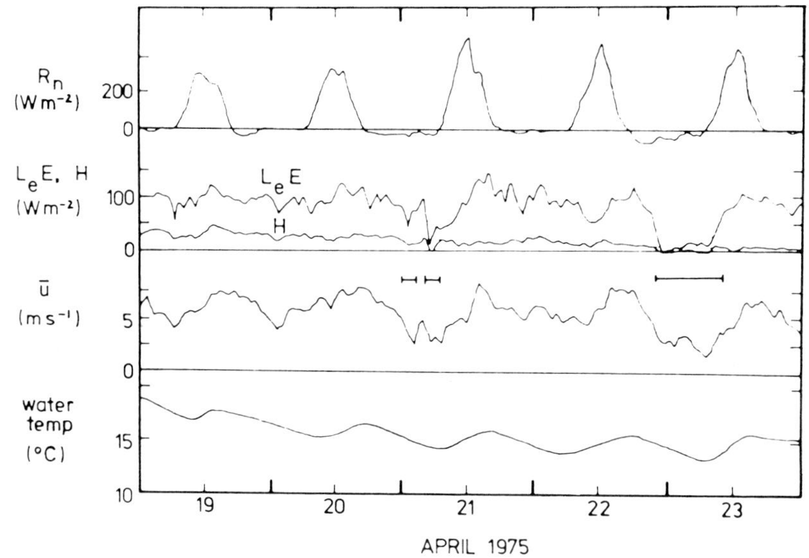

Fig. 6.4. Example of diurnal variation of some terms of the energy budget at the surface of a shallow lake (mean depth 2m) in Southern Australia (35° 35′S, 130° 15′E) with over-water fetches of between 4 to 15 km. The bars show short periods during which the wind strayed outside the direction limits for over water measurements and the fetch consisted of mud flats, so that the measured (with eddy correlation) fluxes probably do not represent the lake evaporation (from Raupach, 1978).

$$R_n = R_s(1 - \alpha_s) + \varepsilon_s R_{ld} - R_{lu} \tag{6.2}$$

where R_s is the (global) short-wave radiation, α_s the albedo of the surface, R_{ld} the downward long-wave or atmospheric radiation, ε_s the emissivity of the surface and R_{lu} the upward long-wave radiation. Whenever possible, the net radiation should be measured, and at present, fairly reliable instruments are available for this purpose. In the absence of direct measurements, R_n can be obtained from its components on the right-hand side of (6.2). When these measurements are not available, the components can be obtained by theoretical methods or simpler empirical formulae.

a. Global Short-Wave Radiation

The short-wave radiation is the radiant flux resulting directly from the solar radiation. This incoming short-wave radiation has most of its energy contained in the wavelength range from 0.1 to 4 μm. At the outside of the atmosphere this flux, i.e., the solar constant, is of the order of 1395 W/m^2 (or 2 cal min^{-1} cm^{-2}). As it passes through the atmosphere, the solar radiation is modified by scattering, absorption, and reflection by different types of molecules and colloidal particles; thus at the earth's surface the global short-wave radiation consists of direct solar radiation and diffuse sky radiation.

The short-wave radiation can easily be measured (e.g., Robinson, 1966; Kondratyev, 1969; Coulson, 1975) and data are readily available from most national weather services and agricultural organizations. In the event that suitable data are not available, it may be necessary to make an estimate by means of one of several theoretical models or simpler empirical formulae that relate short-wave radiation with other physical factors such as extra-terrestrial radiation, optical air mass, turbidity, water vapor content of the air, amount and type of cloud cover, etc. However, these are still under development and should be exercised with caution.

A simple equation for daily averages was proposed by Ångström (1924) in terms of the short-wave radiation under clear skies R_{sc} and of the fraction of sunshine hours (n/N), in which n is the actual number of hours of bright sunshine and N the number of daylight hours; it can be written as

$$R_s = R_{sc}[a + (1 - a)n/N] \tag{6.3}$$

where a is a constant which was found to be of the order of 0.235 at Stockholm. However, a appears to depend on the location, the season and the state of the atmosphere, so it is best determined by local calibration. At many locations totally cloudless days are so scarce, that this calibration may be impossible. This difficulty is avoided in the equation proposed by Prescott (1940) in terms of the extraterrestrial radiation R_{se}, that is, the solar radiation which would reach a horizontal surface in the absence of the atmophere; this equation, which may be useful for weekly to monthly averages, is

$$R_s = R_{se}[a + b(n/N)] \tag{6.4}$$

where a and b are constants which, again, depend on the location, the season and the state of the atmosphere. R_{se} can readily be calculated for a given latitude and time of the year when the solar constant is known; the daily totals given in Table 6.1 are based on the calculations of Milankovitch (List, 1971) applied here to a solar constant of 2.0 cal min^{-1} cm^{-2} (see also Figure 6.5). Values of a and b have been determined for many locations, some examples of which are shown in Table 6.2. The averages of the values in the table are $a = 0.25$ and $b = 0.50$. Attempts have been made to relate a and b to latitude and/or climate by, among others, Glover and McCulloch (1958), Löf *et al.* (1966) and Linacre (1967).

Instead of the relative duration of sunshine (n/N), the mean fractional cloud cover m_c has also been used in empirical relationships. Kimball (1927) proposed an equation analogous to (6.3), viz.

$$R_s = R_{sc}[1 - am_c] \tag{6.5}$$

where a is a constant which he found to be of the order of 0.71; under a slightly different form, (6.5) is also known as the Savinov–Ångström equation and the dependence of a on latitude has been determined by T. G. Berliand (e.g., Budyko, 1974, p. 45 ff.; Kondratyev, 1969, p. 467). The analog of (6.4), in terms of m_c instead of (n/N), has been used by Pochop *et al.* (1968) and a quadratic extension of it has been studied by Black (1956).

The analogy between (6.3) and (6.5) and the fact that (n/N) is the portion of the day during which clouds are not obstructing the sun, lead to the supposition that

$$m_c + \frac{n}{N} = 1 \tag{6.6}$$

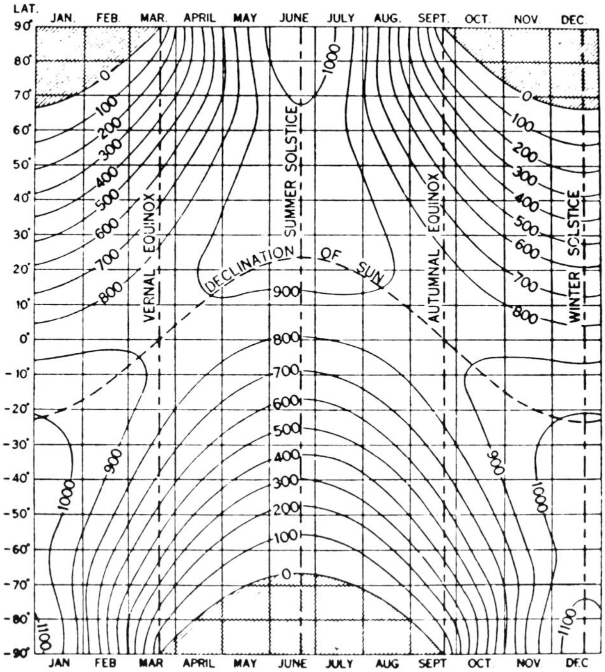

Fig. 6.5. Solar radiation on a horizontal plane without atmosphere in (cal day^{-1} cm^{-2}). The solar constant is taken as 1.94 cal cm^{-2} min^{-1} (from J. B. Leighly in List, 1971).

approximately. This may be satisfactory when no other information is available. Actually, it has been observed that the right-hand side of (6.6) is not quite unity, but that it is larger in summer and smaller in winter. For example, in the Netherlands DeVries (1955) found

$$a\frac{n}{N} + bm_c = 1 \tag{6.7}$$

where $a = 1.12$, $b = 0.88$ in summer and $a = 1.29$, $b = 1.00$ in winter; Kondo (1967) derived $a = 1.11$, $b = 0.78$ from data in Japan.

Several other correlation equations like (6.3)–(6.5) have been proposed in the literature, but a more extensive review would lead too far here. Clearly, such simple equations can only be poor substitutes for direct measurements. Nevertheless, it is possible to obtain fairly accurate radiation estimates by better empirical and partly theoretical methods, which are, however, more difficult to apply. Examples of such methods, which have given satisfactory results, are those presented by Kondo (1967; 1976), Paltridge and Platt (1976, p. 137) and Satterlund and Means (1978).

b. Albedo

The surface albedo is the ratio of the global short-wave reflected radiative flux and the flux of the corresponding incident radiation; in contrast to the term reflectivity, the albedo also includes the diffuse portion of the radiation. In energy budget studies,

TABLE 6.1

Solar radiation on horizontal plane without atmosphere [cal day^{-1} cm^{-2}]

Longtitude of Sun	0°	22.5°	45°	67.5°	90°	112.5°	135°	157.5°	180°	202.5°	225°	247.5°	270°	292.5°	315°	337.5°
Date (day/month)	21/3	13/4	6/5	29/5	22/6	15/7	8/8	31/8	23/9	16/10	8/11	30/11	22/12	13/1	4/2	26/2
Latitude																
90 N		436	796	1030	1110	1025	789	431								
80 N	160	436	784	1014	1093	1010	777	431	158	7						7
70 N	316	541	772	968	1043	963	765	535	312	133	25				25	135
60 N	461	655	834	963	1009	957	826	648	456	281	151	74	51	75	151	285
50 N	593	755	894	988	1020	984	886	747	586	427	295	210	181	211	298	432
40 N	707	832	938	1002	1022	997	929	823	698	562	442	359	327	361	447	570
30 N	799	892	958	997	1005	990	949	882	789	684	581	507	480	509	586	691
20 N	867	922	952	964	964	959	944	911	857	784	706	646	624	649	712	793
10 N	909	925	921	908	900	904	913	914	898	861	813	771	756	775	820	871
0	923	900	863	829	814	825	856	890	912	913	897	877	869	881	905	924
10 S	911	849	784	729	708	726	776	839	898	938	956	960	962	965	965	949
20 S	867	773	680	611	585	608	674	764	857	935	989	1019	1030	1024	998	946
30 S	799	674	560	479	449	477	555	666	789	904	994	1052	1073	1057	1003	915
40 S	707	555	426	339	306	338	422	549	698	844	973	1059	1092	1064	982	854
50 S	593	421	285	199	170	198	282	416	586	766	929	1045	1089	1049	937	775
60 S	461	277	144	70	48	70	143	274	456	664	866	1018	1078	1023	873	672
70 S	316	131	24				24	130	312	548	802	1024	1114	1029	809	556
80 S	160	7						7	158	442	814	1073	1167	1078	821	447
90 S										442	826	1089	1185	1095	834	447

TABLE 6.2
Some values of the constants in Prescott's Equation (6.4) obtained for different locations

Location	Latitude	Period	*a*	*b*	Reference
Accra (Ghana)	6°	monthly	0.30	0.37	Davies (1965)
Kano (Nigeria)	12°	monthly	0.26	0.54	Davies (1965)
Kunumura (W. Australia)	16°S	daily	0.334	0.431	Fitzpatrick and Stern (1965)
Delhi (India)	29°	weekly	0.31	0.46	Yadav (1965)
Tateno (Japan)	36°	monthly	0.25	0.54	Kondo (1967)
Dodge City (Kansas, USA)	38°	daily	0.230	0.542	Baker and Haines (1969)
Cleveland (Ohio, USA)	41°	daily	0.188	0.539	Baker and Haines (1969)
Madison (Wisc. USA)	43°	daily	0.208	0.530	Baker and Haines (1969)
DeBilt (Netherlands)	52°	daily	0.22 ± 0.01	0.50 ± 0.02	W. Kohsiek (1971) – unpublished
Rothamsted (England)	52°	monthly	0.18	0.55	Penman (1948)
Matanuska-Anchorage (Alaska, USA)	61°	daily	0.261	0.465	Baker and Haines (1969)

the albedo usually refers to an integral value over all wave lengths; however, sometimes, to distinguish it from the spectral albedo, it is called the integral albedo. In the case of an ideal rough surface, the albedo should be independent of the direction of the primary beam.

For most natural surfaces the fraction of the directly and diffusely reflected radiation depends on the direction of the incoming beam. Therefore, on days with sunshine, the albedo of most surfaces depends on the altitude of the sun, but this dependence decreases with increasing cloudiness.

For example, for the water surface of Lake Hefner, Anderson (1954) proposed the following empirical equation

$$\alpha_s = aS_A^b \tag{6.8}$$

TABLE 6.3
Values of the parameters ($\frac{a}{b}$) in Anderson's (1954) Equation (6.8)

Cloudiness m_c	0	0.1–0.5	0.6–0.9	1.0
Cloud Type	Clear	Scattered	Broken	Overcast
High Clouds	1.18	2.20	1.14	0.51
	−0.77	−0.98	−0.68	−0.58
Low Clouds	1.18	2.17	0.78	0.20
	−0.77	−0.96	−0.68	−0.30

TABLE 6.4
Approximate mean albedo values for various natural surfaces

Nature of surface	Albedo
Deep Water	0.04–0.08
Moist dark soils; plowed fields	0.05–0.15
Gray soils; bare fields	0.15–0.25
Dry soils; desert	0.20–0.35
White sand; lime	0.30–0.40
Green grass and other short vegetation (e.g., alfalfa, potatoes, beets)	0.15–0.25
Dry grass; stubble	0.15–0.20
Dry prairie and savannah	0.20–0.30
Coniferous forest	0.10–0.15
Deciduous forest	0.15–0.25
Forest with melting snow	0.20–0.30
Old and dirty snow cover	0.35–0.65
Clean, stable snow cover	0.60–0.75
Fresh dry snow	0.80–0.90

where S_A is the sun altitude in degrees and a and b are constants having values of 1.18 and −0.77, respectively, for clear skies. Values of a and b obtained for different types and amounts of cloud are presented in Table 6.3. Similar but more detailed results were obtained by Payne (1972). The albedo's of other surfaces obey relationships similar to (6.8). However, for calculations of daily radiation totals, it is common practice to use a mean value of the albedo. In the case of a water surface, the albedo is then taken to be of the order of 0.06. In Table 6.4 a brief summary is presented of mean albedo values for various surfaces obtained from different summaries of available data (e.g., Van Wijk and Scholte Ubing, 1963; Kondratyev, 1969; List, 1971; Budyko, 1974).

c. Long-Wave or Terrestrial Radiation

The long-wave radiation is the radiant flux resulting from the emission of the atmospheric gases and the land and water surfaces of the earth. All materials on earth and around it have a much lower temperature than the sun, so that the radiation they emit has much longer wavelengths than the global radiation. There is practically no overlap, since most of the radiation emitted by earth and atmosphere is contained in the range from 4 to 100 μm.

Although good instruments are becoming available, it is still not as easy to measure long-wave radiation in nature as global radiation; one of the reasons for this is that any instrument for this purpose emits radiation of comparable wavelengths and intensity as that which it is supposed to measure. Therefore, in many practical situations in meteorology, it is still expeditious to calculate it on the basis of measurements of more easily-measured variables. Accurate methods for this purpose are described in the general works of Goody (1964), Kondratyev (1969) and Paltridge and Platt (1976).

It is convenient to consider two components of the terrestrial radiation at the earth's surface separately, namely the component of downward radiation from the atmosphere, R_{ld}, and that of upward radiation from the surface, R_{lu}.

Upward Long-Wave Radiation

This term R_{lu} is usually obtained by assuming that the ground, the canopy or the water surface under consideration is equivalent to an infinitely deep gray body of uniform temperature and emissivity ε_s, which is close to unity. This allows the following formulation

$$R_{lu} = \varepsilon_s \sigma T_s^4 \tag{6.9}$$

in terms of the surface temperature; σ ($= 5.6697 \times 10^{-8}$ W m^{-2} K^{-4} $= 1.354\ 10^{-12}$ cal cm^{-2} s^{-1} K^{-4}) is the Stefan–Boltzmann constant.

For a water surface ε_s can usually be taken as 0.97 (e.g., Anderson, 1954; Davies *et al.*, 1971). For other surfaces ε_s is not as well known, but it is usually also quite close to unity. Some values which have been compiled in the literature (e.g., Van Wijk and Scholte Ubing, 1963; Kondratyev, 1969) are summarized in Table 6.5. In

TABLE 6.5
Values of the emissivities ε_s of some natural surfaces

Nature of surface	Emissivity
Bare soil (mineral)	0.95–0.97
Bare soil (organic)	0.97–0.98
Grassy vegetation	0.97–0.98
Tree vegetation	0.96–0.97
Snow (old)	0.97
Snow (fresh)	0.99

many practical applications it is simply assumed that $\varepsilon_s = 1$. Furthermore, since T_s is rarely known, especially over land, (6.9) is often applied by using the air temperature T_a instead of T_s.

Downward Long-Wave Radiation Under Clear Skies

The more accurate methods for calculating the atmospheric radiation under clear skies, R_{ldc}, require, in general, vertical profile data of humidity and temperature. Such data are rarely available where the long-wave radiation is needed; as a result, simpler methods have been developed, which are mostly based on an equation of the type

$$R_{ldc} = \varepsilon_{ac} \sigma T_a^4 \tag{6.10}$$

where T_a is the air temperature near the ground, usually taken at shelter level and ε_{ac} is the atmospheric emissivity under clear skies.

Several expressions have been published for ε_{ac}. Most of these are strictly empirical, but it is also possible to derive ε_{ac} on physical grounds. In one such derivation (Brutsaert, 1975d) the starting point is the equation for radiative transfer in a plane stratified atmosphere. For the downward radiation at the surface, the specific flux can be shown to be (e.g., Goody, 1964)

$$R_{ld} = \int_{a(z)=0}^{\infty} \sigma T^4 \frac{\partial \varepsilon_{sl}(a, T)}{\partial a} \, \mathrm{d}a \tag{6.11}$$

where $T = T(z)$ is the temperature, ε_{sl} the slab emissivity, relating the emission of a slab of gas to that of a black body. The quantity $a = a(z)$ is the scaled amount of matter or, in the case of a cloudless sky, mainly the water vapor in the air column from the surface up to the level z, scaled for the pressure effect. This pressure effect is often taken as a square root, so that the scaled amount of water vapor is

$$\mathrm{d}a = \rho_v (p/p_a)^{1/2} \, \mathrm{d}z \tag{6.12}$$

where p_a is the pressure at the surface.

Note that (6.11) and (6.12), or very similar equations, constitute the basis of the more accurate methods of calculation mentioned earlier, and of the radiation charts developed from them.

It is possible to obtain a closed-form solution of (6.11) and (6.12) by assuming first a power function relationship for the slab emissivity

$$\varepsilon_{sl} = A\, a^m \tag{6.13}$$

where A and m are constants (when the CO_2 in the air is included these are typically 0.75 and 1/7, respectively) and second, a near-Standard Atmosphere in the lowest 15 km. This last assumption allows the following for T, p and ρ_v as first approximations,

$$T = T_a \exp\left[-(\gamma/T_a) z\right], \tag{6.14}$$

$$p = p_a \exp(-k_p z), \tag{6.15}$$

$$\rho_v = (0.622\, e_a/R_d T_a) \exp(-k_w z) \tag{6.16}$$

where e_a is the vapor pressure near the surface; the attenuation parameter (γ/T_a) is of the order of 2.26×10^{-2} km^{-1} to approximate a standard atmosphere, and k_p and k_w are typically of the order of 0.13 km^{-1} and 0.44 km^{-1}, respectively. Substitution of (6.12) through (6.16) in (6.11), and integration yield

$$R_{ldc} = \sigma T_a^4 m A \left[\frac{0.622\, e_a}{k_2 R_d T_a}\right]^m B\left(\frac{k_1}{k_2}, m\right) \tag{6.17}$$

where $B(\ )$ is the complete Beta function (e.g., Abramowitz and Stegun, 1964), $k_1 = [k_2 + (4\gamma/T_a)]$ and $k_2 = [k_w + (k_p/2)]$. Substitution of the above-mentioned typical values of the constants immediately yields for the atmospheric emissivity under clear skies

$$\varepsilon_{ac} = 1.24\left(\frac{e_a}{T_a}\right)^{1/7} \tag{6.18}$$

if e_a is in mb and T in K.

Because the right-hand side of (6.18) is quite insensitive to changes in T_a, it can be further approximated with the near-surface temperature of the Standard Atmosphere $T_a = 288$ K, as

$$\varepsilon_{ac} = 0.552\, e_a^{1/7} \tag{6.19}$$

where e_a is in mb. Both (6.18) and (6.19) have been found (e.g., Mermier and Seguin, 1976; Aase and Idso, 1978) to yield satisfactory results with daily means at intermediate latitudes and at temperatures above 0°C; these are probably conditions which, on average, are fairly well described by a Standard Atmosphere.

In the literature, also purely empirical equations based on correlations with measurements have been proposed. One of the better-known formulae used in practice is due to Brunt (1932), namely

$$\varepsilon_{ac} = a + b\, e_a^{1/2} \tag{6.20}$$

where a and b are constants to be determined from observational data. In Table 6.6 some values are given which were obtained by different investigators and in Figure 6.6 the clear-sky emissivities obtained with (6.20) are compared with (6.19).

A second empirical formula, which has been used in practice, is that obtained by Swinbank (1963), $\varepsilon_{ac} = 0.398 \times 10^{-5}\, T_a^{2.148}$, which he rounded off to

$$\varepsilon_{ac} = 0.92 \times 10^{-5}\, T_a^2. \tag{6.21}$$

Albeit the basis of (6.21) appears to be purely empirical, it can be reconciled with the theoretical result (6.18). Indeed, it has been found (Deacon, 1970) from monthly means at locations covering a considerable climatic range that the amount of precipitable water is proportional to the 16.8 power of T_a. Thus, according to (6.16), one has $e_a \sim T_a^{17.8}$, and substitution in (6.18) shows $\varepsilon_{ac} \sim T_a^{2.4}$, which is not very different from $T_a^{2.148}$. Paltridge and Platt (1976) noted that, since (6.21) is based on night-time data, it is biased towards inversion conditions. Therefore, they suggested that on average day-time calculations with (6.21) should be decreased by about 2 mW cm^{-2}.

TABLE 6.6
Some values of the constants in Brunt's formula (6.19) (with e_a in mb)

Reference	Location	Latitude	Altitude (m)	a	b	Correlation	Period
Brunt (1932)	Benson (England)	52°N	6	0.52	0.065	0.97	Monthly
Yamamoto (1950)	From fit with theoretical calculation and various data sets			0.51	0.066		
Goss and Brooks (1956)	Davis, Central Valley (Calif.)	38°N	14	0.66	0.039	0.89	Monthly
Anderson (1954)	Lake Hefner (Oklahoma)	36°N	369	0.68	0.036	0.92	Monthly
DeCoster and Schuepp (1957)	Kinshasa (Zaire)	4.8°N	321	0.645	0.048		Daily

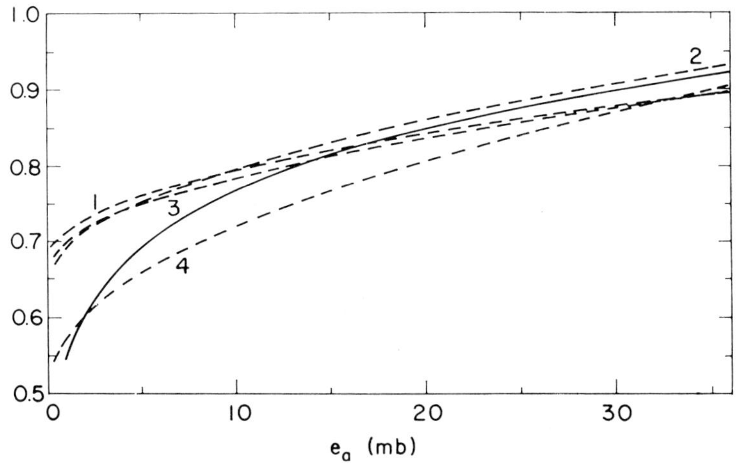

Fig. 6.6. The dependence of the effective atmospheric emissivity, ε_{ac}, on the vapor pressure near the surface at screen level, e_a (in mb), as calculated by means of Brunt's Equation (6.20) (- - -) and Brutsaert's (6.18) for $T_a = 288°$ (———). Curve 1 is based on the Brunt parameters obtained by Anderson (1954); Curve 2 on those of DeCoster and Schuepp (1957); Curve 3 on those of Goss and Brooks (1956); Curve 4 on those of Yamamoto (1950).

Another empirical equation based only on temperature is that of Idso and Jackson (1969), viz.

$$\varepsilon_{ac} = \{1 - 0.261 \exp[-7.77 \times 10^{-4} (273 - T_a)^2]\} \tag{6.22}$$

which appeared to be valid over a wider temperature range than (6.21); however, Aase and Idso (1978) concluded that (6.22) was less accurate at temperatures below freezing. More recently, Satterlund (1979) proposed an empirical equation which, just like (6.18) takes both temperature and atmospheric humidity into account, viz.

$$\varepsilon_{ac} = 1.08[1 - \exp(-e_a^{T_a/2016})] \tag{6.23}$$

where again e_a is in mb and T_a in K. The main point of interest here is the comparison Satterlund (1979) made of the three Equations (6.18), (6.22) and (6.23). This comparison, which is shown in Figure 6.7, gives some idea of the scatter that may be expected with equations of the type of (6.10) on the basis of daily mean temperature.

The question remains which among (6.18) through (6.23) is preferable. It is clear from the derivation of (6.18) from (6.11), which after all is the basis of more accurate methods and radiation charts, that the effect of the atmospheric humidity should be included, when possible. Brunt's (6.20) has been tested most; but the variability of its constants is a serious disadvantage and it may be an indication that (6.20) does not have the proper functional form. As shown in Figure 6.7 for daily means, (6.18) which is based on physical grounds, produces results which are apparently at least as good as the empirical equations. Equation (6.18) has the advantage that it may be possible to adjust it to unusual conditions of humidity or temperature stratification by adopting more appropriate functions for (6.13) through (6.16). But (6.23) is in better agreement with the data presented for below freezing.

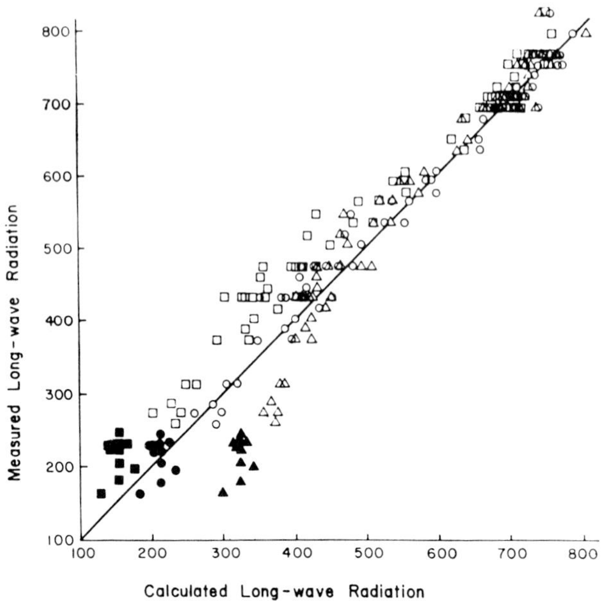

Fig. 6.7. Measured versus calculated atmospheric radiation (in cal cm^{-2} d^{-1}) using Idso-Jackson's (6.22) (triangles); Satterlund's (6.23) (circles); and Brutsaert's (6.18) (squares). The data are based on clear sky measurements of Aase and Idso (1978) (open symbols) and Stoll and Hardy (1955) (solid symbols) (adapted from Satterlund, 1979).

The Effect of Cloudiness

In contrast to R_{lu}, R_{ld} is affected by the cloudiness. Among the empirical methods of taking this effect into consideration, one can distinguish two types of procedures. The first consists of applying an adjustment to R_{ldc} or ε_{ac} as defined in (6.10); in the second, the adjustment is applied to the calculated net long-wave radiation under clear skies.

Several adjustment equations of the first type can be combined in one expression as follows

$$R_{ld} = R_{ldc}(1 + a\, m_c^b) \tag{6.24}$$

where m_c is the fractional cloud cover of the sky and a and b are constants; a is usually made to depend on the cloud type. An example is Bolz's (1949) version of (6.24) with $b = 2$ and a, which is listed in Table 6.7, dependent on cloud type. Equation (6.24), but with different a and b, was also applied by Kuzmin and Kirillova, respectively (e.g., Budyko, 1974, p. 59).

The net long-wave radiation is

$$R_{nl} = \varepsilon_s R_{ld} - R_{lu}. \tag{6.25}$$

By analogy with (6.24) several adjustment equations, which have been proposed in the literature, can be expressed as

$$R_{nl} = R_{nlc}(1 - a' m_c^{b'}) \tag{6.26}$$

TABLE 6.7
Values of the Bolz parameter a with $b = 2$ in (6.24) for various sky conditions

Cloud type	La
Cirrus (Ci)	0.04
Cirrostratus (Cs)	0.08
Altocumulus (Ac)	0.17
Altostratus (As)	0.20
Cumulonimbus (Cb)	0.20
Cumulus (Cu)	0.20
Stratocumulus (Sc)	0.22
Nimbostratus (Ns)	0.25
Fog	0.25
Average	0.22

where

$$R_{nlc} = \varepsilon_s R_{ldc} - R_{lu} \tag{6.27}$$

is the net long-wave radiation under clear skies, and a' and b' are constants. Clearly, if one can assume that $T_s = T_a$ in (6.9), and if $b = b'$ one would have the following approximate relationship between the two methods of adjusting for cloud effect

$$a = a'\left(\frac{1 - \varepsilon_{ac}}{\varepsilon_{ac}}\right). \tag{6.28}$$

Since ε_{ac} is of the order of 0.75 (cf., Figure 6.4), this would yield that a' in (6.26) is roughly of the order of three times a of (6.24).

Most applications of (6.26) have been made with $b' = 1$. Already, some 60 years ago, Ångström concluded with $b' = 1$ that $a' = 0.9$; Unsworth and Monteith (1975) derived (6.26) with $b' = 1$ and $a' = 0.84$ on the basis of a simple cloud radiation model with $T_a - T_c = 11$ K (where T_c is the cloud base temperature). On the other hand, in 1960, Barashkova (e.g. Kondratyev, 1969) found that $b' = 2$ and $a' = 0.7$. Several attempts have been made to relate a' (mostly with $b' = 1$) to the amount of clouds at the lower, middle and upper levels, viz. m_l, m_m and m_u, respectively; a' is then weighted as follows

$$a' = (a_l' m_l + a_m' m_m + a_u' m_u)/m_c \tag{6.29}$$

where a_l', a_m' and a_u' are constants. As an example, values of these constants obtained in 1952 by Berliand and Berliand (e.g., Kondratyev, 1969) are shown in Table 6.8.

TABLE 6.8
Empirical coefficients for (6.29) to determine cloud effect on net long-wave radiation in (6.26) with $b' = 1$ (obtained by Berliand and Berliand)

Latitude	Season	a_l'	a_m'	a_u'	Average a'
>60°	Cold	0.90	0.77	0.28	0.82
	Warm	0.86	0.72	0.27	0.80
60–50	Cold	0.86	0.74	0.27	0.77
	Warm	0.80	0.67	0.24	0.70
50–40	Cold	0.82	0.69	0.24	0.71
	Warm	0.78	0.65	0.19	0.69

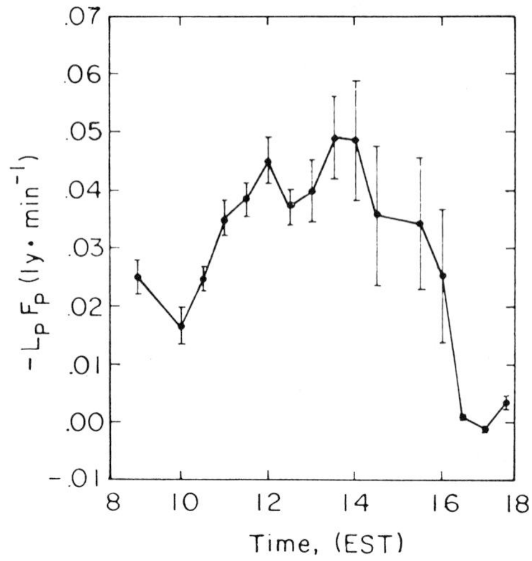

Fig. 6.8. CO_2 flux density $-L_pF_p$ into a stand of maize cal (cm^2 min)$^{-1}$ estimated by means of (10.8) of the energy budget method (EBBR). The experiment took place near Ithaca, NY on August 13, 1970. The values shown were about 8 percent of the measured latent heat flux L_eE, which displayed a similar diurnal variation. The time is Eastern Standard Time. The vertical bars are error estimates of the flux values (from Sinclair *et al.*, 1975).

M. E. Berliand (Budyko, 1974; Table 9) has also calculated mean values of a' (for $b' = 1$) for different latitudes by considering the frequency of clouds at different levels for each latitude. A somewhat different method of adjusting R_{nlc}, taking account not only of cloud amounts at different levels, but also of the mean humidity of the air, was developed by Kondo (1967, 1976), and applied to the calculation of the energy budget over the ocean.

When data on cloud cover are not available, it may be adequate as an approximation to substitute (n/N) in (6.24) and (6.26) for m_c by means of (6.6) or (6.7). Actually, this is implicit in an equation used by Penman (1948)

$$R_{nl} = R_{nlc}[a + (1-a)(n/N)] \tag{6.30}$$

where a is a constant which he took to be 0.10. Other studies have yielded estimates of 0.23 (Impens, 1963) and 0.30 (Fitzpatrick and Stern, 1965), so that $a = 0.2$ may be taken as an average value for practical calculations.

6.2. ENERGY ABSORPTION BY PHOTOSYNTHESIS

In energy budget studies the flux of CO_2, F_p, is usually neglected, except when its determination is the main objective (e.g., Sinclair *et al.*, 1975). Under favorable conditions over vegetation, L_pF_p can be of the order of 5 percent of the global radiation, but usually it is smaller than 1 percent. An example of the daily variation of the photosynthetic energy flux density abovze maie is shown in Figure 6.8. The thermal conversion for fixation of CO_2, L_p, is approximately 1.05×10^7 J kg^{-1} of CO_2 (or 2500 cal g^{-1}).

6.3 ENERGY FLUX AT LOWER BOUNDARY OF THE LAYER

The nature of G and the optimal method of its determination depend on the type of substrate to which the energy budget equation is applied.

For a thin layer of soil, for a vegetational canopy or for a whole lake or stream, the term G in (6.1) represents the heat flux into the ground. For a water surface, G is the heat flux into the underlying water body.

a. Land Surfaces

Over land covered with vegetation the daily mean value of G, that is the soil heat flux, is often one or more orders of magnitude smaller than the major terms in the energy budget, R_n, H and L_eE. However, over shorter periods it can be quite important. Several methods are available to determine G, which warrant a brief discussion of basic concepts.

Heat Transfer in the Soil

Since conduction is the main mechanism, even though convection and radiation also play a role, the most important features of soil heat transfer can be described by casting it in the form of a conduction phenomenon. Thus, the specific flux of heat is given by Fourier's law. For the vertical downward flux, which is of main concern here, with z pointing down, this can be written as

$$Q_H = -K_T \frac{\partial T}{\partial z} \tag{6.31}$$

where K_T is called the thermal conductivity. Because the heat transfer under a temperature gradient does not involve only conduction but other mechanisms as well (primarily vapor movement) K_T is also referred to as the apparent thermal conductivity. Substitution of (6.31) in the equation of conservation of heat produces

$$C_s \frac{\partial T}{\partial t} = \frac{\partial}{\partial z}\left(K_T \frac{\partial T}{\partial z}\right) \tag{6.32}$$

where $C_s = \rho_s c_s$ is the volumetric heat capacity of the soil, c_s, the specific heat and ρ_s the density of the soil. In certain situations when K_T can be assumed to be independent of z, (6.32) can be simplified to

$$\frac{\partial T}{\partial t} = D_T \frac{\partial^2 T}{\partial z^2} \tag{6.33}$$

where $D_T = (K_T/C_s)$ is the thermal diffusivity.

(*i*) *Specific heat.* A knowledge of the volume fractions of mineral soil θ_m, organic matter θ_c, water θ, and air θ_a, allows the determination of the volumetric heat capacity C_s, as follows

$$C_s = \rho_m \theta_m c_m + \rho_c \theta_c c_c + \rho_w \theta c_w + \rho_a \theta_a c_a \tag{6.34}$$

where the c terms are the specific heats and the ρ terms the densities as indicated by

TABLE 6.9
Properties of soil components at 293 K

	Specific heat c(J kg^{-1} K^{-1})	Density ρ (kg m^{-3})
Soil minerals	733	2650
Soil organic matter	1926	1300
Water	4182	1000
Air	1005	1.20

the subscripts. Values of these properties of the soil components, obtained from a compilation by De Vries (1963) are presented in Table 6.9. Accordingly, the volumetric heat capacity in (J m^{-3} K^{-1}) is

$$C_s = (1.94\ \theta_m + 2.50\ \theta_c + 4.19\ \theta)\ 10^6. \tag{6.35}$$

(*ii*) *Thermal conductivity*. Not only are field soils rarely homogeneous, but it is usually quite difficult to reproduce their structure in small samples required for laboratory measurements. For hydrological purposes the soil thermal properties are preferably measured *in situ*. Field methods to determine the thermal conductivity involving special probes have been described in the literature (e.g., De Vries and Peck, 1958a, b; Janse and Borel, 1965; Fritton *et al*., 1974). However, these are not without difficulties; errors may result from air entrapment (Nagpal and Boersma, 1973) and from imperfect probe-soil contact (Hadas, 1974).

Since the main role of K_T is to serve as a parameter in (6.31) and (6.32), it can also be determined by inverse calculations from known values of Q_H and T. For example, such a method has been used by Kimball and Jackson (1975) as follows; for the times of the day when a zero thermal gradient exists somewhere in the soil profile, it is possible to calculate the soil heat flux Q_H at any level by means of the calorimetric method (see (6.40)); K_T at that level is then computed by dividing Q_H by the local temperature gradient.

In addition, methods have been proposed to calculate K_T. De Vries (1963) has developed a theoretical model for heat conduction, by making use of earlier ideas of Burger (1915); the basic assumption is that the soil is a suspension of soil particles and small air pockets in water. The effect of the vapor transport, involving distillation due to the temperature gradient, is included in the model by means of the formulation of Krischer and Rohnalter (1940). DeVries showed that values of K_T calculated with this model agreed usually within less than 10 percent with experimental data. The method was tested with field data on a loam soil by Kimball *et al.* (1976a). They concluded that heat transfer by pure conduction was far more important than any other mechanism, as the method yielded better results by ignoring the transfer term due to the water vapor movement. In a similar vein, the laboratory experiments of Moench and Evans (1970) showed that the apparent conductivity is usually only a few (around 5) percent different from the real thermal conductivity.

As an illustration of their magnitude, in Figure 6.9 experimental data are presented of the thermal conductivity as function of water content for a quartz sand, a sandy loam, a loam, and a peat soil.

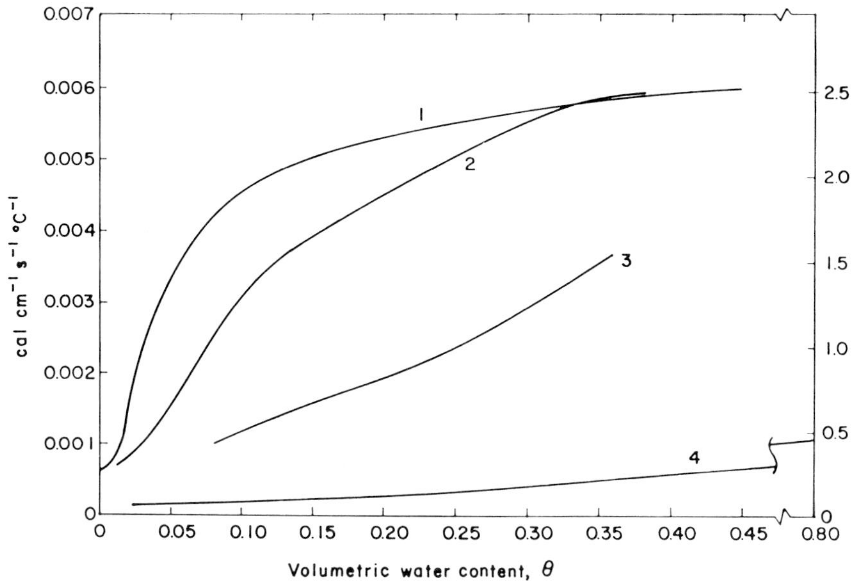

Fig. 6.9. Thermal conductivities K_T of four types of soils as functions of water content. Curve 1 represents laboratory results on a quartz sand of porosity 0.43 (De Vries, 1963); 2 laboratory data on a sandy loam of porosity 0.38 (Moench and Evans, 1970); 3 is the median line through field data on a loam with mean porosity 0.41 (Kimball *et al.*, 1976a); 4 laboratory data on a peat soil with porosity around 0.8 (Kersten, 1949).

(*iii*) *Thermal diffusivity*. With C_s and K_T known as functions of depth and water content, (6.32) can be solved to calculate the soil temperature and heat flux profiles. However, since the soil heat flux is usually considerably smaller than other more important terms in the surface energy budget, extreme precision in the calculation of the soil heat flux may not always be necessary. Hence, for certain problems involving the energy budget or the propagation of annual temperature waves into the ground, it may be adequate to make use of solutions of the linear Equation (6.33) with a constant diffusivity D_T. Since both K_T and C_s, and thus also D_T may be quite variable, it is not obvious what this constant or characteristic value of D_T should be to obtain optimal results.

Methods proposed in the past to determine D_T consist mainly of fitting observed temperature measurements to an appropriate solution of (6.33). For example, use can be made of the solution for a sinusoidal temperature wave at the surface (e.g., Chudnovskii, 1962; Van Wijk and De Vries, 1963). This solution can be written as

$$T = T_{sa} + a_0 \exp\left[-\left(\frac{\omega}{2\,D_T}\right)^{1/2} z\right] \sin\left[\omega t - \left(\frac{\omega}{2\,D_T}\right)^{1/2} z + b\right] \qquad (6.36)$$

where T_{sa} is a constant average soil surface temperature, a_0 the amplitude of the sine wave at $z = 0$, $\omega = (2\pi/p)$ the radial frequency, p the period in the same units as t, and b a phase constant. This solution shows that the maxima of a temperature wave occur at later and later times with increasing depth. The diffusivity can be calculated from observations of when (say $t_1, t_2, \ldots$) the maxima of a sinusoidal temperature wave, applied at the surface, penetrate into the soil profile at different depths, (say $z_1, z_2, \ldots$). It can also be determined from determinations of the amplitude of the wave at different depths z. Methods based on (6.36) are rather simple in principle, and they have been used extensively during the past 100 years (e.g., Chudnovskii, 1962). However, in the field they are of limited applicability, because the surface temperature wave is rarely sinusoidal.

Some methods have been proposed, based on solutions of (6.33), for more general, i.e., non-sinusoidal, temperature variations at the surface. For example, Van Wijk (1963) suggested a method when temperature records are available at two depths z_1 and z_2. The method can be summarized as follows. The Laplace transform of the solution of (6.33) for a constant initial temperature T_i is

$$\mathscr{L}\{T - T_i\} = a \exp\left[-z\left(\frac{s}{D_T}\right)^{1/2}\right] \tag{6.37}$$

where a is a constant and s is the Laplace transform variable. The temperature records are Laplace-transformed by multiplying $(T - T_i)$ by $\exp(-st)$ and integrating over time for any suitably-chosen value of s. The diffusivity D_T is obtained by taking the ratio of the two integrals which equals $\exp[-(z_1 - z_2)(s/D_T)^{1/2}]$. The value of s should be chosen, so that it gives proper weight to the important portions of the record; if s is large, only the record for small t contributes to the record and *vice versa*. In his example with a record of one hour at 1 cm depth, Van Wijk (1963) used $s = 0.001\ s^{-1}$.

A method, based on the Green's function solution of (6.33) for arbitrary, but known temperature $T(z_1, t)$ at some level z_1 (e.g., at or near the surface) was suggested by Laikhtman and Chudnovskii (1962). For an initially linear temperature distribution $T(z, 0) = T_i(z) = az + b$ (where a and b are constants), this solution can be written as

$$T(z, t) - T_i(z) = \frac{x}{2(\pi D_T)^{1/2}} \int_0^t [T(z_1, \tau) - T_i(z_1)] \frac{\exp[-x^2/(4D_T(t-\tau))]}{(t-\tau)^{3/2}}\, \mathrm{d}\tau \tag{6.38}$$

where $x = (z - z_1)$. Multiplying by $\mathrm{d}x$ and integrating from $x = 0$ to $x = \infty$, one obtains

$$(D_T)^{1/2} = \frac{(\pi)^{1/2} \int_{z=z_1}^{\infty} [T(z, t) - T(z, 0)]\, \mathrm{d}z}{\int_0^t [T(z_1, \tau) - T(z_1, 0)](t-\tau)^{-1/2}\, \mathrm{d}\tau}\,. \tag{6.39}$$

The thermal diffusivity can be calculated if measured temperature profiles are available for a certain time period t. The integral in the numerator of (6.39) is obtained as the area between an initial temperature-depth curve and that at the end of the period; the depth of integration is that to which the temperature changes penetrate. The denominator is a convolution operation; if the time period is subdivided into n small time intervals Δt, during which T_j is the averaged temperature at $z = z_1$ (near the surface), the denominator can be calculated from its discrete form $\sum_{j=1}^{j=n} [T_j - T(z_1, 0)](\Delta t)^{1/2}(n - j + 1/2)^{-1/2}$.

Methods to Determine Soil Heat Flux

The heat flux at the soil surface can be measured directly by means of calibrated heat flux plates or it can be deduced from temperature and moisture content measurements or from theoretical calculations.

(*i*) *Measurement with heat flux plates.* This device usually consists of a thin plate or sheet of insulating material, which is placed in the soil normal to the direction of the heat flow; the temperature difference, which is measured across the plate, is a direct measure of the heat conduction through the plate, so that it can be related to the heat flux in the surrounding soil by a proper calibration. Although heat flux plates are simple to use, their construction, calibration and installation require great care (e.g., Deacon, 1950; Philip, 1961; Fuchs and Tanner, 1968; Idso, 1972). The thermal properties of the plate material are likely to be different from those of the soil, which vary with moisture content. If this difference is large, or if the plate is placed too close to the surface, the soil heat flux pattern may be distorted considerably. In addition, proximity to the soil surface may cause sampling difficulties resulting from surface heterogeneity. Therefore, soil heat flux plates should probably be placed at least 5 to 10 cm below the soil surface. If the soil thermal properties vary over too wide a range, the calibration may be invalid, or it may have to be adjusted for moisture content. Problems may also arise from poor contact between the plate and the soil, and from possible interference of the plate with the soil water movement.

(*ii*) *Temperature gradient method.* The soil heat flux at a given depth can, in principle, be calculated by means of (6.31) from measurements of the soil temperature gradient, provided the thermal conductivity of the soil is known. However, the gradient obtained from differences involves often large errors and, as seen above, it is not easy to determine K_T. Therefore, this method is not usually suitable for direct calculation of the heat flux at the surface, but only to determine the heat flux at some larger depth where the moisture content and the temperature gradients do not vary as drastically.

(*iii*) *Soil calorimetry.* The soil heat flux can be determined from changes in soil heat storage. The procedure is based on the integral of (6.32) with (6.31), viz.

$$Q_{H1} - Q_{H2} = \int_{z_1}^{z_2} C_s(z) \frac{\partial T}{\partial t}\, dz \tag{6.40}$$

where Q_{H1} and Q_{H2} are the heat flux densities at levels z_1 and z_2, respectively. Thus, if z_1 refers to the soil surface and z_2 to some lower level where Q_{H2} is known, the surface heat flux $G = Q_{H1}$, during a certain time interval, may be calculated by numerical integration of (6.40) for measured soil temperature and moisture content profiles at the beginning and at the end of the time interval. The volumetric heat capacity can be calculated from (6.35). If the depth z_2 is sufficiently large, Q_{H2} can be assumed to be negligible. If z_2 is not large enough to allow this assumption, the heat flux Q_{H2} can be determined by one of the following methods.

One possibility is the temperature gradient method applied at z_2. As mentioned, at greater depths $(\partial T/\partial z)$ and θ tend to be more uniform and more stable with time, so that the measurements are probably more reliable. However, at such depths the accurate determination of K_T is likely to be difficult.

The null-alignment method, described by Kimball and Jackson (1975), resolves this difficulty as follows. Null points in the temperature gradient are used to provide known zero heat fluxes at known depths in the soil profile. In other words, for the times of day when a zero-gradient is observed at some level z_2 in the T-profile, the soil heat flux, Q_{H1}, can be determined by means of (6.40), not just at the surface but at any level z_1 both above and below z_2, where $Q_{H2} = 0$. This information then allows the determination of the thermal conductivity $Q_{H1}/(\partial T/\partial z)$ at some reference depth, e.g., 20 cm, where the moisture content θ does not vary too much so that K_T remains approximately constant through the day. For those times of day when no zero-gradient exists, the soil heat flux at any level can be determined by means of (6.40) with Q_{H2} as the heat flux at the reference depth calculated by the usual temperature gradient method.

In the combination method, suggested by C. B. Tanner of Wisconsin, Q_{H2} is measured by means of a heat flux plate placed at a depth of 5 to 10 cm below the surface. The integral in (6.40) is then determined from successive temperature profiles measured above the level of the heat flux plate. This combination of two types of measurements eliminates some of the undesirable features of both the heat flux plate and of the calorimetric method. Since the plate is installed at greater depth, there is less interference with the heat and moisture flow pattern near the surface. It also removes some of the uncertainties of the standard calorimetric method because, especially for computations over periods shorter than a day (e.g., Hanks and Tanner, 1972), it may sometimes require accurate temperatures to depths of 1 m or more.

(*iv*) *Empirical relationships.* When necessary measurements are not available, the surface soil heat flux may be estimated on the basis of empirical relationships. The simplest assumption here is that the surface flux G is proportional to some other term in the energy budget equation. An obvious choice is the sensible heat flux into the air; thus

$$G = c_H H \tag{6.41}$$

where c_H is a constant; for bare soil, Kasahara and Washington (1971) have taken $c_H = 1/3$, which was suggested by Sasamori's (1970) numerical simulation of the field data of Lettau and Davidson (1957). The soil heat flux can also be assumed to be proportional to net radiation, or

$$G = c_R R_n \tag{6.42}$$

where, again, c_R is an empirical constant. For a bare soil, Fuchs and Hadas (1972) found that, on average, $c_R = 0.3$, approximately; as shown in Figure 6.10, (6.42) appeared to be best satisfied for the moist soil, and it displayed some hysteresis for the dry soil. Nickerson and Smiley (1975) concluded from the data of Lettau and Davidson (1957), that c_R is 0.19 for $R_n > 0$ during the day time, and 0.32 for $R_n < 0$. For a stand of maize, Perrier (1975b) estimated that $c_R = 0.2$ when R_n is taken at the soil surface. The data of Idso *et al.* (1975) showed some variation in c_R with moisture content; but inspection shows that, combining all their data, one would obtain a

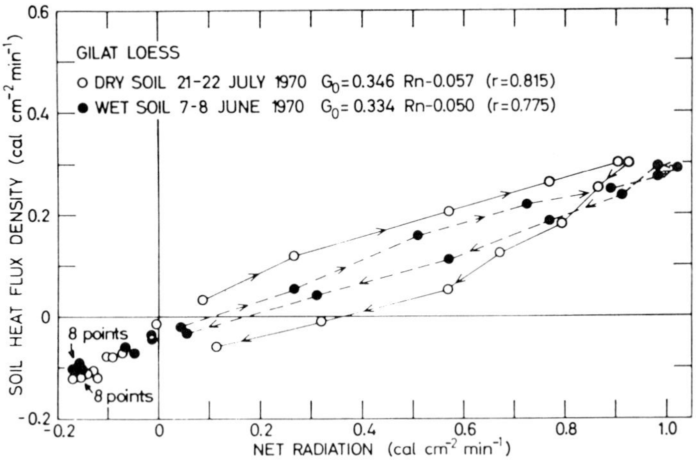

Fig. 6.10. Relationship between soil heat flux and net radiation on a bare loess (from Fuchs and Hadas, 1972).

c_R value of approximately 0.4. In the light of these different studies it seems that (6.42) with $c_R = 0.3$ may be used as a good compromise value for bare soils, or when R_n is taken at the soil surface. However, for surfaces covered by vegetation and with R_n taken at the top of the canopy c_R is likely to be considerably smaller so that it is often negligible; this is illustrated in Figure 6.1 for grass and in Figure 6.2 for maize.

Of course, both (6.41) and (6.42) are oversimplifications since G is related not to one but to all the terms in (6.1); therefore, such simple relationships must be calibrated anew for each given problem, and the values given for the constants c_H and c_R are accurate only for specific conditions. Nevertheless, in certain practical applications (6.41) and (6.42) can be quite useful.

(*v*) *Analytical solutions for simplified cases.* For a know temperature or heat flux variation at or near the surface, and with known $K_T = K_T(z, \theta)$ and $C_s = C_s(z, \theta)$, it is, in principle, possible to calculate $T = T(z, t)$ and $Q_H = Q_H(z, t)$ from (6.31) and (6.32). However, since $\theta = \theta(z, t)$, this type of calculation involves at least also a solution of the water flow problem [cf. Section 11.1a]. Although water and heat transport in the soil are actually coupled phenomena (e.g., Philip and De Vries, 1957; De Vries, 1958), in many applications it is quite adequate to treat them separately. But even so, the available information on the soil and on the boundary conditions is usually inadequate to allow a very rigorous simulation.

The mathematical formulation is simplified considerably when use is made of the linearized Equation (6.33); in this approach, solutions may be obtained which illustrate, or give an order of magnitude of, certain features of the soil heat flux phenomenon. Examples of such solutions are those implicit in (6.36)–(6.38); the simplest among these is the solution (6.36) for an harmonic temperature variation at the surface; the heat flux in the soil is by virtue of (6.31),

$$Q_H = (K_T C_s \omega)^{1/2} a_0 \exp\left[-\left(\frac{\omega}{2 D_T}\right)^{1/2} z\right] \sin\left[\omega t - \left(\frac{\omega}{2 D_T}\right)^{1/2} z + \frac{\pi}{4} + b\right] \tag{6.43}$$

or, at the surface,

$$G = a_0 (K_T C_s \omega)^{1/2} \sin\left(\omega t + \frac{\pi}{4} + b\right). \tag{6.44}$$

When an average thermal conductivity, an average specific heat and the soil surface temperature amplitude a_0 are known, (6.44) can be readily applied for a diurnal, or for an annual cycle. While this solution is derived for idealized conditions, it can be used to obtain rough estimates of G and of the amount of heat taken up or released from storage. Equation (6.43) gives a rough estimate of the penetration of the temperature wave: for example, 95 percent of the wave is damped at a depth of $z = 3(2D_T/\omega)^{1/2}$.

b. Whole Water Bodies

In the case of shallow ponds or streams, with a depth of at most a few meters, G can be important; in past studies it has been measured (e.g., Brown, 1969) and calculated (e.g., Jobson, 1977; Jirka, 1978) usually by temperature gradient methods based on Fourier's law, as given in (6.31). In the case of a deep lake, G is usually negligible.

c. Water Surfaces

When the energy budget (6.1) is calculated for a very thin layer or film of water at the surface of the ocean, or of a lake, G comprises several heat transport mechanisms; these are the conduction and vertical convection of sensible heat to deeper layers, and the penetration of radiation below the surface.

The heat flux to the water layers, below the thin surface layer under consideration, can be determined experimentally by measuring water temperature profiles. Since the specific heat of water is known (see Table 3.4), temperature changes provide a direct indication of heat storage changes. Thus, an equation analogous to (6.40) with $\rho_w c_w$ instead of C_s can be applied by means of successive temperature profiles to calculate G, provided the heat flux is known or negligible at the lower boundary of the water column.

The heat transfer into a water body can also be calculated. However, a treatment of this problem, which is so complex as to deserve a book of its own, is well beyond the scope of this one. Therefore, the reader is referred to reviews by Niiler and Kraus (1977) and Sherman *et al.* (1978); the penetration of radiation in water bodies has been treated by Jerlov (1968). A more recent example of a calculation of temperature profiles in the ocean is the paper by Kondo *et al.* (1979).

6.4. REMAINING TERMS

a. Energy Advection

This term comprises all the energy which is advected by water flowing in or out of the system to which (6.1) is applied. Precipitation is a source of vertical advection at the upper surface of the layer; rainfall may be important in the case of a snow cover, and snowfall may affect the energy balance of a warm lake. Lateral advection is probably only to be considered in the case of the energy budget for a whole lake. The total advection rate per unit area of a lake can be written approximately as follows

$$A_h = \rho_w c_w (q_i T_i + P T_p - q_o T_o) \tag{6.45}$$

in which c_w is the specific heat of water, ρ_w its density, q_i the total rate of inflow per unit area of lake into the lake during the period, q_o the corresponding outflow, P the rate of precipitation, T_i, T_p and T_o the weighted temperature of the inflow, precipitation and outflow water, respectively. For energy budget calculations the accuracy of the inflow and outflow terms in (6.45) is not very important. For this reason, other terms, such as evaporation and ground water seepage, are usually negligible.

b. Rate of Change of Energy Stored in the Layer

In the case of a thin layer of water, soil or canopy, the term $(\partial W/\partial t)$ is omitted from the energy budget Equation (6.1). In the case of a tall vegetation, however, it may have to be considered; it has been observed (e.g., Stewart and Thom, 1973) that this term can be especially significant after sunrise and near sunset, when it may be of the same order of magnitude as the net radiation R_n. Still, on a daily basis it is usually neglected. When the layer under consideration is a snow pack, this term is generally rather important (e.g., McKay and Thurtell, 1978), since as formulated in (6.1) it includes the energy utilized in fusion; however, it is very difficult to determine it directly from the properties of the snow.

In the case of a lake $(\partial W/\partial t)$ can be determined by the same methods described above in Section 6.3c for G below a thin water layer at the surface of the ocean. Thus, experimentally, this term is determined from successive temperature profile surveys. Crow and Hottman (1973) have studied the influence of network density of the water temperature profile stations on the accuracy of the evaporation from Lake Hefner, calculated by means of the energy budget. The optimal number of stations was found to be five, or one station per 2.1 km^2. Increasing this to 19 resulted in an accuracy increase of only 1 percent.

CHAPTER 7

Advection Effects Near Changes in Surface Conditions

7.1. THE INTERNAL BOUNDARY LAYER

The concepts reviewed up to this point are applicable to study local evaporation from surfaces, which are sufficiently uniform and large, so that edge effects involving horizontal advection by the mean wind are relatively unimportant. The assumption of a horizontally homogeneous and steady boundary layer allows a one-dimensional treatment of the transport phenomena near the surface. However, under natural conditions, this assumption is often invalid. In the case of evaporation from surfaces of limited extent, such as finite-size lakes or irrigated areas surrounded by arid land, the horizontal inhomogeneity can be very important.

As an illustration, consider a partly saturated air mass which moves from a uniform, dry land surface, over to a water surface. All the lower boundary conditions change abruptly. Not only is the surface humidity greatly increased, but also the surface roughness and temperature are, in general, likely to be different from their upwind value. As a result, immediately behind the leading edge of the water surface the water vapor flux is suddenly considerably larger than its value over the land. Likewise, the discontinuity in the thermal and water vapor stratification produces a rapid change in the stability of the atmosphere and the concomitant sensible heat transfer. Due to this change in stratification and to the change in surface roughness, the mean wind velocity profile and the turbulence structure of the air must readjust. But the turbulence structure controls the transport mechanisms and, thus, in turn the temperature and water vapor profiles. In other words, behind the surface discontinuity the vertical profiles are no longer in equilibrium, as over the upwind uniform land surface, nor are the horizontal gradients equal to zero. It is only further downwind that the profiles of specific humidity, wind and temperature tend to a new equilibrium with the changed surface conditions and that gradually the mean gradients again become vertical.

The region of the atmosphere affected by such a step change in surface conditions is referred to as an internal boundary layer. In the case of evaporation it has been called the 'vapor blanket'. The lower portion of this region, in which a new equilibrium is established, may be called the internal equilibrium sublayer. The horizontal transport near the discontinuity is often referred to as local advection.

Relatively few experimental results have been published on the structure of internal boundary layers. The observations of Rider *et al.* (1963), Millar (1964), Dyer and Crawford (1965), Davenport and Hudson (1967a, b), Lang *et al.* (1974) and Brakke

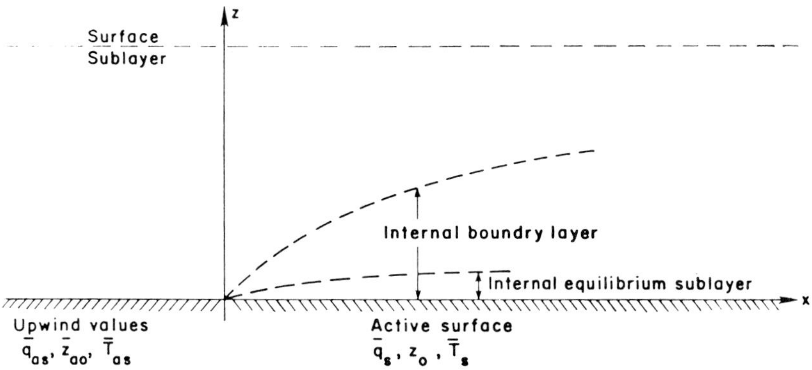

Fig. 7.1. Definition sketch of two-dimensional internal boundary layer problem.

et al. (1978) provide examples and descriptions of this type of situation over land, involving water vapor and heat transfer. Data on the evaporation from water surfaces of different sizes have been summarized and discussed by Harbeck (1962) and Brutsaert and Yu (1968). Experimental results related to the local advection of momentum in the atmosphere have been published and discussed by Bradley (1968; Mulhearn, 1978), Lettau and Zabransky (1968), Panofsky and Petersen (1972), Petersen and Taylor (1973), Munro and Oke (1975).

a. Equations for the Mean Field

To simplify the analysis of the internal boundary layer, consider a situation where a fully-turbulent steady air flow is normal to a surface discontinuity at $x = 0$. The horizontal component of the wind is in the x-direction; z is the vertical and y the lateral coordinate (Figure 7.1). As long as the fetch x is not too large, the internal boundary layer is sufficiently thin, so that it is submerged within the atmospheric surface sublayer, and the Coriolis effect may be assumed negligible. For this type of boundary layer problem one can write $\bar{u} \gg \bar{w}$, $\partial/\partial z \gg \partial/\partial x$ and $\bar{v} = \partial/\partial y = 0$. The governing equations are not as simple as for the homogeneous boundary layer of Section 3.4; therefore, they are now briefly reviewed.

Under these circumstances (3.44), for the mean specific humidity $\bar{q} = \bar{q}(x, z)$, can be written as

$$\bar{u}\frac{\partial \bar{q}}{\partial x} + \bar{w}\frac{\partial \bar{q}}{\partial z} = -\left[\frac{\partial}{\partial x}\overline{(u'q')} + \frac{\partial}{\partial z}\overline{(w'q')}\right] \tag{7.1}$$

The analogous equations of the mean motion, describing momentum transfer, can be obtained from the Reynolds Equations (3.62), namely to a close approximation

$$\bar{u}\frac{\partial \bar{u}}{\partial x} + \bar{w}\frac{\partial \bar{u}}{\partial z} = -\frac{1}{\rho}\frac{\partial p}{\partial x} - \frac{\partial}{\partial x}\overline{(u'u')} - \frac{\partial}{\partial z}\overline{(w'u')}, \tag{7.2}$$

$$\bar{u}\frac{\partial \bar{w}}{\partial x} + \bar{w}\frac{\partial \bar{w}}{\partial z} = -\frac{1}{\rho}\frac{\partial p}{\partial z} - g - \frac{\partial}{\partial x}\overline{(u'w')} - \frac{\partial}{\partial z}\overline{(w'w')} + g\frac{T_{VD}}{T_a} \tag{7.3}$$

where ρ is the mean density of the air at the reference temperature T_a and $T_{VD} = T_V - T_{VS}$ is the mean deviation of the virtual temperature (3.9) from the hydrostatic reference profile T_{VS} defined by (3.54); under neutral conditions this term vanishes.

The equation for the mean potential temperature (3.67), when radiative flux divergence is negligible, becomes

$$\bar{u}\frac{\partial \bar{\theta}}{\partial x} + \bar{w}\frac{\partial \bar{\theta}}{\partial x} = -\left[\frac{\partial}{\partial x}(\overline{u'\theta'}) + \frac{\partial}{\partial z}(\overline{w'\theta'})\right]. \tag{7.4}$$

Finally, because $\bar{u}$ varies in the downwind direction, there is also the equation of continuity for bulk air (3.48)

$$\frac{\partial \bar{u}}{\partial x} + \frac{\partial \bar{w}}{\partial z} = 0. \tag{7.5}$$

Often the formulation of the internal boundary-layer problem is simplified even further by the assumption that the effects of downwind gradients of turbulent fluxes and pressure are also negligible. As regards (7.1), analytical and numerical solutions (Yeh and Brutsaert, 1970; 1971a, b) have dealt with the importance of the first term of the right-hand side in the calculation of evaporation in the case of a step change in surface specific humidity $\bar{q}$, but with uniform roughness and surface temperature, so that $\bar{w} = 0$. By using an eddy diffusivity with the Reynolds analogy, it was found that the effect of this term is practically imperceptible for fetches larger than, say, 1 m. Peterson (1972) considered the importance of various terms in (7.2) and (7.3) in the solution of the surface roughness change problem. By using a closure method based on the turbulent kinetic energy equation, he concluded that, except in the immediate vicinity of the roughness step-change, the first and second terms on the right of (7.2) are relatively unimportant; because the effect of the pressure change is negligible, the vertical equation of motion (7.3) can be omitted. If these assumptions are adopted, (7.1), (7.2) and (7.4) assume the same form, viz.

$$\bar{u}\frac{\partial \bar{q}}{\partial x} + \bar{w}\frac{\partial \bar{q}}{\partial z} = -\frac{\partial}{\partial z}(\overline{w'q'}), \tag{7.6}$$

$$\bar{u}\frac{\partial \bar{u}}{\partial x} + \bar{w}\frac{\partial \bar{u}}{\partial z} = -\frac{\partial}{\partial z}(\overline{w'u'}), \tag{7.7}$$

$$\bar{u}\frac{\partial \bar{\theta}}{\partial x} + \bar{w}\frac{\partial \bar{\theta}}{\partial z} = -\frac{\partial}{\partial z}(\overline{w'\theta'}). \tag{7.8}$$

In general, a change of surface conditions involves changes in evaporation, surface shear stress and sensible heat flux. Thus (7.1) through (7.5), or (7.5) through (7.8), must be solved simultaneously for a set of suitable boundary conditions, that specify a particular problem. In Section 3.3c, it was indicated how the problem of closure arises for any turbulent flow. Even for the simplified formulation of (7.5) through (7.8), there are only four equations with seven unknowns, namely the four mean variables $\bar{q}$, $\bar{u}$, $\bar{v}$, $\bar{\theta}$ and the three fluxes $\overline{w'q'}$, $\overline{w'u'}$, $\overline{w'\theta'}$. If (7.1) through (7.5) are used, the situation is even more complicated. Thus, to close the system, some appropriate assumption must be made for the turbulent fluxes, or some additional equations must be constructed.

b. Methods of Closure for Disturbed Boundary Layers: A Brief Survey

Numerous theoretical studies have been devoted to the problem of a step-change of one or more surface properties, i.e., roughness, specific humidity, temperature, their respective fluxes or some combination. Although one of the earliest dealt with a surface discontinuity of specific humidity (Sutton, 1934), most attention has gone to the effect of a surface roughness change. Because the turbulent transport of humidity, momentum and sensible heat are usually quite related, the methods that have been used can be reviewed and classified together.

Solutions With Assumed Self-Preservation

In the internal boundary layer the number of relevant variables affecting the flow and the transport phenomena is much larger than for a uniform boundary layer. This makes it quite difficult to consider all the important dimensionless variables and to find general similarity relationships between them. However, it is possible to simplify the problem of local advection somewhat by assuming 'self-preservation'. Self-preservation means that certain similarity features of the flow are maintained while the turbulence changes. Specifically, in the case of local advection, it refers to the situation where vertical profiles of mean velocity, shear stress, mean specific humidity, vertical vapor flux, etc. can be represented by functional forms which are independent of the fetch x, and where only the scales of these variables are functions of fetch. In other words, the profiles of these variables can be assumed to be invariant with fetch when they are nondimensionalized with local scales. One of the main similarity features in a dynamic or surface sublayer under non-advective conditions, is the relationship between the vertical fluxes and the vertical gradients. Although this is not an essential element in self-preservation, most studies based on this approach have also assumed that such flux-gradient relationships are valid under conditions of local advection. Thus, for example, for neutral conditions the relationship between shear stress and gradient of mean wind speed was mostly assumed to be given by (4.1) but with u_*^2 replaced by the local shear stress $-\overline{u'w'}$.

The assumption of self-preservation has been applied in different ways. Several studies of local advection have been based on the Kármán – Pohlhausen (e.g., Schlichting, 1960) approach, or some extension of it. In the case of the change-of-roughness problem, this approach consists of assuming the form of the wind profile *a priori*. This wind profile is assumed to have some self-preservation and the unknown parameters are obtained from suitable boundary conditions and by imposing the integral form of the momentum equation [cf., (7.7) with (7.5)] as a constraint, viz.

$$\frac{d}{dz}\int_0^{\delta_m}\bar{u}^2\,dz - \bar{u}(\delta_m)\frac{d}{dx}\int_0^{\delta_m}\bar{u}\,dz = u_{*a}^2 - u_*^2 \tag{7.9}$$

where $\delta_m = \delta_m(x)$ is the thickness of the internal momentum boundary layer, u_{*a}^2 the upwind ($x < 0$) friction velocity and $u_* = u_*(x)$ the friction velocity downwind of the change in roughness. This approach was introduced by Elliott (1958) and further developed by Panofsky and Townsend (1964), Plate and Hidy (1967), Lettau and Zabransky (1968), and others to analyze the change-of-roughness problem. Itier and

Perrier (1976) extended the Kármán – Pohlhausen method to the problem of local advection of a scalar admixture of the flow. They assumed *a priori* that the vertical flux profile $\overline{w'q'} = \overline{w'q'}(x, z)$ can be described by the lowest-order polynomial in (z/δ_v), which satisfies the boundary conditions on the flux in the boundary layer. The thickness of the internal boundary layer for a scalar, $\delta_v = \delta_v(x)$ is a function of x, so that the assumed profile is self-preserved. This thickness can be obtained from the scalar analog of (7.9); in the case of water vapor, one has from (7.6) with $\bar{w} = 0$ (if the local advection of momentum can be neglected)

$$\int_0^{\delta_v} \bar{u}(z) \frac{\partial \bar{q}}{\partial x} \, dz = E - E_a \tag{7.10}$$

where $E = E(x)$ is the rate of evaporation for $x > 0$, and E_a is the constant rate of evaporation upwind for $x < 0$. Equation (7.10) can be written as

$$\frac{d\delta_v}{dx} = (E - E_a)\left(\int_0^{\delta_v} \bar{u}(z) \frac{\partial \bar{q}(x, z)}{\partial \delta_v}\right) dz^{-1} \tag{7.11}$$

for which Itier and Perrier (1976) obtained numerical solutions after assuming, in addition, that $\bar{u}(z)$ is logarithmic [cf. (4.3)] and that $(\partial \bar{q}/\partial z)$ is related to $\overline{w'q'}$ by (7.15) with K_v a linear function of elevation. One of the main approximations in the Kármán – Pohlhausen method is that the conservation equations are not satisfied at every x and z but only in an average sense, since they are integrated over the whole boundary layer.

A different way of applying self-preservation was proposed by Townsend (1965a b; 1966). In an analysis of the change-of-roughness problem, he postulated self-preservation in the wind profile and in the shear-stress. The momentum equation (7.7) was not approximated by its integral form; rather, its solution was derived, by satisfying self-preservation and by making certain assumptions suggested by similarity features of the non-advective boundary layer. The method was subsequently further developed by Blom and Wartena (1969) and Mulhearn (1977). The latter also applied it to the local advection of a scalar admixture. Mulhearn (1977) showed that this approach is in satisfactory agreement with the temperature profiles of the experiments of Rider *et al.* (1963) and Dyer and Crawford (1965) quoted above. However, the results of this method are in implicit form, which somewhat complicates their application.

First-Order Closure: Turbulent Diffusion Approach

This approach is based on a direct approximation of the turbulent fluxes in the conservation Equations (7.6)–(7.8); it is often referred to also as mixing-length or K theory. This approach is based on flux-gradient assumptions which are quite related to, and in some cases the same as, those used in most applications of the self-preservation approach in the previous section. Nevertheless, the two approaches should be considered separately. The turbulent diffusion approach depends more explicitly and critically on a conceptual model for the turbulence involving diffusion or gradient transport. The main, and often only *a-priori* assumption required is the functional form of the eddy diffusivity or eddy viscosity. Thus, the conservation equations are transformed into a type of convective diffusion equation and solved as such. As mentioned in Chapter 2, the K-theory approach was initiated by Boussinesq (1877) for momentum, and later by Schmidt (1917) for a scalar admixture. In general the approach is based on the assumption that the turbulent flux is linearly related to the mean concentration gradients. In the case of a scalar admixture, such as water vapor, both are vectors so that the proportionality factor, which is called eddy diffusivity, must be a tensor. For the present problem of an internal boundary layer, the turbulent fluxes of (7.1) can be written as

$$\overline{u'q'} = -\left(K^v_{xx}\frac{\partial \bar{q}}{\partial x} + K^v_{xz}\frac{\partial \bar{q}}{\partial z}\right), \tag{7.12}$$

$$\overline{w'q'} = -\left(K^v_{zx}\frac{\partial \bar{q}}{\partial x} + K^v_{zz}\frac{\partial \bar{q}}{\partial z}\right) \tag{7.13}$$

in which the superscript v refers to water vapor. The justification of these equations rests on empirical grounds and seems to have no physical value except an analogy between the mean free path of molecular motion and the scale of turbulence. It is clear that the same idea, relating fluxes to gradients, was already implicit in the similarity formulations for a uniform boundary layer in Chapter 4. The general development of gradient transport models such as (7.12) and (7.13) has been treated by Monin and Yaglom (1971) and their limitations to describe turbulent transfer have been discussed by Corrsin (1974). Very little is known about the degree of anisotropy of the K-tensor, but various models have been proposed. For example, on the basis of a simple hypothesis, the author suggested (Brutsaert, 1970)

$$\begin{aligned} K_{xx} &= [\overline{(u')^2}]^2/\varepsilon, \\ K_{xz} &= K_{zx} = -(\overline{u'w'})^2/\varepsilon, \\ K_{zz} &= [\overline{(w')^2}]^2/\varepsilon \end{aligned} \tag{7.14}$$

in which ε is the turbulent energy dissipation rate per unit mass. In other words, it was assumed that the tensor is symmetrical, and that $(K_{xx}/K_{zz}) = [\overline{(u')^2}/\overline{(w')^2}]^2$ and $(K_{xz}/K_{zz}) = -[\overline{u'w'}/\overline{(w')^2}]^2$. Although this result was not inconsistent with some experimental results, other evidence has indicated (Yaglom, 1972, 1976) that the matter is more complicated.

Fortunately, for many practical problems related to internal boundary layers this may not be very important. In the case of evaporation with local advection as described by (7.1) with (7.12) and (7.13), it has been found (Yeh and Brutsaert, 1970; 1971a, b) that the effect of terms containing K_{xz}, K_{zx} and K_{xx} is negligible when the fetch exceeds a few meters. Also, all the studies of the change of roughness problem, that have appeared in the literature, have considered only K_{zz} in the analysis. Thus, most of the local-advection studies based on this approach have closed (7.6)–(7.8) with the following

$$\overline{w'q'} = -K_v\frac{\partial \bar{q}}{\partial z}, \tag{7.15}$$

$$\overline{w'u'} = -K_m\frac{\partial \bar{u}}{\partial z}, \tag{7.16}$$

$$\overline{w'\theta'} = -K_h \frac{\partial \bar{\theta}}{\partial z} \tag{7.17}$$

in which the subscripts v, m and h refer to vapor, momentum and heat respectively; the subscripts zz of K can be omitted when only this one component is used. Invariably, the form of these K's has been taken as a direct generalization or extension of the flux-profile similarity formulations for the uniform boundary layer of Chapter 4.

Many analytical studies of fetch-dependent evaporation, in the absence of a roughness change, have been carried out with an equilibrium-wind profile given by (4.18) and with a power function for the eddy diffusivity

$$K_v = bz^n \tag{7.18}$$

where b and n are constants; usually, these constants were determined by applying Reynolds's analogy that $K_v = K_m$, and by combining (7.16), (4.18) and $-u_*^2 = \overline{w'u'}$, so that

$$n = 1 - m, \qquad b = a_v u_*^2/(ma) \tag{7.19}$$

where a and m are the parameters of (4.18). The underlying ideas of this formulation were developed and discussed, for a uniform turbulent boundary layer, by Schmidt (1917), Prandtl and Tollmien (1924), Ertel (1933) and others. Sutton's (1934) treatment of the advection problem was pioneering; later studies are reviewed below in Section 7.2a. Dimitriev and Sokolova (1954; see Panchev *et al.*, 1971) obtained a solution of the change-of-roughness problem by using a similar power function approximation. The power profile is not based on similarity theory; however, as noted, it can be made to fit the similarity based profile functions by a proper choice of the parameters a and m. Its main advantage is that it greatly simplifies the mathematical formulation of problems of turbulent diffusion.

The earlier applications of similarity of a uniform boundary layer in first-order closure modeling were directly obtained from the logarithmic profile (4.1) for the dynamic sublayer. For example, the solutions of the change-of-roughness problem for neutral conditions by Gandin (1952: see Panchev *et al.*, 1971) and Nickerson (1968) made use of the following eddy viscosity in (7.16)

$$K_m = ku_*z \tag{7.20}$$

where u_* is the friction velocity at the surface downwind from $x = 0$; u_* was assumed to be known and constant. However, most of the more recent K models for the internal boundary layer are in terms of the local fluxes; they can be written in the common form

$$K_v = l(\overline{-u'w'})^{1/2}/\phi_{sv}, \tag{7.21}$$

$$K_m = l(\overline{-u'w'})^{1/2}/\phi_{sm}, \tag{7.22}$$

$$K_h = l(\overline{-u'w'})^{1/2}/\phi_{sh} \tag{7.23}$$

where l is referred to as mixing length. Taylor (1969a) studied the change-of-roughness problem under neutral conditions with $\phi_{sm} = 1$ and $l = k(z + z_0)$; subsequently

(Taylor, 1969b, c) he studied deeper layers by retaining the y-component for the momentum equation and by using a mixing length proposed by Blackadar (1962)

$$l = kz(1 + kz\lambda^{-1})^{-1} \tag{7.24}$$

in which he replaced z by $(z + z_0)$ for numerical convenience, and $\lambda = aG/|f|$, where a is a constant of the order of 0.0004 and G is defined in (3.73). Taylor (1970; 1971) studied airflow changes resulting from a change in surface temperature and roughness with (7.22) and (7.23) in (7.16), (7.17), (7.7) and (7.8) in which he took $l = k(z + z_0)$ and the ϕ_{sm} and ϕ_{sv}-expressions from the equilibrium surface layer; however, he redefined the Obukhov length in terms of local fluxes, i.e., (7.26) but without $(\overline{w'q'})$. Similarly, Estoque and Bhumralkar (1970) used the closure assumption of (7.21)–(7.23) as an extension of the work of Onishi and Estoque (1968) to study the behavior of the whole planetary boundary layer over nonhomogeneous terrain with step-changes in roughness, temperature and humidity. They included in the model the horizontal equation of motion in $\bar{v}$ and $\overline{w'v'}$ and they retained the Coriolis terms and the pressure gradient $\partial p/\partial x$. All three eddy K's were assumed to be the same, viz.

$$K = l^2\left[\frac{\partial}{\partial z}(\bar{u}^2 + \bar{v}^2)^{1/2}\right](1 - \beta\,\mathrm{Ri})^{\alpha} \tag{7.25}$$

in which α and β are constants, and l was given by an expression similar to (7.24). Huang and Nickerson (1974a) studied stratified flow over a nonhomogeneous surface by using a similar model with functions $\phi_{sm} = \phi_{sm}(\mathrm{Ri})$ and $\phi_{sh} = \phi_{sh}(\mathrm{Ri})$ which were developed for the uniform surface layer. Although various boundary conditions were considered, the studies of Taylor (1970, 1971), Estoque and Bhumralkar (1970) and Huang and Nickerson (1974a) did not deal with evaporation.

Weisman and Brutsaert (1973) studied evaporation from a warm lake without roughness change. The basic model was similar to Taylor's (1970), but the specific humidity transport was included, and l was given by (7.24); in this model, which is discussed further in Section 7.2b, the expressions for the ϕ_s-functions in (7.21)–(7.23) were given by (4.42)–(4.44) for the equilibrium surface layer, but with a 'local' Obukhov length $L_a = L_a(x, z)$, [cf. (4.25)],

$$L_a = \frac{-(-\overline{u'w'})^{3/2}}{kg[(\overline{w'\theta'}/T) + 0.61\,\overline{w'q'}]} \tag{7.26}$$

Higher-Order Closure Models

The essence of higher-order closure models is that the second moments, namely the Reynolds stresses and Reynolds fluxes, are not approximated by eddy diffusion approximations like (7.12), (7.13), (7.15), (7.16), (7.17), but that they remain as unknown variables of the problem. To close the problem, field equations are introduced for these second moments. The third moments and other third-order turbulence variables, which these equations (e.g., (3.47), (3.64)) contain, are then approximated on the basis of some higher-order similarity assumptions. The models currently in use are mostly based on turbulence formulations that were pioneered by Kolmogorov (1942), Prandtl and Wieghardt (1945) and Rotta (1951). It is beyond the scope of the present treatment to go into the details of the available models, and only the main closure features will be touched upon.

Most closure assumptions in the literature, that have been tried, appear reasonable from the physical point of view. However, just like those for first-order closure, their theoretical justification is not easy. A very common assumption, found in higher-order approaches, is that the third moments are linear functions of the gradients of appropriate second moments. This assumption is evidently a generalization of the principle underlying the eddy diffusion models; in other words, it is similarity and K theory extended to the higher moments. For example, for the present simple case of an internal boundary layer, if c represents $u'q'$, $w'q'$, $(q')^2$, $u'w'$, $u'\theta'$, etc. then, in the respective field equations for the second moments $\bar{c}$, one might consider the following [cf. (7.12), (7.13)] as a possibility

$$\overline{u'c} = -\left(K^c_{xx}\frac{\partial \bar{c}}{\partial x} + K^c_{xz}\frac{\partial \bar{c}}{\partial z}\right), \tag{7.27}$$

$$\overline{w'c} = -\left(K^c_{zx}\frac{\partial \bar{c}}{\partial x} + K^c_{zz}\frac{\partial \bar{c}}{\partial z}\right) \tag{7.28}$$

in which the superscript refers to the particular c in question. Again, the main problem lies, of course, in the determination of the K terms. Moreover, from the physical point of view, the limitations of the gradient transport assumptions of (7.27) and (7.28) are probably no less serious than those of (7.12) and (7.13). The implicit justification for higher-order closure is that the mean field and the lower turbulence statistics are presumably relatively insensitive to a poor approximation or even incorrect modeling of higher moments. Wyngaard (1973) has discussed the implications of higher-order closure in the light of experimental turbulence data; Mellor and Yamada (1974) have compared several alternative simplifications to determine the sensitivity of this type of modeling to the completeness of the equations. A more general discussion of the subject was given by Lumley (1978).

(*i*) *With the equation of turbulent kinetic energy*. This approach, sometimes referred to as 1.5-order closure, has only been applied to solve change-of-roughness problems, and no attempts have been made to use it in the solution of evaporation problems. In this method, the surplus of unknowns in the mean field Equations (7.5) and (7.7) is remedied by introducing the field equation for a special second moment, namely the turbulent kinetic energy Equation (3.64); for the present problem of steady local advection this may be written approximately as

$$\bar{u}\frac{\partial \bar{e}_t}{\partial x} + \bar{w}\frac{\partial \bar{e}_t}{\partial z} = -\overline{u'w'}\frac{\partial \bar{u}}{\partial z} + \frac{g}{T_a}(\overline{w'\theta'} + 0.61\, T_a\, \overline{w'q'}) - \frac{\partial}{\partial z}\left(\overline{w'e_t} + \frac{\overline{w'p'}}{\rho}\right) - \varepsilon \tag{7.29}$$

where, under neutral conditions, the second term on the right is zero. Equation (7.29) contains, however, new unknowns, namely the turbulent energy $\bar{e}_t$, a third moment $\overline{w'e_t}$, a related term $\overline{w'p'}/\rho$, and the viscous dissipation of turbulent kinetic energy ε. The system is then closed by means of additional simple relationships which are assumed to be valid even under advective conditions. In the studies that have been published, the third moment $\overline{w'e_t}$ is approximated with the usual gradient transport model [cf. (7.27), (7.28)]

$$\overline{w'e_t} = -K_e \frac{\partial \overline{e_t}}{\partial z} \qquad (7.30)$$

and the related pressure term is neglected. Peterson (1969a), in a numerical model to simulate neutral conditions in the atmospheric surface layer, made use of (7.30) and assumed that K_e equals K_m as given in (7.16). In addition, he introduced the hypothesis, used also by Bradshaw *et al.*, (1967) that shear stress is proportional to turbulent kinetic energy,

$$-\overline{u'w'} = a\overline{e_t} \qquad (7.31)$$

where a is a constant of the order of 0.16; he assumed further that the viscous dissipation rate, ε, can be determined as usual (cf. Taylor, 1935; Monin, 1959)

$$\varepsilon = (-\overline{u'w'})^{3/2}/l_e \qquad (7.32)$$

where $l_e = kz$. Shir (1972) also used the closure assumption of (7.30) and (7.31); however, in contrast to Peterson (1969a) he did not neglect the pressure gradient term in the equation of motion, but he included p implicitly by transforming the equation of motion to a vorticity equation. The length l_e in (7.32) was assumed to be given by (7.24) for $z > 10$ m. Huang and Nickerson (1974b) applied (7.30) in the turbulent kinetic energy equation with $K_e = (\overline{e_t})^{1/2} l_1$; instead of (7.31) they used (7.16) with $K_m = K_e$, and instead of (7.32)

$$\varepsilon = (\overline{e_t})^{3/2}/l_2. \qquad (7.33)$$

The two length parameters were assumed to be given by $l_1 = a_1 z$ and $l_2 = a_2 z$, where a_1 and a_2 are empirical constants. According to Panchev *et al.* (1971), the use of the turbulent kinetic energy equation was suggested already in 1966 by Nadejdina to analyze simultaneous local advection of momentum and sensible heat; the idea was expressed mathematically in 1969 by Novikova. An approximate analytical solution for neutral conditions has been given in 1969 by Nadejdina.

None of the studies just mentioned dealt specifically with local advection of atmospheric moisture, but their models have produced interesting information on the applicability of first-order closure models to describe turbulent transport. For example, Peterson (1969b, 1971) and Huang and Nickerson (1974b) obtained the result that $[kz/(-\overline{w'u'})^{1/2}]\,(\partial \bar{u}/\partial z)$ is not equal to unity under neutral conditions in accelerating or decelerating flow of an internal boundary layer; they considered this to be clear evidence of the limitations of the mixing length models. Peterson (1972) found that, except very close to the discontinuity, the effect of pressure is likely to be small. On the other hand, Shir (1972) and Peterson and Taylor (1973) felt that the neglect of the pressure term may introduce serious errors in the model. Surprisingly, the latter authors also obtained slightly more accurate results with the first-order closure model than with the model using the turbulent kinetic energy. Huang and Nickerson (1974b) concluded from their calculations that a in (7.31) is not a constant in non-equilibrium flow.

(*ii*) *With equations for second moments.* In this approach, which can also be referred to as second-order closure or mean Reynolds-flux approach, equations for the second moments are introduced in addition to the mean field equations. The system of equations is closed by describing the higher-order turbulence statistics by some plausible approximations.

Rao *et al.* (1974a) have modeled the internal boundary layer resulting from a sudden change of roughness under neutral conditions. In a second paper (Rao *et al.*, 1974b), they have treated the problem of the local advection of momentum, heat and moisture due to a horizontal inhomogeneity in surface conditions. In the latter study, which is further discussed in Section 7.2b, their model consisted of a set of 16 coupled parabolic partial differential equations; these included four mean field equations in $\bar{u}$, $\bar{w}$, $\bar{\theta}$ and $\bar{q}$ ((7.5) through (7.8)), four Reynolds stress equations for $\overline{(u')^2}$, $\overline{(v')^2}$, $\overline{(w')^2}$, $\overline{(u'w')}$, two heat flux equations for $\overline{(\theta'w')}$, $\overline{(\theta'u')}$, two water vapor flux equations for $\overline{(q'w')}$, $\overline{(q'u')}$ and one equation each for temperature variance $\overline{(\theta')^2}$, specific humidity variance $\overline{(q')^2}$, temperature-humidity covariance $\overline{(q'\theta')}$, and energy dissipation rate ε. The third moments, contained in the equations for the second moments, where approximated by gradient transport equations like Equations (7.27) and (7.28); the K tensor was taken to be of the following form

$$
\begin{aligned}
K_{xx} &= a\overline{(u')^2}\,\overline{e_t}/\varepsilon,\\
K_{xz} &= K_{zx} = a\overline{(u'w')}\,\overline{e_t}/\varepsilon,\\
K_{zz} &= a\overline{(w')^2}\,\overline{e_t}/\varepsilon
\end{aligned}
\tag{7.34}
$$

in which a is a constant of the order of 0.3. It is of interest to note that (7.34), for the third moments, are quite similar to (7.14), proposed for the second moments.

Rao *et al.* (1974a) obtained good agreement between their calculations and the widely quoted experimental results of Bradley (1968). They found, however, that most previously used lower-order closure assumptions are invalid in the transition layer caused by a roughness change; this means that for non-equilibrium flow l in (7.22), l_e in (7.32) and l_2 in (7.33) are not simply proportional to z, and that a in (7.31) is not constant. The two studies of Rao *et al.* (1974a, b) showed that higher-order closure methods can become a powerful tool in the numerical analysis of evaporation problems with local advection. However, several aspects of the available models, in particular the formulation of proper boundary conditions in the case of plant canopies over a drying soil, will require further investigation.

c. Some General Features of Local Momentum Advection: Fetch Requirement

An important conclusion from the above review of closure methods is that solutions obtained at a given order of closure often disprove the similarity assumptions used in closure at the next level down. For example, the model with the turbulent kinetic energy equation of Peterson (1969b, 1971) disproved the eddy diffusion or mixing length assumption as used by Taylor (1969a), Weisman and Brutsaert (1973) and others in first-order closure models. However, the assumptions of Bradshaw *et al.*

(1967), Peterson (1969a) and Shir (1972) given in (7.31) and (7.32) were disproved in turn by the results of Rao *et al.* (1974a).

Nevertheless, in spite of the scarcity of experimental evidence, and the wide diversity of physical models used in theoretical analyses, certain features of internal boundary layers are now fairly well established. Consider, as a preamble to the treatment of evaporation, the local advection of momentum due to an abrupt change in surface roughness.

The growth of the internal boundary layer can be estimated by the following simple argument. The same idea was already used by Monin (1959) in connection with the propagation of a smoke plume. It is first assumed that the vertical rate of propagation of the disturbance due to the change in roughness at the leading edge is proportional to $\sigma_w = [\overline{(w')^2}]^{1/2}$; however, it is known that σ_w is proportional to u_* and that the proportionality depends on ζ. Hence, one has

$$\frac{\mathrm{d}z}{\mathrm{d}t} = a\, u_* \tag{7.35}$$

where a is a constant. On the other hand, the horizontal rate of propagation $\mathrm{d}x/\mathrm{d}t$ equals the mean wind speed, so that with (7.35), the slope of the disturbance trajectory is

$$\frac{\mathrm{d}z}{\mathrm{d}x} = \frac{a\, u_*}{\bar{u}}. \tag{7.36}$$

This can be immediately integrated for any wind profile function $(\bar{u}/u_*)$. For example, for the logarithmic law (4.3) with (4.4), one obtains, approximately,

$$\delta_0(\ln \delta_0 - 1) = b x_0 \tag{7.37}$$

where b is another constant, and $\delta_0 = \delta_m/z_0$ and $x_0 = x/z_0$; the symbol δ_m is the height z up to which the disturbance has propagated, that is, the thickness of the internal boundary layer for momentum. Equation (7.37) with $b = 1$ was introduced by Panofsky and Townsend (1964) after earlier work by M. Miyake. Later Panofsky (1973) recommended that (7.37) be used with $b = 0.6$ and z_0 as the roughness over the rougher terrain. Equation (7.36) can, of course, also be integrated with the power profile (4.19). For the simple case $d_0 = 0$, one obtains

$$\delta_0 = [a(m+1)/C_p]^{(m+1)^{-1}} x^{(m+1)^{-1}} \tag{7.38}$$

or, for neutral conditions when $m = 1/7$ and $C_p \simeq 6$, and with $a \simeq 1.5$,

$$\delta_0 = 0.334 x_0^{0.875} \tag{7.39}$$

which agrees roughly with the previous estimates of the growth of internal boundary layers. In fact, in all previous studies, it has been found that, except very close to the roughness discontinuity at $x = 0$, the thickness of the internal boundary layer can be approximated by a power function of fetch, viz.

$$\delta_m = c_m x^{b_m} \tag{7.40}$$

where c_m and b_m depend on the stability and c_m also on the upwind and downwind roughness heights. Under neutral conditions $b_m = 0.7 \sim 0.8$, which was first obtained

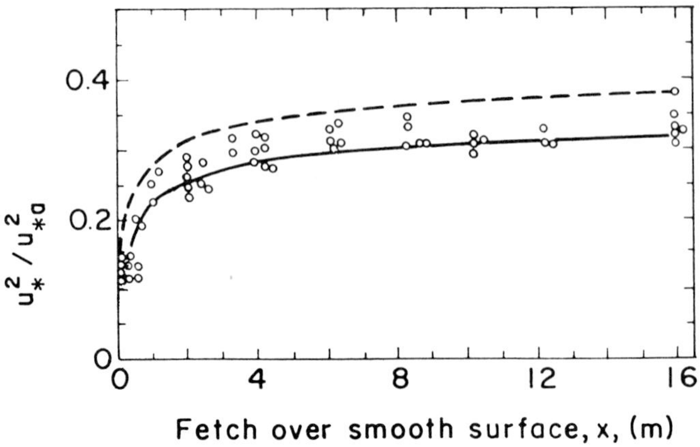

Fig. 7.2. Non-dimensional surface shear stress as a function of fetch for a transition rough-smooth with $z_{a0} = 0.25$ cm and $z_0 = 0.002$ cm. The experimental data are those of Bradley (1968). The solid line represents (7.41) with (7.37) in which $b = 0.6$ or with (7.39) (the results cannot be distinguished). The numerical result of Rao *et al.* (1974a) by means of a higher-order closure model is shown as a dash curve.

by Elliott (1958). This growth rate was confirmed experimentally by Bradley (1968) and it is the same as that of a turbulent boundary layer on a flat plate in a wind tunnel (Schlichting, 1960, p. 537). There are indications (Huang and Nickerson, 1974a) that b_m is slightly smaller for the transition from rough to smooth, than for smooth to rough. Equation (7.38) suggests that b_m increases with increasing instability of the atmosphere. Actually, Rao (1975) with a second-order closure model found that $b_m = 0.77$, 0.88, 1.39 for $L = -\infty$, -20 m, -2 m, respectively; according to Miyake's (Panofsky, 1973) work integration of (7.36) yields $b_m = \frac{3}{2}$ under extremely unstable conditions or free convection.

Several higher-order closure models (e.g., Peterson, 1969a; Shir, 1972; Rao *et al.*, 1974a) have produced the result that the internal equilibrium sublayer occupies the lower 10 percent of the internal boundary layer for a smooth-rough transition and the lower 5 percent for a rough-smooth transition. This equilibrium sublayer is defined as the region where the shear stress is calculated to be within 10 percent of its surface value. Expressions such as (7.37) and (7.38) suggest that for fetches of the order of 100 to 200 m x/δ_m is roughly of the order of 5 to 20. Thus, as a practical rule of thumb, it is sometimes suggested that equilibrium conditions are established within a height-fetch ratio of roughly 0.01 for smooth-rough flow and 0.005 for rough-smooth flow; naturally, this is only valid provided the difference in roughnesses is not too severe. A grass-forest transition is not merely a roughness change, but rather a tall obstruction. From the practical point of view, it is clear that any method, to determine evaporation over uniform surfaces, can also be used over a non-uniform surface provided the fetch requirement is satisfied. Thus, a surface may be considered uniform whenever the measurements are taken at elevations which are smaller than approximately 1/100 to 1/300 of the distance from an upwind discontinuity in surface conditions.

The surface shear stress appears to approach relatively rapidly to a new equilibrium value. In the experiments of Bradley (1968) (see Figure 7.2) this occurred within a few meters behind the roughness change. Again. a simple argument for neutral conditions can be used to describe the surface shear stress adjustment as follows. It is assumed that the wind profile behind the roughness change can also be described by a logarithmic function; the upwind and the downwind profile intersect at the boundary of the internal boundary layer $\delta_m = \delta_m(x)$. Thus one has

$$\frac{u_*}{u_{*a}} = \frac{\ln(\delta_m/z_{a0})}{\ln(\delta_m/z_0)}$$

or, denoting $\ln(z_{a0}/z_0)$ by M_0,

$$\frac{u_*}{u_{*a}} = 1 - \frac{M_0}{\ln(\delta_m/z_0)} \tag{7.41}$$

which can be applied with anyone of (7.37), (7.39) or (7.40). Equation (7.41) has also been used by Logan and Fichtl (1975) and Jensen (1978); as shown in Figure 7.2, (7.41) compares at least as favorably with Bradley's experimental data as those obtained with a second-order closure model.

The rapid adjustment of u_* shown in Figure 7.2 has been obtained with most other theoretical models as well. This suggests that a step-change in u_* may be a satisfactory approximation for large fetches.

Finally, except for certain details or local 'kinks', most models appear to give an adequate representation of the wind profiles, which is rarely very different from the experimental profiles of Bradley (1968). Even the simple self-preservation models, with an assumed form for the profiles, yield velocity distributions that are rather close. Although the shear stress distribution of Townsend (1965b, 1966) may not have been correct, it was found by Rao *et al.* (1974a) that his assumption of self-preservation of the vertical shear stress profile is probably valid.

7.2. EVAPORATION WITH LOCAL ADVECTION

Few closure assumptions have been able to stand scrutiny by means of a more complete or higher-order closure model. However, as illustrated in the previous section, certain features of an internal boundary layer are relatively insensitive to the order or the type of similarity assumptions that are used. Any given characteristic requires, by virtue of its own order, a minimal order of closure; but this does not mean that a more complicated or complete model will always produce better results. For example, Petersen and Taylor (1973) found that a mixing length model produced slightly more accurate results for the wind profiles than a model with the turbulent kinetic energy equation. Mellor and Yamada (1973) concluded that it is not necessary to use the most complete second-order model to obtain satisfactory results. A similar conclusion can be drawn from Figure 7.2. All this is an indication that relatively simple models can be very useful to obtain certain lower-order features.

In what follows, several simple evaporation problems will be considered. Although they probably represent rather special situations, the results provide insight and information for the solution of practical problems, and suggestions for their parameterization.

a. Analytical Solutions with Power Laws

One major advantage of the analytical treatment of a problem is that it may yield a solution in a concise and well-ordered form; this allows an easier interpretation of the results and it also provides a suitable framework for the parameterization of the phenomenon under study. Unfortunately, the logarithmic and related similarity functions for the turbulent boundary layer are not always easy to handle in an analytical context. For this reason, power functions such as (4.18) have been used in the past. The power law does not appear to have any theoretical basis, and it must be considered an approximation. Still, in integrals or other averaging expressions it can provide a rather accurate description. It has been found (Yeh and Brutsaert, 1971a) that evaporation under neutral conditions could be analyzed by means of a power function for the wind profile equally as well as by means of the logarithmic law. Because the power law is an algebraic function without a singularity near $z = 0$, it is well suited for the analytical solution of certain turbulent transport problems.

Known Step-Change in Surface Humidity

This section deals with the evaporation from a uniformly-moist surface of limited size, such as a lake or an irrigated field, whose temperature and roughness are uniform and the same as the surrounding presumably drier land surface. Under these simple conditions, the water vapor is a passive admixture that does not affect the dynamics of the motion or the stability. Hence, the problem can be considered non-dynamic and non-energetic and only the passive transfer of water vapor is of concern. This means that the wind profile $\bar{u} = \bar{u}(z)$, $\bar{v} = \bar{w} = 0$, remains unchanged as the air travels past the humidity discontinuity.

(*i*) *Large wet surface: Sutton's problem.* If the surface is sufficiently large, the effects of the horizontal gradients of the turbulence are negligible, and the boundary conditions may be taken as those of a strip of a width or downwind fetch, x_f, extending to infinity in both lateral directions. The governing equation, obtained from (3.44) or (7.1), is

$$\bar{u}\frac{\partial \bar{q}}{\partial x} = -\frac{\partial}{\partial z}(\overline{w'q'}). \tag{7.42}$$

The term on the left represents the longitudinal change in vapor transport due to local advection by the mean wind. This is balanced by the term on the right, which is the vertical change in vertical vapor transport due to turbulent transport. This turbulent flux can be expressed as diffusion in terms of the specific humidity gradient by means of (7.15) and one obtains from (7.42)

$$\bar{u}\frac{\partial \bar{q}}{\partial x} = \frac{\partial}{\partial z}\left(K_v \frac{\partial \bar{q}}{\partial z}\right). \tag{7.43}$$

As shown in Figure 7.3, the boundary conditions are to describe the following situation: at the wet surface the specific humidity is given and it is constant; far

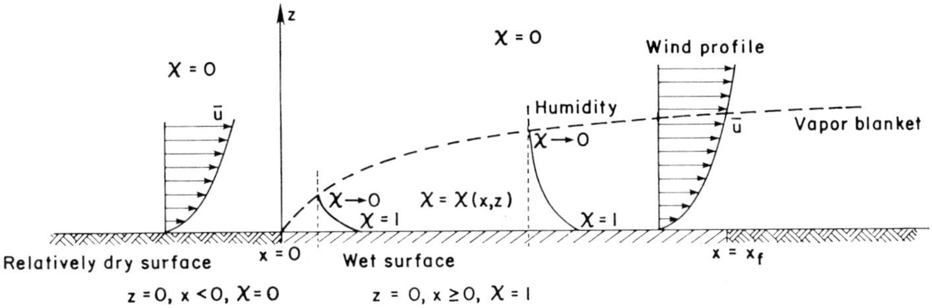

Fig. 7.3. Internal boundary layer of specific humidity in a dynamically uniform boundary layer.

away from the wet surface and upwind from the leading edge the specific humidity is known and it is unaffected by the wet surface. These conditions can be written as

$$\begin{aligned} z &= 0, & x &\geq 0, & \bar{q} &= \bar{q}_s, \\ z &\to \infty, & x &\geq 0, & \bar{q} &= \bar{q}_a, \\ z &> 0, & x &= 0, & \bar{q} &= \bar{q}_a, \end{aligned} \tag{7.44}$$

where $\bar{q}_a = \bar{q}_a(z)$ is the specific humidity in the part of the atmosphere unaffected by the presence of the moist surface. Assuming a uniform evapotranspiration rate E_a from the surface surrounding the active moist surface, one can impose on q_a the condition

$$E_a = -\rho K_v \frac{\partial \bar{q}_a}{\partial z} = \text{const.} \tag{7.45}$$

As formulated, this problem can be readily solved numerically by means of functional relationships for $\bar{u}$ and K_v that are derivable on the basis of the logarithmic law and the other functions for the surface sublayer given in Sections 4.1 and 4.2. Analytical solutions with these functions appear to be rather difficult. However, it is easy to obtain an analytical solution if use is made of the power law approximations of (4.18) and (7.18).

Such a solution with $E_a = 0$ was first given by Sutton (1934); but because of its mathematical form and because of the particular model used for a and b in (4.18) and (7.18), which were restricted to smooth flow, this solution as presented by Sutton was difficult to use. More suitable forms were published later (Jaeger, 1945; Frost, 1946; Calder, 1949; Sutton, 1943; Yih, 1952; Philip, 1959) and the mathematical aspects of the problem have been analyzed (Sutton, 1943). The historical comments made by Frost and Calder are also of interest.

In most derivations the evapotranspiration from the surrounding land, viz. E_a, was neglected; but as will now be shown (cf. also Laikhtmann, 1964), it is straightforward to include it, by defining a normalized specific humidity

$$\chi = \frac{\bar{q} - \bar{q}_a}{\bar{q}_s - \bar{q}_{as}} \tag{7.46}$$

where q_{as} is the value of q_a at the surface, $z = 0$ for $x < 0$. This allows the combination of (7.43) with (7.45) and (4.18), (7.18) to yield

$$\frac{\partial \chi}{\partial x} = \frac{b}{a} z^{-m} \frac{\partial}{\partial z}\left(z^n \frac{\partial \chi}{\partial z}\right). \tag{7.47}$$

The nature of the boundary conditions (7.44) allows the introduction of a similarity variable, as suggested by Frost (1946),

$$\xi = \frac{a}{b(2+m-n)^2} \frac{z^{2+m-n}}{x}. \tag{7.48}$$

This reduces (7.47) to an ordinary differential equation (provided $2 + m - n > 0$)

$$(2 + m - n)\xi \frac{d^2\chi}{d\xi^2} + [(2 + m - n)\xi + (1 + m)] \frac{d\chi}{d\xi} = 0 \tag{7.49}$$

and the boundary conditions

$$\chi = 0 \quad \text{for } \xi \to \infty$$

$$\chi = 1 \quad \text{for } \xi = 0. \tag{7.50}$$

Integrating twice, and applying these conditions one obtains

$$\chi = \int_\xi^\infty y^{-(m+1)/(2+m-n)} e^{-y} \, dy / \Gamma\left(-\frac{m+1}{2+m-n} + 1\right) \quad \text{or} \quad \chi = 1 - P(\nu, \xi) \tag{7.51}$$

where $P(a, x)$ denotes the incomplete gamma function (e.g., Abramowitz and Stegun, 1964), $\Gamma(n)$ the complete gamma function, and where $\nu = (1 - n)/(2 + m - n)$. Note that the gamma function is subject to the limitation that $\nu > 0$ or $n < 1$ which is clearly satisfied in the atmospheric surface layer. Provided a proper effective roughness $\hat{z}_0$ is chosen such that $\hat{z}_0 = z_0 = z_{0v}$, approximately, the existence of the interfacial sublayer can be neglected. Note that although, in general, $z_0 \neq z_{0v}$, in view of the closeness of (5.11) with the values given in Table 5.3, this is probably an adequate approximation for average conditions over open water. The local vertical vapor flux in the proximity of the wet surface can thus be assumed to be given by

$$E = -K_v \rho \left.\frac{\partial \bar{q}}{\partial z}\right|_{z \to 0} = E_a - (q_s - q_{as}) K_v \rho \left.\frac{d\chi}{d\xi} \frac{\partial \xi}{\partial z}\right|_{z \to 0} \tag{7.52}$$

and the mean evaporation from the strip of unit lateral width and of downwind fetch x_f, is

$$\bar{E} = \int_0^{x_f} E \, dx / x_f. \tag{7.53}$$

This gives the final result

$$\bar{E} = E_a + \rho b \left(\frac{a}{b x_f}\right)^\nu \frac{(1 - \nu)^{2\nu - 2}(m + 1)^{1 - 2\nu}}{\Gamma(\nu)} (\bar{q}_s - \bar{q}_{as}). \tag{7.54}$$

With (4.18), (4.19) and (7.19), with $a_v = 1$, and $d_0 = 0$, so that

$$a = 5.5\, u_* / \hat{z}_0^m \quad \text{and} \quad b = u_* \hat{z}_0^m / (5.5\, m) \tag{7.55}$$

Equation (7.54) can be expressed in a more convenient form. For an air temperature of 20°C, a pressure of 1013.2 mb and an air density $\rho = 1.2\ 10^{-3}$ g/cc, one can write (7.54) as follows

$$\bar{E} = E_a + N\bar{u}_2(\bar{e}_s - \bar{e}_{as}) \qquad (7.56)$$

where $\bar{u}_2$ is the mean wind speed at 2 m elevation above the surface, and N is a mass transfer coefficient given by (Brutsaert and Yeh, 1970a)

$$N = \frac{7.36\ 10^{-7}\ [1.62m(1+2m)^2\ \hat{z}_0{}^{2m}]^{(m+1)/(1+2m)}}{200^m\Gamma[m/(1+2m)](1+m)}\ x_f^{-m/(1+2m)} \qquad (7.57)$$

if $\bar{E}$ and E_a are in the same units as $\bar{u}_2$, $\hat{z}_0$ and x_f are in cm, and $\bar{e}_s$ and $\bar{e}_{as}$, the vapor pressures, respectively at the wet surface and the upwind dry surface, are in millibars. This equation can be used to calculate $\bar{E}$ after a proper choice is made for the roughness of the wet surface, and the wind profile parameter, m.

This solution of Sutton's problem has several features which are of some interest. First there is the growth of the internal boundary layer. The height of the internal boundary layer of water vapor can be defined as the locus of points where χ of (7.46) has a certain small constant value, say 0.05 or 0.01. Since χ is a function only of ξ, a constant χ also implies a constant ξ. Hence, (7.48) immediately yields the thickness of the vapor blanket as a function of fetch x, viz.

$$\delta_v = c_v x^{b_v} \qquad (7.58)$$

where $b_v = (2 + m - n)^{-1}$ and c_v is another constant which can be calculated from (7.51) for a given choice of χ. Under near-neutral conditions one has $m = (\frac{1}{7})$ to $(\frac{1}{8})$, so that by virtue of the first of (7.19), one obtains $b_v = 0.78$ to 0.80; this is the same as the power b_m in (7.40), which was confirmed in most experimental and theoretical studies of the internal boundary layer for momentum.

Other features of the solution, namely the values of m and $\hat{z}_0$ required to make it physically plausible, can be brought out by comparing it with two available empirical formulae for lake and pan evaporation. As a result of extensive measurements of means, usually over a week or longer, at numerous selected reservoirs in the Western United States, Harbeck (1962), proposed the following empirical mass transfer formula

$$\bar{E} = N\ \bar{u}_2(\bar{e}_s - \bar{e}_a) \qquad (7.59)$$

in which

$$N = 3.367\ 10^{-9} A^{-0.05} \qquad (7.60)$$

where $\bar{u}_2$ has the same units as $\bar{E}$, $\bar{e}_s$ and $\bar{e}_a$ are in mb, and A is the water surface area in m^2. From Figure 7.4 it appears that the surface areas of the investigated reservoirs varied between approximately 4 10^{-3} and 1.2 10^2 km^2. Nevertheless, a comparison with data obtained by a water budget method, has suggested that (7.59) with (7.60) produces reasonable results with monthly mean data even for Lake Ontario, which has a surface area of approximately 19 700 km^2 (Yu and Brutsaert, 1969a, b). For lakes located in a relatively arid climate, the evapotranspiration from the surrounding land, E_a, is neglibible, This was probably the case for most of the reservoirs that provided the data for (7.60).

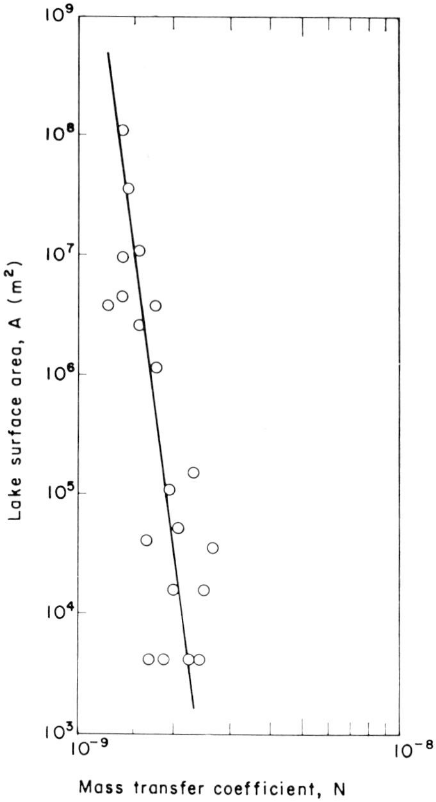

Fig. 7.4. Relation between the mass transfer coefficient N in (7.59) and the lake surface area, derived from experimental data by Harbeck (1962). $\bar{E}$ is in the same units as $\bar{u}_2$ and e in millibars (adapted from Harbeck, 1962).

The approximate value of m to make the theoretical result (7.57) agree with Harbeck's empirical (7.60) can be deduced by equating powers of $A^{1/2}$ and x_f, that is $m/(1 + 2m)$, which yields $m = \frac{1}{8}$. This value is very nearly the same as that obtainable from the velocity profile in a pipe on the basis of Blasius's expression (Prandtl and Tollmien, 1924), viz. $m = \frac{1}{7}$. It indicates that, roughly speaking, Harbeck's experimental data, for lakes and reservoirs represent on the average atmospheric conditions close to neutral, or slightly unstable. Further solution between (7.57) and (7.60) for the roughness yields a value $\hat{z}_0 = 0.0213$ cm. It is remarkable that this value is so close to $z_{0v} = z_0 = 0.0228$ cm, which corresponds to $Ce_{10} = Cd_{10} = 1.4\ 10^{-3}$, a typical value for sea surfaces, as seen in (5.11) and in Table 5.3. All this means that the solution of Sutton's problem with the power law, as given here in (7.57), is in good agreement with Harbeck's empirical formula (7.60) for water with fetches roughly between 50 m and 10 km, and for weekly or monthly mean data.

A second set of experimental data is available to check the suitability of (7.57). It was obtained by measuring evaporation from shallow square pans at ground level, whose sizes varied between 1.0 and 64.0 ft^2. Regression analysis of these data yielded,

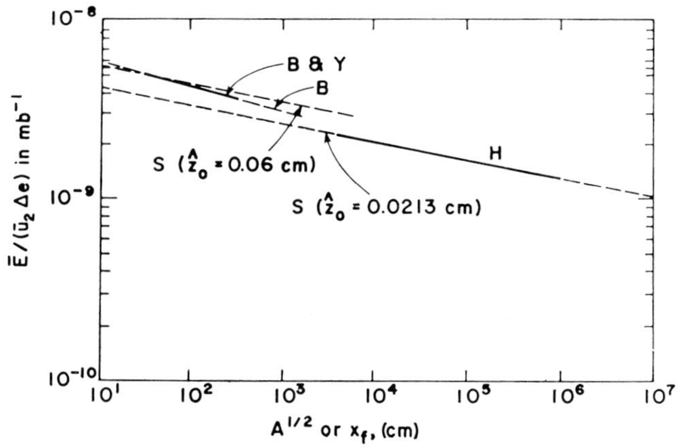

Fig. 7.5. Comparison of the empirical formulae (solid lines) (7.60) of Harbeck (H), (7.61) of Brutsaert and Yu (B & Y), and the theoretical equations (7.57) of Sutton's model (S) and (7.67) of Brutsaert (B). S coincides with H for $m = 1/8$ and $\hat{z}_0 = 0.0213$ cm; B coincides with B&Y for $m = 1/7.6$ and $\hat{z}_0 = 0.05$ cm.

among several other possible alternative forms, (7.59) with the following mass transfer coefficient (Brutsaert and Yu, 1968)

$$N = 7.70\ 10^{-9}(A^{1/2})^{-0.132} \tag{7.61}$$

where now A is in cm^2. The range of validity of this result was for values of A between 10^3 and $6\ 10^4$ cm^2, approximately. The value of m needed to make (7.57) agree with (7.61) and obtained by equating powers of x_f and $A^{1/2}$ turns out to be $\frac{1}{5.6}$. This is somewhat large, since the exponents of the power functions made to fit observed (Yu and Brutsaert, 1967) wind profiles lay between $\frac{1}{7}$ and $\frac{1}{8}$ which would result in 0.11 and 0.10 in (7.57) instead of 0.132 as in (7.61). The discrepancy is probably due to the effect of the neglect of the longitudinal and lateral gradients of the turbulence in (7.42), the effect of the roughness of the surrounding land, and the related effect of the edges of the pans; another factor may be the omission of E_a which is not negligible during the humid summers in Up-state New York. If, in accordance with the observed wind profiles, m is chosen to be $\frac{1}{8}$, it is found that (7.57) matches (7.61) if $\hat{z}_0$ is taken as 0.06 cm for the smaller, and 0.04 cm for the larger pans.

A comparison of (7.60) and (7.61) with (7.57) is shown in Figure 7.5. A different way of modeling evaporation from very small surfaces is presented in the following Section (ii).

(ii) Solutions for small water surfaces. One of the assumptions used to obtain (7.43) is that the gradients $\partial(\overline{u'q'})/\partial x$ and $\partial(\overline{v'q'})/\partial y$ of (3.44) are negligible for large surfaces. Conversely for small surfaces these gradients of the turbulence are relatively more important, and the terms representing advection by the mean wind are relatively less important than for large surfaces. This can be seen by dimensional inspection of the governing equation. In the framework of the turbulent diffusion approach, (3.44) or (7.1) with the assumption of (7.12) and (7.13), is for the present problem of flow over a surface with uniform roughness and temperature, including lateral turbulence,

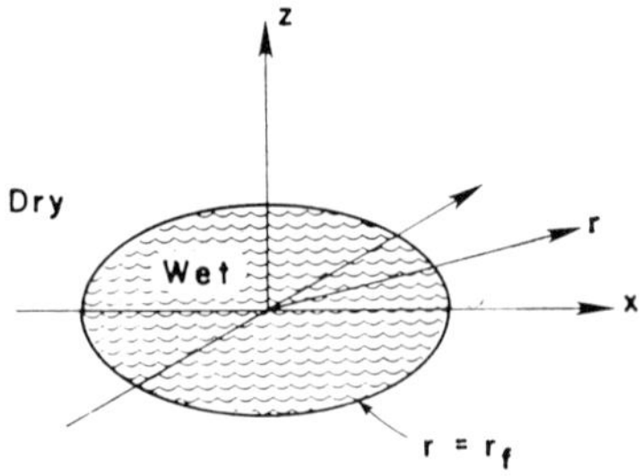

Fig. 7.6. Definition sketch for situation described by boundary conditions (7.65).

$$\bar{u}\frac{\partial \bar{q}}{\partial x} = \frac{\partial}{\partial x}\left(K^v_{xx}\frac{\partial \bar{q}}{\partial x} + K^v_{xz}\frac{\partial \bar{q}}{\partial z}\right) + \frac{\partial}{\partial y}\left(K^v_{yy}\frac{\partial \bar{q}}{\partial y}\right) + \frac{\partial}{\partial z}\left(K^v_{zx}\frac{\partial \bar{q}}{\partial x} + K^v_{zz}\frac{\partial \bar{q}}{\partial z}\right) \tag{7.62}$$

If K^v_{zz} and the fetch x_f are taken as characteristic diffusivity and length, respectively, to normalize the variables, on the left-hand side the term $\bar{u}x_f/K^v_{zz}$ appears as a dimensionless parameter. This suggests that decreasing the dimension x_f of the problem, may be equivalent with decreasing the wind speed $\bar{u}$, that is the relative importance of the left-hand side in (7.62). In what follows, three solutions are presented for special cases, which allow an assessment of the importance of the terms that were neglected in (7.42).

Although for small surfaces, the term on the left of (7.62) is not likely to be ever totally negligible, some features resulting from the turbulence gradients can be brought out by simply omitting the left-hand side. This 'extreme' case of evaporation from a very small surface can thus be studied by means of the equation

$$\frac{\partial}{\partial x}\left(K^v_{xx}\frac{\partial \bar{q}}{\partial x}\right) + \frac{\partial}{\partial y}\left(K^v_{yy}\frac{\partial \bar{q}}{\partial y}\right) + \frac{\partial}{\partial z}\left(K^v_{zz}\frac{\partial \bar{q}}{\partial z}\right) = 0 \tag{7.63}$$

in which it can be assumed that K^v_{zz} is given by (7.18) and, similarly that

$$K^v_{xx} = K^v_{yy} = c\, z^p \tag{7.64}$$

in which c and p are constants. The boundary conditions for a circular surface of radius r_f and radial coordinate r, are, as shown in Figure 7.6

$$\begin{aligned} z &= 0, & 0 \le r \le r_f, && \bar{q} &= \bar{q}_s, \\ z &\to \infty, & r \ge 0, && \bar{q} &= \bar{q}_a, \\ z &> 0, & r \to \infty, && \bar{q} &= \bar{q}_a, \\ z &= 0, & r > r_f, && E_a &= 0. \end{aligned} \tag{7.65}$$

The last condition, in which E_a is the evaporation from the land surrounding the water surface, is not very important; it could be replaced by $E_a = \text{const}$, as in (7.45), without making the problem more difficult. It would merely add the term E_a to the final solution (7.66), just like in (7.54). After transforming (7.63) in cylindrical coordinates, one can derive the result for the mean evaporation rate (Brutsaert, 1967)

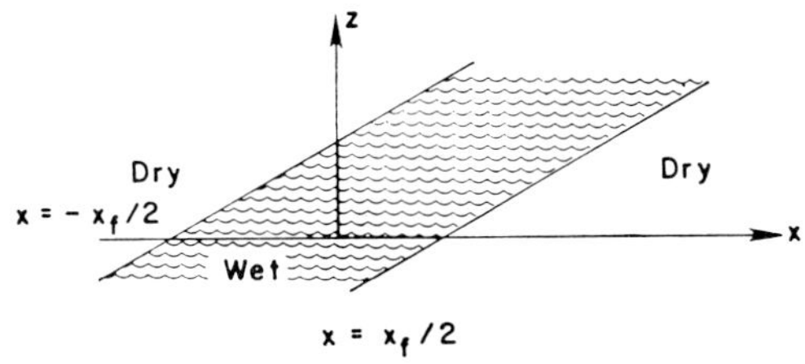

Fig. 7.7. Definition sketch for situation described by boundary conditions (7.69).

$$\bar{E} = 2^{2\mu} b(c/b)^{\mu}\left(\frac{\mu}{1-n}\right)^{2\mu-1} \sin(\mu\pi)\, r_f^{-2\mu} \rho(\bar{q}_s - \bar{q}_a)/\pi(1 - \mu) \tag{7.66}$$

in which $\mu = (1-n)/(p-n+2)$. If it is assumed that $n = p = (1-m)$(see (7.19)) this parameter is simply $\mu = m/2$, or approximately $(\frac{1}{14})$ under near-neutral conditions.

Equation (7.66) can be put in a more practical form for actual calculations, by using (7.55) for b. Putting $(c/b) = d$, as a measure of the degree of anisotropy, $r_f = (A/\pi)^{1/2}$, and $\sin(\mu\pi) \simeq \mu\pi$, for an air temperature of 20°C, a pressure of 1013.2 mb and an air density $\rho = 1.2\ 10^{-3}$ g^{-1} cc^{-1}, one obtains from (7.66), a mass transfer equation like (7.56) with

$$N = \frac{4.87\ 10^{-8} (\pi d)^{m/2}\, \hat{z}_0^{2m}}{200^m(2-m)} (A^{1/2})^{-m} \tag{7.67}$$

where $\hat{z}_0$ and $A^{1/2}$ are in cm. The solution is not very sensitive to the degree of anisotropy d of the K tensor, so that for the present purpose it is probably sufficiently accurate to assume (cf. (7.14)) that d is of the order of 20. Suitable values of m and $\hat{z}_0$ can be obtained by comparison with the empirical pan evaporation formula (7.61). The powers of $A^{1/2}$ in (7.61) and (7.67) yield $m = 0.132 = \frac{1}{7.6}$ which is quite realistic and slightly smaller than the neutral value $\frac{1}{7}$. The roughness value obtained in this way is $\hat{z}_0 = 0.05$ cm; this result is also remarkable, since it is intermediate between the values of z_{0_v} (cf. Figure 4.24) and z_0 (cf. Table 5.1), one would expect for the type of water surface surrounded by grass in the experiments of (7.61). It can be concluded that, albeit, (7.63) represents an extreme case, it leads to a result which is not inconsistent with experimental data. Its solution, (7.66) (or (7.67)) appears to be slightly better than (7.54) (or (7.57)) of Sutton's model to describe evaporation from small pans at ground level.

The relative importance of the longitudinal and lateral turbulent flux terms $\partial(\overline{u'q'})/\partial x$ and $\partial(\overline{v'q'})/\partial y$ in the computation of evaporation can be studied in a similar way. The lateral turbulent flux gradient $\partial(\overline{v'q'})/\partial y$ is most important for small water surfaces; therefore, the extreme case can be considered of a very narrow strip, that extends laterally to infinity, in order to compare it with the previous case of an extremely small circular surface. The equation for this geometry is

$$\frac{\partial}{\partial x}\left(K_{xx}^v \frac{\partial \bar{q}}{\partial x}\right) + \frac{\partial}{\partial z}\left(K_{zz}^v \frac{\partial \bar{q}}{\partial z}\right) = 0 \tag{7.68}$$

in which K_{xx}^v and K_{zz}^v can be taken the same as for (7.63). The boundary conditions, as shown in Figure 7.7, are

$$
\begin{array}{lll}
z = 0, & 0 \le |x| \le x_f/2, & \bar{q} = \bar{q}_s, \\
z \to \infty, & |x| \ge 0, & \bar{q} = \bar{q}_a, \\
z > 0, & |x| \to \infty, & \bar{q} = \bar{q}_a, \\
z = 0, & |x| > x_f/2, & z^n \dfrac{\partial \bar{q}}{\partial z} = 0 \text{ (or } E_a = 0), \\
z > 0, & |x| = 0, & z^p \dfrac{\partial \bar{q}}{\partial x} = 0.
\end{array}
\tag{7.69}
$$

The solution for the mean rate of evaporation can be written as (Brutsaert and Yeh, 1969)

$$
\bar{E} = \pi b \left(\frac{c}{b}\right)^{\mu} \left(\frac{2\mu}{1-n}\right)^{2\mu - 1} \frac{\Gamma(1-\mu)}{\Gamma(\mu)\Gamma[(\frac{1}{2}) - \mu]\Gamma[(\frac{3}{2}) - \mu]} \times \left(\frac{x_f}{2}\right)^{-2\mu} \rho(\bar{q}_s - \bar{q}_a)
\tag{7.70}
$$

in which $\mu = (1-n)/(p-n+2)$. By using (7.55) and taking $\mu = m/2$, one can also express it as

$$
\bar{E} = \frac{2\pi d^{m/2}\,\Gamma(1-m/2)}{(5.5)m(1-m)\{\Gamma[(1-m)/2]\}^2\Gamma(m/2)}\, \hat{z}_0^m (x_f/2)^{-m}\, \rho(\bar{q}_s - \bar{q}_a)\, u_*.
\tag{7.71}
$$

The effect of the lateral gradient of the turbulent flux $\overline{(v'q')}$ can be determined by taking the ratio of (7.66) and (7.70) with $x_f^2 = \pi r_f^2$; this shows that when m lies in the neighborhood of $(\frac{1}{7})$, this ratio is approximately 1.13. In other words, even in the extreme case of a very small water surface, when the advection by the mean wind is neglected, the lateral turbulence v' is not likely to ever account for more than 13 percent of the total evaporation rate. Hence, for larger surfaces, this effect is bound to be totally negligible. This supports the common assumption that the surface flux in an internal boundary layer may be considered a two-dimensional problem without lateral effects, as shown in Figures 7.1 and 7.3.

The relative importance of the terms $\bar{u}\partial\bar{q}/\partial x$ and $\partial\overline{(u'q')}/\partial x$ of (3.44) or (7.1) in the two-dimensional evaporation problem can be evaluated by considering the following equation,

$$
\bar{u}\frac{\partial \bar{q}}{\partial x} = \frac{\partial}{\partial x}\left(K_{xx}^v \frac{\partial \bar{q}}{\partial x}\right) + \frac{\partial}{\partial z}\left(K_{zz}^v \frac{\partial \bar{q}}{\partial z}\right)
\tag{7.72}
$$

with $\bar{u}$, K_{xx}^v and K_{zz}^v as power functions, given in (4.18), (4.19), (7.18) and (7.64), with $n = p = 1 - m$. The boundary conditions are the same as those shown in Figure 7.3

$$
\begin{array}{lll}
z = 0, & 0 \le x \le x_f, & q = \bar{q}_s, \\
z \to \infty, & -\infty < x < \infty, & q = \bar{q}_a, \\
z = 0, & x \to \pm\infty, & q = \bar{q}_a, \\
z = 0, & x > x_f \text{ and } x < 0 & \rho K_{zz}^v \dfrac{\partial \bar{q}}{\partial z} = 0.
\end{array}
\tag{7.73}
$$

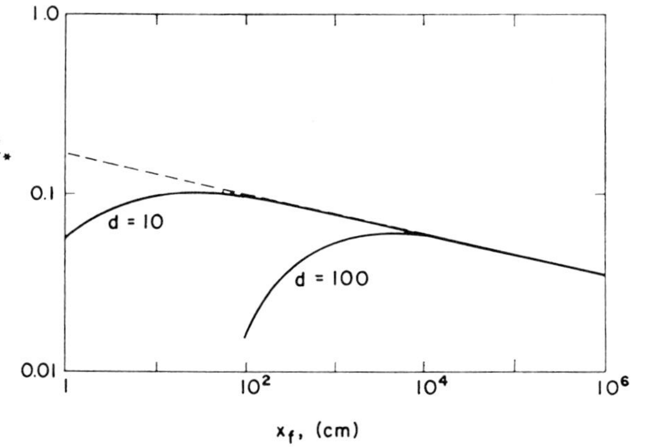

Fig. 7.8. Dimensionless mean evaporation rate $E_* = \bar{E}/[\rho u_*(\bar{q}_s - \bar{q}_a)]$ versus fetch, as given by (7.75) for zero-th order (dashed line) [cf., (7.54)] and up to the first order (solid line). The parameters were taken as typical values, $m = 1/7$, $(C_p/\hat{z}_0^m) = 9$, and $(c/b) = d = 10$ and 100 (from Yeh and Brutsaert, 1970).

A solution can be obtained by means of a regular perturbation method with small parameter ε_p, taken as the coefficient of the first term on the right of the dimensionless form of (7.72); this parameter can be written as

$$\varepsilon_p = (c/b)(m^m x_f^m C_p/\hat{z}_0^m)^{-4/(1+2m)} \tag{7.74}$$

where b and c are given in (7.18) and (7.64). The zero-th order solution is the solution free from the effect of longitudinal diffusion, as obtained in Sutton's problem (7.51). The first order solution, which is in terms of the hypergeometric function, is more complicated. The average evaporation rate up to the order of ε_p is (Yeh and Brutsaert, 1970)

$$\bar{E} = \bar{E}_0 + \varepsilon_p \bar{E}_1 \tag{7.75}$$

in which $\bar{E}_0$ is the solution of Sutton's problem, that is, the second term on the right of (7.54) or (7.56), and $\bar{E}_1$ is the first-order term which can be written as

$$\begin{aligned}\bar{E}_1 = &\frac{-\rho u_*(\bar{q}_s - \bar{q}_a)}{(mx_f)^{m/(1+2m)}(C_p/\hat{z}_0^m)^{1/(1+2m)}} \frac{(\mu-\nu)^2 + \nu - \nu^2/4}{(2\mu - 1 - \nu)(\mu - 1 - 3\nu/2)} \times \\ &\times \frac{\nu^{1+4\nu-2\mu}\,\Gamma(\mu+\nu/2)\,\Gamma(\mu-\nu/2)\,\Gamma(\mu-1-\nu/2)}{\Gamma(\nu)\,\Gamma(1+\nu)\,\Gamma(2\mu-1-\nu)}\end{aligned} \tag{7.76}$$

in which here $\nu = m/(1+2m)$ and $\mu = (4+m)/(2+4m)$. Equation (7.75) is plotted in Figure 7.8 for typical values of the parameters. It can be seen that $\bar{E}_1$ is negative; this means that the longitudinal turbulent flux gradient term reduces the evaporation rate. However, Figure 7.8 also shows that this effect is small and that it is likely to be negligible for water surfaces with a fetch larger than say, a few meters.

In summary, a comparison of the solutions given in (7.54), (7.66), (7.70), (7.75) and (7.76), allows an evaluation of the relative importance of the longitudinal and lateral turbulent diffusion terms under certain special conditions. For most practical problems, involving evaporation from lakes and reservoirs, but not small evaporation pans, these terms may probably be neglected like in Sutton's formulation (7.43). Moreover, the error introduced by neglecting one of the two is apparently at least partly compensated by also neglecting the other.

Note, that in (7.63), (7.68) and (7.72) the off-diagonal terms of the diffusivity K^v_{zx} and K^v_{xz} were not included. However, their effect was considered in a numerical study (Yeh and Brutsaert, 1971b). It was found that these terms probably decrease the rate of evaporation from small surfaces by, at most 10 percent. Thus, for larger surfaces with fetches exceeding a few meters, this effect is also negligible.

Surface Water Vapor and Heat Transfer Linked by Energy Budget

The physical situation dealt with in this section is again the evaporation from a moist surface of limited size. In contrast to the analyses of the previous section, the surface specific humidity and the temperature are left unspecified but the evaporation and the surface heat flux are linked by the surface energy budget. In the available analytical solutions of this problem, the roughness is uniform and the same as the surrounding land surface. Also, the wind profile $\bar{u} = \bar{u}(z)$ and the eddy diffusivity are assumed to remain in equilibrium as the air moves across the sudden change in surface humidity and temperature. Thus, water vapor and sensible heat are treated as nearly passive admixtures of the flow, in the sense that they affect the dynamics, at most, in an average way. Various aspects of local advection and simultaneous vapor and heat transfer in the lower atmosphere have been studied in this way by Timofeev (1954), De Vries (1959), Rider *et al.* (1963), Laikhtman (1964) and Yeh and Brutsaert (1971c). In what follows, an outline is given of the latter analysis.

Under the assumption that $\bar{u} = \bar{u}(z)$, $\bar{v} = \bar{w} = 0$, and that the fluxes are given by (7.15) and (7.17) for intermediate to large surfaces, the problem is governed by (7.43) and

$$\bar{u}\frac{\partial \bar{T}}{\partial x} = \frac{\partial}{\partial z}\left(K_h \frac{\partial \bar{T}}{\partial z}\right) \tag{7.77}$$

in which the mean temperature $\bar{T}$ is used as a good approximation of $\bar{\theta}$. Although it would not complicate the problem if a_v and a_h were retained, it is assumed that $K_v = K_h = K$.

The boundary conditions are deduced from the following considerations: far away from the wet surface the atmospheric conditions remain unaffected by it; the incoming air has a known humidity profile $\bar{q}_a(z)$, which is a result of equilibrium conditions dictated by the heat flux and the evapotranspiration from the windward land surface; although this is not a necessary assumption for the solution, at the active surface the specific humidity can be assumed to be saturated and, hence, a known function of the temperature; at the surface the net energy flux is zero. In this energy budget, the incoming radiation on the active surface, R_d, and on the upwind land surface, R_{da}, are both constant and independent of surface temperature, though they may be different on account of the different albedo's of the surface. The evaporation rate and sensible heat flux at the surface, E and H, are given by (7.15) and (7.17) in terms of the gradients, so that they are part of the solution of (7.43) and (7.77). The long-wave radiation emitted from the surface (see (6.9)) is obtained by assuming that it is a grey body with emissivity ε_s radiating at its temperature, $\bar{T}_s$; hence, since $\bar{T}_s$ is part of the solution, this outgoing radiation is also allowed to vary along the downwind direction. The rate of change of heat stored below the surface can be formulated in several ways; two possibilities are either

$$G = G_w = \text{const} \tag{7.78a}$$

or

$$G = K_s(\bar{T}_s - T_{rw}) \tag{7.78b}$$

where K_s is a thermal exchange coefficient and T_{rw} is the temperature at a given reference depth below the surface, $z = 0$. The boundary conditions are thus

$$\begin{aligned}
&\bar{q} = \bar{q}_a(z); \qquad \bar{T} = \bar{T}_a(z) \quad \text{at } x = 0,\, z > 0,\\
&\bar{q} = \bar{q}_s(\bar{T}) \quad \text{at } 0 < x < x_f,\, z = 0,\\
&-c_p\rho K\frac{\partial \bar{T}}{\partial z} - L_e\rho K\frac{\partial \bar{q}}{\partial z} + \varepsilon_s\sigma \bar{T}^4 + G = R_d \quad \text{at } 0 < x < x_f,\, z = 0,\\
&-c_p\rho K\frac{\partial \bar{T}}{\partial z} = H_a; \; -\rho K\frac{\partial \bar{q}}{\partial z} = E_a \quad \text{at } x > x_f,\, z = 0
\end{aligned} \tag{7.79}$$

where x_f is the fetch, c_p the specific heat of air at constant pressure, L_e the latent heat of vaporization of water, σ the Stefan–Boltzmann constant, and H_a and E_a are the heat and water vapor fluxes from the land surface upwind and downwind from the active surface. The upwind profiles, $\bar{q}_a$ and $\bar{T}_a$ are in equilibrium, satisfying (7.45) and

$$\frac{\partial}{\partial z}\left(K\frac{\partial \bar{T}_a}{\partial z}\right) = 0 \tag{7.80}$$

with the following conditions

$$\begin{aligned}
&\bar{q}_a = q_{as}, \qquad \bar{T}_a = T_{as} \quad \text{at } z = 0,\\
&-c_p\rho K\frac{\partial \bar{T}_a}{\partial z} - L_e\rho K\frac{\partial \bar{q}_a}{\partial z} + \varepsilon_s\sigma T_a^4 + G_a = R_{da} \quad \text{at } z = 0,\\
&-c_p\rho K\frac{\partial \bar{T}_a}{\partial z} = H_a, \qquad -\rho K\frac{\partial \bar{q}_a}{\partial z} = E_a \quad \text{at } z = 0
\end{aligned} \tag{7.81}$$

where G_a is the heat conducted into the ground upwind from $x = 0$. Note that only one of H_a and E_a can be arbitrary, since, for a given incoming radiation and $\bar{T}_{as}$, they must satisfy the second of (7.81). The value of $\bar{q}_{as}$ need not be that of saturation at $\bar{T}_{as}$.

The system of (7.43), (7.77), (7.78), (7.79), (7.45), (7.80) and (7.81) can be solved when the wind profile and the eddy diffusivity are given by (4.18) and (7.18), respectively. This solution was apparently first obtained by Laikhtman (1964), for the case of (7.78a), by means of a method of integral transforms; later, the same solution was obtained independently by Yeh and Brutsaert (1971c) by means of the method of Green's function for both (7.78a) and (7.78b). Because the method of solution is beyond the scope of the present treatment, it is not given here, but it may be found in those papers.

It can be shown that at the active surface, i.e., for $z=0$ and $0 < x < x_f$, the temperature is

$$\bar{T}(\xi, 0) = \bar{T}_{as} - \frac{L_e(\bar{q}^*_{as} - \bar{q}_{as})}{c_p + \alpha_q L_e} + \frac{\nu^{1-2\nu}\,\Gamma(\nu)\{c_4 + c_2c_3c_6/(c_1 + c_3c_5)\}}{\Gamma(1-\nu)\,\rho b\{a/(bx_f)\}^{\nu}(1-n)^{1-2\nu}(c_p + \alpha_q L_e)} \times \sum_{n=0}^{\infty} \frac{(-\omega)^n\,\xi^{\nu+\nu n}}{\Gamma(1+\nu+\nu n)} \tag{7.82}$$

and similarly the specific humidity

$$\bar{q}(\xi, 0) = \bar{q}_{as} + \frac{c_p(\bar{q}^*_{as} - \bar{q}_{as})}{c_p + \alpha L_e} + \frac{\alpha_q\nu^{1-2\nu}\,\Gamma(\nu)\{c_4 + c_2c_3c_6/(c_1 + c_3c_5)\}}{\Gamma(1-\nu)\rho b\{a/(bx_f)\}^{\nu}(1-n)^{1-2\nu}(c_p + \alpha_q L_e)} \times \sum_{n=0}^{\infty} \frac{(-\omega)^n\xi^{\nu+\nu n}}{\Gamma(1+\nu+\nu n)} \tag{7.83}$$

in which $\xi = x/x_f$, $\bar{q}^*_{as}$ is the saturated specific humidity at $\bar{T}_{as}$ and, as before, $\nu = (1-n)/(2+m-n)$; the other terms in (7.82) and (7.83) are

$$\omega = \frac{c_2\,\nu^{1-2\nu}\,\Gamma(\nu)}{(c_1 + c_3c_5)\,\Gamma(1-\nu)},$$

$$c_1 = c_p\rho b(\bar{T}_m - \bar{T}_{as})(a/bx_f)^{\nu}(1-n)^{1-2\nu},$$

$$c_2 = 4\varepsilon_s\sigma \bar{T}^3_{as}(\bar{T}_m - \bar{T}_{as}) \quad \text{for model (7.78a)},$$

$$c_2 = (4\varepsilon_s\sigma \bar{T}^3_{as} + K_s)(\bar{T}_m - \bar{T}_{as}) \quad \text{for model (7.78b)},$$

$$c_3 = L_e\rho b(\bar{q}_m - \bar{q}_{as})(a/bx_f)^{\nu}(1-n)^{1-2\nu},$$

$$c_4 = R_d - R_{da} - G_w + G_a \quad \text{for model (7.78a)},$$

$$c_4 = R_n - R_{na} + K_s(T_{rw} - \bar{T}_{as}) + G_a \quad \text{for model (7.78b)},$$

$$c_5 = \frac{\bar{T}_m - \bar{T}_{as}}{\bar{q}_m - \bar{q}_{as}} \left.\frac{dq^*}{dT}\right|_{\bar{T}=\bar{T}_{as}} = \frac{\bar{T}_m - \bar{T}_{as}}{\bar{q}_m - \bar{q}_{as}}\alpha_q,$$

$$c_6 = \frac{\bar{q}^*_{as} - \bar{q}_{as}}{\bar{q}_m - \bar{q}_{as}}$$

in which $\bar{T}_m$ and $\bar{q}_m$ are some representative temperature and specific humidity of the active surface.

The mean rate of evaporation from the strip with unit lateral width and with downwind fetch x_f can be calculated by applying (7.52) and (7.53) to the solution. The result is

$$\bar{E} = E_a + c_p\rho b\left(\frac{a}{bx_f}\right)^{\nu} \frac{(1-\nu)^{2\nu-2}(m+1)^{1-2\nu}}{\Gamma(\nu)} \frac{\bar{q}^*_{as} - \bar{q}_{as}}{(c_p + \alpha_q L_e)} + \frac{\alpha_q}{c_p + \alpha_q L_e}\left[c_4 + \frac{c_2c_3c_6}{(c_1 + c_3c_5)}\right] \sum_{n=0}^{\infty} \frac{(-\omega)^n}{\Gamma(2+\nu n)}. \tag{7.84}$$

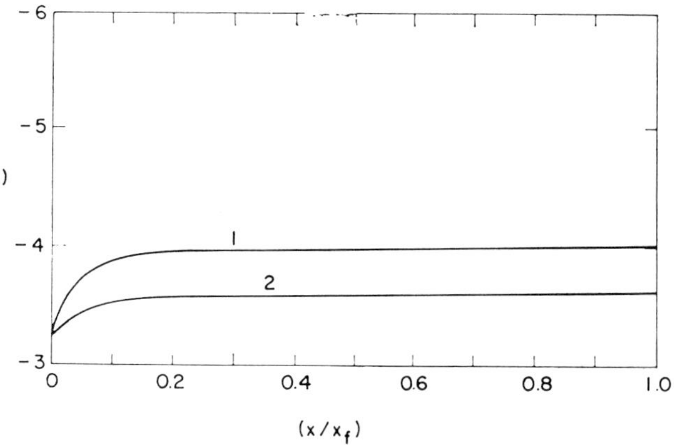

Fig. 7.9. Variation of $\bar{T}(\xi, 0) - \bar{T}_{as}$ with fetch (x/x_f) for a deep water body such as Cayuga Lake, N.Y., in the month of May, calculated by means of (7.82) with (7.78a) (curve 1) and (7.78b) (curve 2) for $x_f = 2.74$ km (from Yeh and Brutsaert, 1971c).

A similar expression can be obtained for the average sensible heat flux.

The result shown in (7.84) is of interest, not so much because of its practical applicability, but because it can give an indication of the type of error that is introduced in some commonly-used assumptions. For example, (7.82) and (7.83) show that the surface specific humidity or temperature can be constant, i.e., independent of x, only if the second term on the right-hand side is zero. In that case, elimination of $(\bar{q}^*_{as} - \bar{q}_{as})/(c_p + \alpha_q L_e)$ between these two equations gives

$$\frac{\bar{T}_s - \bar{T}_{as}}{\bar{q}_s - \bar{q}_{as}} = -\frac{L_e}{c_p} \tag{7.85}$$

where $\bar{q}_s$ and $\bar{T}_s$ are the uniform specific humidity and temperature of the evaporating surface. Equation (7.85) is similar to the equation for the wet-bulb thermometer. Hence, the assumption of a mean surface temperature in the third of boundary conditions (7.79) leads to the result that the latent heat flux of evaporation and the sensible heat flux balance each other. In the case of an irrigated field in an arid environment, which was studied by Rider *et al.* (1963) this assumption is probably quite satisfactory; but in the case of a deep lake, where considerable quantities of heat are taken up or released from storage, this is unrealistic. For such a case, (7.85) implies that the surface temperature of an evaporating lake must always be lower than the air temperature, which is a limitation. Likewise, the last term on the right of (7.84) reflects the effect of nonuniformity of the surface specific humidity on the evaporation. When the specific humidity is uniform, this term does not appear (it is also absent in (7.83)); thence by substitution of $c_p(\bar{q}^*_{as} - q_{as})/(c_p + \alpha_q L_e)$ by means of (7.83), in which $\bar{q}_s$ replaces $\bar{q}(\xi; 0)$, (7.84) becomes exactly the same as (7.54), the solution of Sutton's problem discussed earlier. As an example, Figure 7.9 shows the surface temperature variation in the downwind direction of a deep lake at mid-latitudes in the Northern hemisphere, calculated by means of (7.82) for May. The turbulence parameters a and b were calculated by means of (7.55) with $m = \frac{1}{8}$ and $\hat{z}_0 = 0.02$ cm. The heat flux G_w was assumed to be 5.05 10^{-3} (cal cm^{-2} s^{-1}). The value K_s was calculated by

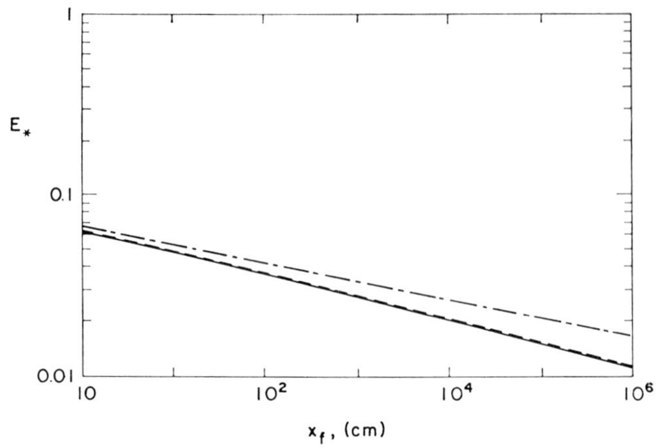

Fig. 7.10. Dimensionless mean evaporation rate $E_* = (\bar{E}-E_a)/[\rho u_*(\bar{q}^*_{as}-\bar{q}_{as}]$ versus fetch x_f, for a deep water body in May, as given by (7.84) with (7.78a) (solid line); by (7.54) with uniform temperature obtained from (7.85) (dash-dot line); and from mean of (7.82) (dash line) (from Yeh and Brutsaert, 1971c).

using experimental mean-water surface temperatures and $T_{rw} = 4°C$. The other necessary parameters and variables were taken as typical values for a lake such as Cayuga Lake near Ithaca, N.Y. Figure 7.9 shows that both phyiscal models of (7.78) give similar temperature distributions, although that of (7.78b) is slightly more uniform. Figure 7.10 shows the variation of the mean evaporation rate from the lake with fetch, as calculated with (7.84) for the model of (7.78a). Also shown are the results of the calculation by means of (7.54), i.e., (7.84) with omission of the last term on the right. In one case, the uniform temperature was obtained by assuming that this term is zero in accordance with the assumption leading to (7.85). In the other case the uniform specific humidity was obtained by integration of (7.83) over x, so that this would be equivalent with an average observed value. As is to be expected for a deep lake, the second case shows better agreement with the complete solution, namely within 1 percent. Similar calculations are possible for heat transfer $\bar{H}$, which would produce similar results to those for evaporation $\bar{E}$. Figure 7.11 shows the variation of the normalized Bowen ratio $\mathrm{Bo}_n=[(H(x) - H_a)/L_e(E(x) - E_a)]$ with distance from the leading edge for the same situation described in Figure 7.9.

The main conclusion that can be drawn on the basis of this solution, is that the error introduced by the use of an experimentally-obtained average surface temperature in the analysis of the advection problem of evaporation or turbulent sensible heat transfer, is often probably quite small, and negligible for practical purposes. It should be added that, in the case of lakes and other water bodies, due to the normal unsteadiness of the wind speed and direction, the presence of lake currents and horizontal mixing in the water, and also the cooling or warming effect of shallow off-shore zones, the water temperatures predicted by the solution (7.82) will only rarely agree with actual observations. In other words, the heat and vapor fluxes are often occasioned by pre-existing differences of temperature and humidity between surface and air. Over land, however, G of (7.78) is usually small and the major

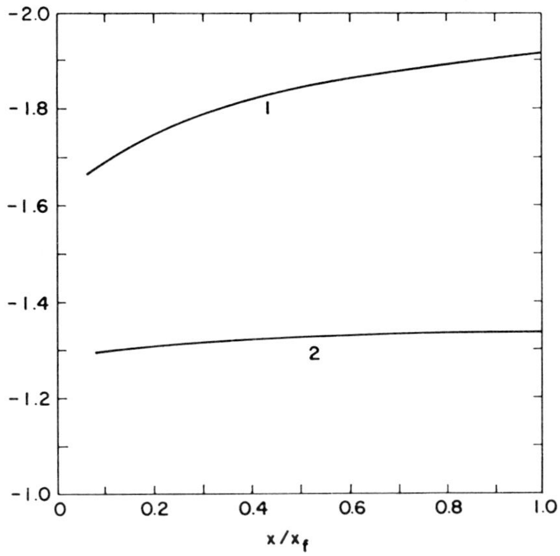

Fig. 7.11. Normalized Bowen ration $[H(x)-H_a]/L_e[E(x)-E_a]$ versus distance from the leading edge with model (7.78a) (curve 1) and model (7.78b) (curve 2) for the same situation as Figure 7.9 (from Yeh and Brutsaert, 1971c).

components of the energy balance are the net radiation and the turbulent fluxes. Hence, the solution may be useful to study certain problems of evaporation from a moist land area within arid surroundings. But for this type of problem the formulation of the boundary condition at the vegetated surface can usually not be treated in the same simple way as in the third of (7.79) (see also Section 5.2).

b. Numerical Studies

Lake Evaporation by Turbulent Diffusion Approach

This approach has been used by Weisman and Brutsaert (1973) in a numerical study of evaporation and cooling of a relatively-warm lake or reservoir. The problem, involving unstable conditions over the water, is of some practical importance; it is frequently encountered, for example, over natural deep lakes in fall and early winter, or over water bodies subject to thermal pollution. For the sake of simplicity the roughness was taken to be constant and the same over the land as over the water, so that the advection of momentum was assumed to be the result solely of the step-change in surface temperature and humidity at the leading edge. It was also assumed implicity that $z_{0v} = z_{0h} = z_0 = \hat{z}_0$ where $\hat{z}_0$ is an effective roughness; in the light of the general similarity of the values of Cd, Ce and Ch for water, given by (5.11) and in Table 5.3, this is probably adequate to model 'average' lake surface conditions. The mathematical model was formulated in terms of (7.5) through (7.8) with the turbulent diffusion assumption of (7.15), (7.16), (7.17) and (7.21) through (7.24); the ϕ-functions were taken as (4.45) but with the local L_a of (7.26). Values of $\lambda = 5 \times 10^4$ and 7×10^4 cm were used in (7.24). The boundary conditions imposed were specified from the following considerations: (i) The incoming air is neutral and it has a known humidity profile $q_a(z)$, resulting from equilibrium dictated by evapotranspiration from the upwind land surface. This neutral wind profile upwind is a log-linear function resulting from integration of (7.16) with (7.22) and (7.24). (ii) High above the water surface and well above the internal boundary layer, conditions are the same as upwind, and unaffected by the water surface. (iii) At the water surface the temperature is known and uniform; the specific humidity is saturated and a function of this temperature. These conditions are the same as those for Sutton's problem, (7.44), except that here a log-linear profile is used for the incoming air stream instead of a power function.

In the non-dimensionalization of this formulation, two dimensionless parameters appear in the stability length, L_a, namely

$$A_* = -\frac{(\bar{T}_s - \bar{T}_{as})}{\bar{T}_{as}} \frac{kg\hat{z}_0}{u_{*a}^2}$$

and (7.86)

$$B_* = -0.61(\bar{q}_s - \bar{q}_{as}) \frac{kg\hat{z}_0}{u_{*a}^2}$$

where u_{*a} is the upwind friction velocity and all the other variables are as defined in Section 7.2a. These stability parameters must be specified to solve the problem, and they represent a measure of the discontinuity at the leading edge.

In the present context, the result of main interest is the mean rate of evaporation. The numerical calculations were carried out for the case of neutral and dry conditions upstream. The results were obtained in terms of a dimensionless rate of evaporation averaged over the fetch x_f

$$E_* = \frac{\bar{E} - E_a}{\rho u_{*a}(\bar{q}_s - \bar{q}_{as})} \tag{7.87}$$

where $\bar{q}_s$ is the specific humidity at the water surface, $\bar{q}_{as}$ the specific humidty at the upwind surface, E_a the evapotranspiration from the upwind surface, and $\bar{E}$ the mean evaporation from the water surface. For practical purposes, one can define a water vapor transfer coefficient in terms of upwind reference values for wind at some level $\bar{u}_{ar}$ and humidity at the surface, namely (cf., (4.114))

$$\overline{\mathrm{Ce}}_r = \frac{\bar{E} - E_a}{\rho \bar{u}_{ar}(\bar{q}_s - \bar{q}_{as})} \tag{7.88}$$

where r refers to the reference level z_r of the wind measurement. This mass transfer coefficient can be expressed in terms of E_* and the upwind drag coefficient Cd_{ar} as

$$\overline{\mathrm{Ce}}_r = \mathrm{Cd}_{ar}^{1/2} E_*. \tag{7.89}$$

Figures 7.12 and 7.13 show some calculated results of E_* versus dimensionless fetch $(x_f/\hat{z}_0)$ for various values of A_* and B_*. These results show that the effect of A_* and B_* may be considerable, especially the former. Nevertheless, Figure 7.13 illustrates how, even for isothermal conditions, i.e., $A_* = 0$, the humidity discontinuity at $x = 0$ may cause instability in the air and local advection of momentum and thus affect the mass transfer coefficient. When $A_* = B_* = 0$ the velocity profile remains

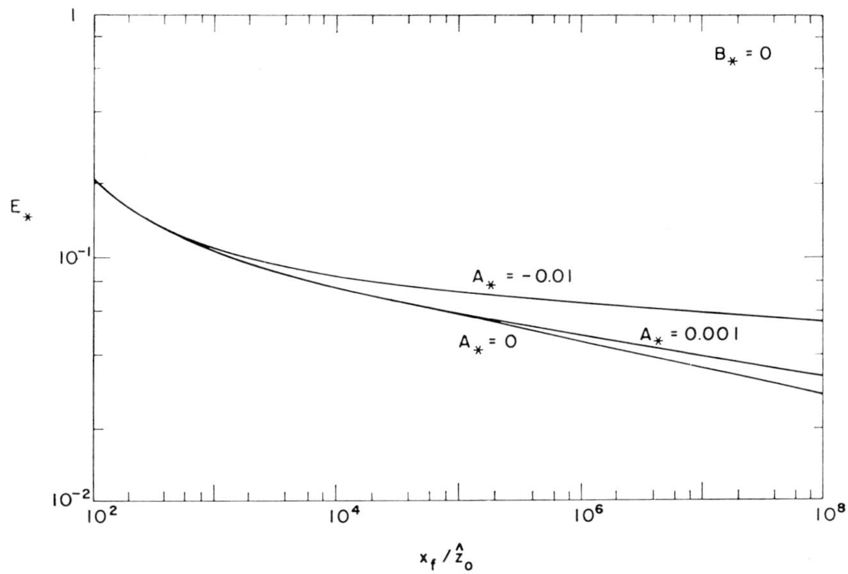

Fig. 7.12. Dimensionless evaporation rate E_* defined in (7.87) versus dimensionless fetch $(x_f/\hat{z}_0)$ under unstable conditions for different values of the stability parameter A_* and constant B_* (adapted from Weisman and Brutsaert, 1973).

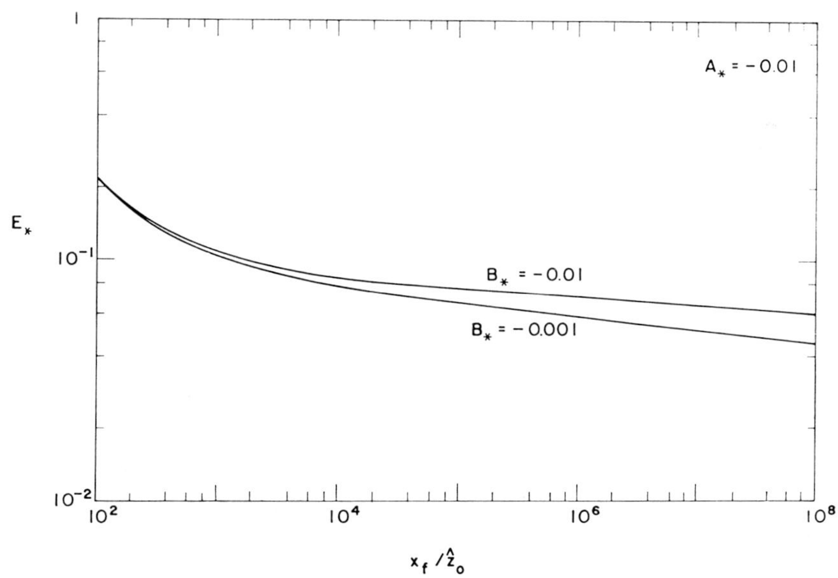

Fig. 7.13. Same as Figure 7.12 for two values of B_*, but constant A_*.

neutral and unchanged downwind. This case is the Sutton problem, except that here the log-linear wind profile is used instead of the power law. The related Sutton problem with the logarithmic profile was treated numerically by Yeh and Brutsaert (1971a). The value of m in (7.57), or (7.54) with (7.55), required to make the power of x_f coincide with the slope of the curve for $A_* = B_* = 0$ and large x_f in Figure 7.12 is $m = \frac{1}{7}$. Again, this is exactly the generally-accepted value for (4.18) under neutral conditions. The numerical solution for large fetches is also in agreement with Harbeck's (1962) empirical (7.60); this agreement requires very slightly unstable conditions with A_* and B_* between 0 and -0.001, and an effective roughness around 0.04 cm; this is not very different from the value $\hat{z}_0 = 0.02$ cm obtained earlier in the comparison between Harbeck's (1962) formula and (7.57).

It is possible, on the basis of the results of the eddy diffusion approach, to obtain a first approximation of the effect of the temperature and humidity discontinuity on lake evaporation. As can be seen in Figures 7.12 and 7.13, for fetches x_f larger than a few meters, E_* appears to be a simple power function of fetch, namely

$$E_* = a(x_f/\hat{z}_0)^{-b} \tag{7.90}$$

where a and b are constants for given A_* and B_*; some values are given in Tables 7.1 and 7.2. For large lakes, as suggested by the above comparision with Harbeck's (7.60) and from Table 5.3, it is probably reasonable to use (7.90) with $\hat{z}_0 = 0.03$ cm.

TABLE 7.1
Some values of the coefficient a in (7.90)

	$-A_*$				
$-B_*$	0.1	0.05	0.01	0.001	0
0.01			0.121	0.112	0.122
0.003.	0.150	0.140	0.120	0.135	0.166
0.001			0.112	0.152	0.167
0			0.120	0.167	0.210

TABLE 7.2
Some values of power b in (7.90)

	$-A_*$				
$-B_*$	0.1	0.05	0.01	0.001	0
0.01			0.036	0.042	0.045
0.003	0.025	0.034	0.046	0.081	0.089
0.001			0.042	0.082	0.093
0			0.045	0.093	0.112

An interesting feature of the eddy diffusion approach outlined here, is that the results are not very sensitive to the exact form of the adopted Monin–Obukhov ϕ_s functions or the exactness of Reynolds's analogy; the results were similarly found to be insensitive to the value of von Kármán's constant. Thus, within the range of values in the literature, changes in the assumed values of a_v, a_h, β_{sv}, β_{sm}, β_{sh} in (4.42) through (4.44), or changing the power in (4.43) from $-(\frac{1}{4})$ to $-(\frac{1}{2})$, did not affect the calculated results very much. This insensitivity probably explains why not very

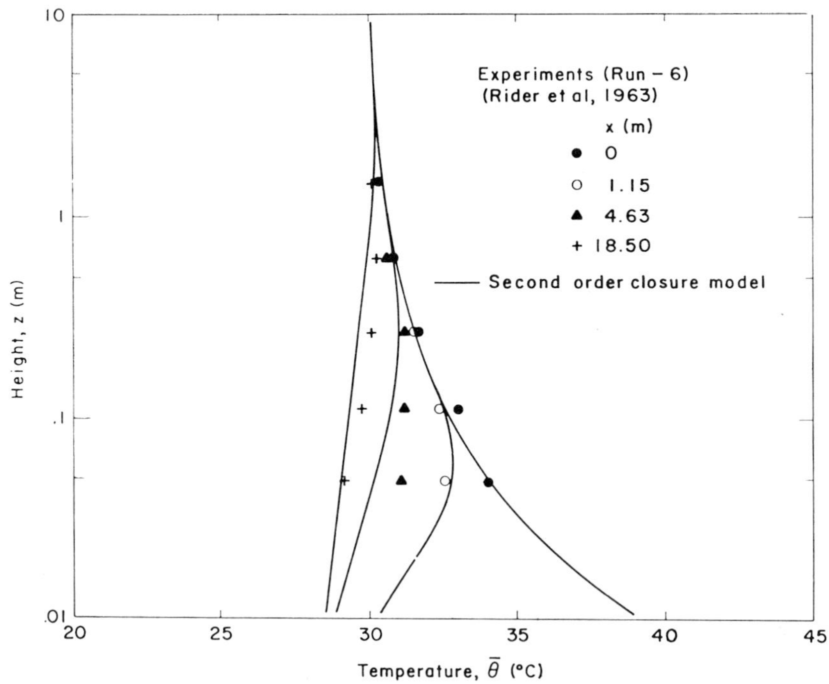

Fig. 7.14. Comparison between experimental mean temperature of Rider *et al.* (1963) and profiles calculated by means of higher order closure model (adapted from Rao *et al.*, 1974b).

dissimilar atmospheric conditions have yielded widely varying experimental ϕ_s-functions. In the formulation of the problem it was also found that it is essential to keep the eddy diffusivities bounded at the value corresponding to the top of the surface sublayer, i.e., about 50 to 100 m above the surface. Blackadar's (1962) mixing length (7.24), provided a suitable way of accomplishing this.

Local Advection by Higher-Order Closure Approach

Simultaneous local advection of water vapor, momentum and sensible heat resulting from an abrupt change in surface conditions has been studied by Rao *et al.* (1974b) by means of a higher-order closure model. As outlined earlier, the governing equations of their model are (7.5) through (7.8) for the mean field, and in addition equations for each of the second-order moments $\overline{(u')^2}$, $\overline{(v')^2}$, $\overline{(w')^2}$, $\overline{(u'w')}$, $\overline{(\theta'w')}$, $\overline{(\theta'u')}$, $\overline{(q'w')}$, $\overline{q'u'}$, $\overline{(\theta')^2}$, $\overline{(q')^2}$, $\overline{(q'\theta')}$ and for the energy dissipation rate ε. The third moments are approximated by gradient transport equations like (7.27) and (7.28) with the K's given by (7.34), and the molecular destruction rates of $\overline{(\theta')^2}$, $\overline{(q')^2}$ and $\overline{(q'\theta')}$ by related similarity expressions. The lower boundary conditions, both upwind and downwind from the discontinuity at $x = 0$, were based on simplified equilibrium flux-profile relations. Thus, it was assumed that at $z = z_0$ the mean variables were $\bar{u} = \bar{w} = 0$, $\bar{\theta} = \bar{\theta}_0$ $\bar{q} = r_0 q^*(\theta_0)$ in which r_0 is the surface relative humidity and q^* the saturated specific humidity; the surface fluxes were assumed to be related to the gradients by (4.5),

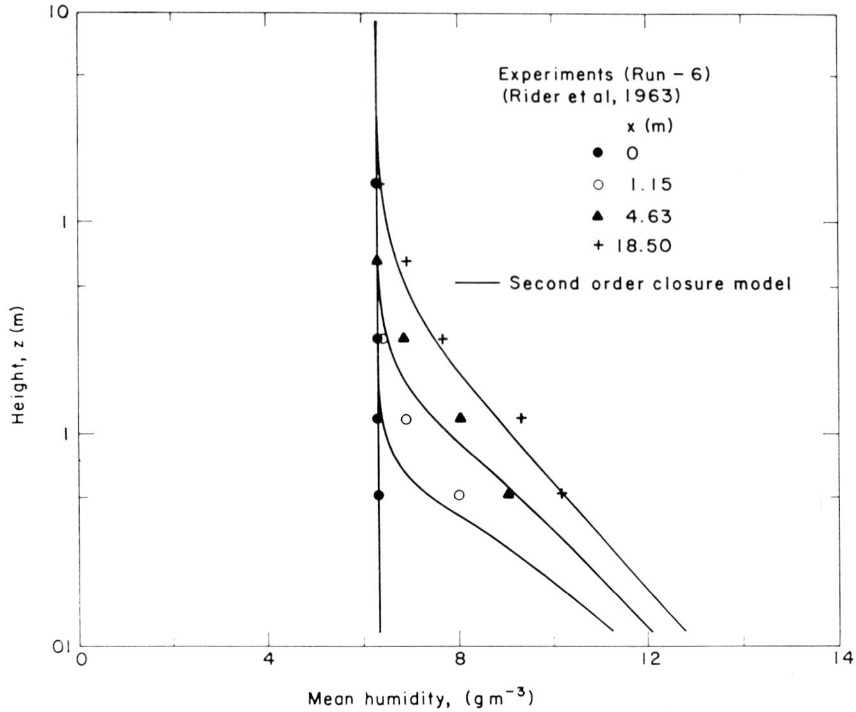

Fig. 7.15. Same as Figure 7.14 for the mean humidity profiles.

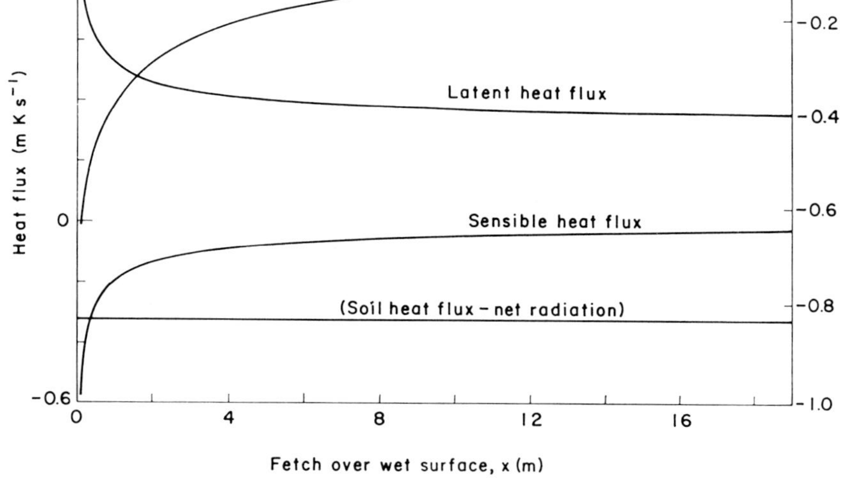

Fig. 7.16. Downwind variation of the energy balance (in units of $\overline{w'\theta'}$) and Bowen ratio for the experimental conditions of Rider *et al.* (1963), as calculated by means of a higher-order turbulence closure model (adapted from Rao *et al.*, 1974b).

(4.11) and (4.15) in which $(z - d_0)$ is replaced by z_0; the other second-order moments (i.e., beside the fluxes) at z_0 were expressed in terms of the surface fluxes E, u^* and H in accordance with available turbulence measurements under equilibrium conditions.

The model was applied to the experimental set-up of Rider *et al.* (1963), in which air moves from an extensive dry, smooth ($z_0 = 0.002$ cm) surface to a grassy ($z_0 = 0.14$ cm) well-irrigated area. The upwind evaporation was assumed to be zero, and the downwind turbulent fluxes at $z = z_0$ were assumed to be related such that the energy and moisture budgets at the surface are satisfied. The calculated profiles of $\bar{q}$ and $\bar{\theta}$ for different x were found to be in good agreement with the data of Rider *et al.* (1963) as shown in Figures 7.14 and 7.15, provided the relative humidity at the wet surface was taken at 60 percent. Since no other experimental data were available, the model was used mainly to determine various advective effects illustrated in Figure 7.16; in addition, the sensitivity of the calculated results was studied for different assumptions in the boundary conditions. Rao *et al.* (1974b) felt that the results of the calculations should not be considered as definitive, but instead, as examples of the type of output that can be generated with higher-order closure models.

In the model of Rao *et al.* (1974b) the higher-order description of the turbulent transport phenomena in the lower atmosphere represents a considerable improvement over the earlier approaches. However, in the context of evaporation and evapotranspiration, it seems that a large degree of uncertainty is still involved in the proper formulation of the lower boundary conditions for surfaces covered with vegetation. In the model it was assumed that the roughness z_0 is the same for the mean wind, temperature and specific humidity. For large water surfaces the assumption that $z_{0v} = z_{0h} = z_0$ (cf. Table 5.3) is probably not very restrictive; but for a grassy surface with $z_0 = 0.14$ cm like in the experiments of Rider *et al.* (1963), this assumption may result in considerable error (cf. Figure 4.24). This may have been one of the reasons why the calculated results were rather sensitive to the change in roughness. Moreover, in the calculations the adopted surface relative humidity r_0 apparently had to be assumed *a priori*, whereas in practical problems it should be left an unknown and a part of the solution. All this suggests that an optimal formulation of the boundary conditions at or near the surface $z = 0$, will require further research.

Nevertheless, higher-order turbulence closure models, such as that of Rao *et al.* (1974b) appear to have a great potential to solve problems involving local advection. Although for certain practical purposes their computational complexity may be thought to be excessive, they should be very useful to guide experimental studies or to check the validity of simpler approaches.

CHAPTER 8

Methods Based on Turbulence Measurements

8.1. DIRECT OR EDDY-CORRELATION METHOD

Equations for the means, such as (3.44), (3.62) and (3.67) constitute the basis for the eddy-correlation method. This method consists of determining the turbulent fluxes of water vapor, momentum, sensible heat, or any other admixture from covariances. Hence, over a uniform surface under steady conditions, the surface fluxes E, H and u_* can be obtained from (3.74), (3.75) and (3.76), respectively. In practice, the flux E is determined by measuring the fluctuations w' and q' and then computing the cross-correlation over a suitable averaging period, and similarly for u_* and H. Equations (3.74) and (3.75) for scalars were first applied by Dyer (1961) and Swinbank (1951), respectively.

a. Instruments

The cross-correlation between the fluctuations w' and q', u', θ' or any other c', can be made by electronic analog or digital computation consisting of a multiplication and an averaging process. Many methods have been applied to measure these fluctuations. For the velocity fluctuations hot wire-type anemometers, which were probably used first, are available commercially. Other instruments for this purpose are the sonic anemometer either with continuous or with pulsed waves (e.g., Kaimal *et al.*, 1968) the pressure-sphere anemometer (e.g., Goltz *et al.*, 1970) the yaw-sphere anemometer (e.g., Yap *et al.*, 1974), various types of propeller anemometers (e.g., Hicks, 1972b), windvanes or trivanes (e.g., Wieringa, 1972) and the thrust anemometer (e.g., Smith, 1974).

The methods for measuring temperature fluctuations are probably better than those of any other admixture because of the availability of different types of thermocouples, thermistors and resistance thermometers, the latter usually with a platinum wire; these sensors have a relatively high accuracy and adequate response characteristics.

Initial attempts to measure humidity fluctuations (Dyer, 1961; see also Hicks, 1970) were made by means of dry-wet bulb thermometer systems. This instrument has, subsequently, become superseded by others with faster response characteristics. An often used instrument is the Lyman-alpha humidiometer with various calibration methods (e.g., Miyake and McBean, 1970; Smith, 1974; Buck, 1976). Some other instruments which have been tested are the dew-point hygrometer (e.g., Miyake and McBean, 1970), the microwave refractometer-hygrometer (e.g., McGavin, 1971;

Martin, 1971), the sensitized quartz-crystal-oscillator hygrometer (e.g., Hicks and Goodman, 1971) and the single beam infrared hygrometer (e.g., Hyson and Hicks, 1975; Raupach, 1978). The reader is referred to these papers for further details.

The measurement of carbon dioxide fluctuations above vegetation has been carried out by Desjardins and Lemon (1974), by means of a modified infrared CO_2 analyzer.

b. Requirements on Instrumentation

The theoretical basis of the eddy-correlation method is straightforward and, in the past one to two decades, considerable progress has been made in its application. Nevertheless, the requirements on the instrumentation are quite stringent and some inherent difficulties remain. These difficulties result mainly from the following requirements: (i) the sensor must have a sufficiently fast response time; (ii) the averaging period must be sufficiently long and (iii) the orientation and placement of the velocity sensors must be precise, if the correlations involving one or more velocity components are to be accurate.

Sensor response

If the response of the sensing instrument is too slow, the higher frequency fluctuations cannot be measured, so that part of the cross-correlation goes undetected and the flux is underestimated. Spectral measurements (e.g., Miyake and McBean, 1970; McGavin *et al.*, 1971; Smith 1974; Wesely and Hicks, 1975) have indicated that the band width required to measure the whole spectrum or cospectrum involving velocity, temperature and specific humidity fluctuations should be at least

$$10^{-3} \leq \frac{nz}{\bar{u}} \leq 5 \text{ to } 10 \tag{8.1}$$

where z is the elevation, $\bar{u}$ the mean velocity and n the frequency in Hz. Slow response characteristics of a sensor can be remedied by placing it higher above the ground surface where the turbulence frequencies are generally lower. However, the higher the sensors, the more difficult it becomes to satisfy the condition of a sufficiently long fetch needed for a constant flux layer. Thus, with fetch limitations, some sensors are less appropriate than others. For example, it is now agreed that mechanical velocity sensors such as the propeller anemometers with a response time of 0.3 s may be of marginal performance to determine an eddy correlation (e.g., Hicks, 1972b; Tsvang *et al.*, 1973). Similarly, among the humidity measuring devices, the dry-wet bulb thermometer system with a response time of 0.3 s appears to be inferior compared to some others listed above (e.g., Hicks and Goodman, 1971). The same probably holds for the dewpoint hygrometer with an upper response limit of 0.1 Hz (Miyake and McBean, 1970).

Averaging Period

In turbulent transport formulations such as (3.44), (3.62) and (3.67) the mean is separated from the short time fluctuations; however, the mean is still subject to longer-term unsteadiness or trend. Consequently, the averaging time should be as short as possible to guarantee a stationary time series without the effect of any trend, but it should also be long enough to cover even the slowest fluctuations of the turbulent spectrum. The lower limit given in (8.1) shows that for a typical elevation of 5 m and a mean wind of 5 m s^{-1} the frequency should be 10^{-3} Hz; this suggests an averaging period of 34 min, since for a given averaging time, all fluctuations with a period of half that value are eliminated. The present consensus is that the averaging period should be at least 15 min and, perhaps, even as long as one hour (Tsvang *et al.*, 1973) to ensure reliable results. In recent applications of the eddy-correlation method the adopted averaging periods appear to have varied over approximately this range with many of them around 30 min. Under some conditions it may be desirable to remove the trend.

Orientation of Velocity Sensors

When the vertical and horizontal velocity sensors are not oriented properly, part of the horizontal component u' is recorded as vertical w'. (Over a sloping but plane surface the velocity components should be taken as parallel and normal.) The resulting error should clearly be largest for the turbulent shear stress estimates $\overline{u'w'}$, but it is still appreciable for the eddy fluxes of scalars as well. As an illustration, consider a system in which the vertical velocity sensor is tilted over an angle α counter-clockwise. The vertical vapor flux appears then to be

$$\overline{w_a'q'} = \cos(\alpha)\,\overline{w'q'} - \sin(\alpha)\,\overline{u'q'} \tag{8.2}$$

whereas, in reality, it is $\overline{w'q'}$; similar expressions can be written for any other scalar. The fractional error is

$$\varepsilon_t = (\cos\alpha - 1) - \sin\alpha\,(\overline{u'q'}/\overline{w'q'}). \tag{8.3}$$

Since α is small and the ratio of the covariances of the order of -2, under slightly unstable conditions, Wesely and Hicks (1975) approximated (8.3) as $\varepsilon_t = 2\sin\alpha$, or 3.5 percent per degree tilt; they also pointed out that when under neutral conditions the ratio decreases to approximately -4, (8.3) yields an error of about 7 percent per degree, whereas under unstable conditions when the ratio tends to -1, (8.3) yields about 1.7 percent per degree. These estimates are roughly in agreement with published experimental errors. For the heat flux, Wieringa (1972) estimated that one degree of instrument misalignment produced 4 ± 2 percent; Yap *et al.* (1974) found that the error was 5 percent per degree of tilt under unstable conditions and that it may be as large as 11 percent under stable conditions. For CO_2 flux Desjardins and Lemon (1974) found errors of a few, but less than 10, percent for 1°, and as large as 25 percent for 3°.

The best method to avoid these errors is to align the instruments carefully. Kaimal and Haugen (1971) recommended that, in practice, the alignment should be accurate to the nearest 0.1°. Dyer *et al.* (1970) proposed to reduce the tilting error by filtering; but it has been suggested that this technique may reduce the measured flux (Kaimal and Haugen, 1971; Wieringa, 1972). For measurements aboard ship Mitsuta and Fujitani (1974) presented a method to correct for the motion of the ship.

8.2. THE DISSIPATION METHOD

This method, also referred to as variance budget method, requires the measurement of the turbulent variables, but it still involves some similarity and other assumptions in its formulation. Because the instrumentation is expensive and complicated, and because some assumptions have not been validated conclusively, the method is still in a developmental stage; nevertheless, recent experience suggests that it may eventually become a practical tool.

The fluxes, which are the covariances, are not determined directly, but they are derived from turbulence measurements through different statistical quantities. Thus, in principle, the method can be used whenever the measuring and data processing equipment are available for the eddy-correlation method. Unlike the eddy-correlation method, however, the dissipation method does not suffer from the stringent requirement of precise orientation and alignment of the sensors. This is probably its main advantage.

The dissipation method is based on the budget equations for the variances. Under steady conditions over a uniform surface these are (3.80) for the specific humidity variance, (3.81) for the turbulent kinetic energy, and an analogous equation for the mean square temperature fluctuations; if needed, a budget equation for the variance of any other scalar can also be added. For the purpose of this method in the surface sublayer by virtue of (4.26), (4.27) and (4.28) these equations can be rewritten as

$$E = \rho[k(z - d_0)u_*\varepsilon_q/\phi_{sv}]^{1/2}, \tag{8.4}$$

$$u_* = [k(z - d_0)\varepsilon/R_m]^{1/3}, \tag{8.5}$$

$$H = \rho c_p[k(z - d_0)u_*\varepsilon_\theta/\phi_{sh}]^{1/2} \tag{8.6}$$

and for any scalar admixture

$$F = [k(z - d_0)u_*\varepsilon_c/\phi_{sc}]^{1/2} \tag{8.7}$$

where ε_θ, ε_q and ε_c are the rates at which molecular diffusion dissipates the quantities $(\overline{\theta'^2}/2)$, $(\overline{q'^2}/2)$ and $(\overline{c'^2}/2)$, respectively, and ε is the rate of dissipation of turbulent kinetic energy $[(\overline{u'^2} + \overline{v'^2} + \overline{w'^2})/2]$. Equations (8.4), (8.6) and (8.7) involve the common assumption that the vertical divergences of the vertical fluxes of a scalar variance, e.g., in the case of specific humidity $\partial(\overline{w'q'^2})/\partial z$ [cf., (3.80)] are negligible; experimental studies by Wyngaard and Coté (1971) for θ' and by Leavitt and Paulson (1975) for q', suggest that these divergence terms are relatively small. In (8.5) the form of the term R_m depends on the assumptions which are made to simplify the turbulent kinetic energy Equation (3.81). If, as suggested by Busch and Panofsky (1968), the vertical divergences of the vertical flux of turbulent kinetic energy and of the pressure transport are simply neglected, this term can be written as

$$R_m = \phi_{sm} - \left(\frac{z - d_0}{L}\right). \tag{8.8}$$

A second form for R_m, which has been used in the dissipation method, is based on the empirical results of Wyngaard and Coté (1971), viz. for $-2 < \zeta \leq 0$

$$R_m = (1 + 0.5|(z - d_0)/L|^{2/3})^{3/2}. \tag{8.9}$$

When the dissipation terms ε_q, ε and ε_θ are known, (8.4) through (8.6) are three equations with three implicit (cf. (4.25)) unknowns, the fluxes E, u_* and H. When the flux of another scalar is required, in addition (8.7) must be included with a fourth unknown, viz. F. The solution of (8.4) through (8.6) can be carried out by iteration. For example, initially neutral conditions are assumed so that $\phi_{sv} = \phi_{sm} = \phi_{sh} = 0$; this provides a first estimate for the fluxes, from which L can be obtained with (4.25). The new values of the ϕ-functions provide a second estimate of the fluxes, and so on. Hicks and Dyer (1972) and Champagne *et al.* (1977) have constructed auxiliary functions to facilitate the solution of this implicit system.

Methods of determing ε_θ, ε_q and ε are discussed in the following two sections.

a. The Direct Variance Dissipation Method

The dissipation by viscosity and molecular diffusion takes place in eddies of the smallest scales, at which the turbulence is generally assumed to be isotropic. This allows (Taylor, 1935) the approximation of (3.82) as follows

$$\varepsilon = 15\,\nu\overline{\left(\frac{\partial u'}{\partial x}\right)^2}. \tag{8.10}$$

If, in addition, Taylor's (1938) hypothesis $t = x/\bar{u}$, relating convective with temporal accelerations, is used, the dissipation terms of (3.80) [cf. (3.47)] and (8.10) can be written as

$$\varepsilon_q = \frac{3\kappa_v}{(\bar{u})^2}\overline{\left(\frac{\partial q'}{\partial t}\right)^2}, \tag{8.11}$$

$$\varepsilon = \frac{15\nu}{(\bar{u})^2}\overline{\left(\frac{\partial u'}{\partial t}\right)^2}, \tag{8.12}$$

$$\varepsilon_\theta = \frac{3\kappa_h}{(\bar{u})^2}\overline{\left(\frac{\partial \theta'}{\partial t}\right)^2}. \tag{8.13}$$

Thus experimentally, the problem consists of measuring the mean wind speed and of calculating the variance of the time derivatives of measured θ', q' and u'. According to Champagne *et al.* (1977), these measurements require spatial resolution of the sensors down to about Kolmogorov's microscale $(\nu^3/\varepsilon)^{1/4}$ which is of the order of 1 mm in the atmosphere, and also a sufficiently high frequency (about 10^3 Hz) response and low noise level of the equipment. They also pointed out that the dissipation rates obtained with Taylor's hypothesis are overestimated in high intensity turbulence; thus, corrections should be applied for which they presented expressions, in accordance with the work of Heskestad (1965) and Lumley (1965). This method of determining ε and ε_θ has also been investigated by Boston and Burling (1972) and Stegen *et al.* (1973); but apparently ε_q has not yet been determined this way.

b. The Inertial Dissipation (or Spectral Density) Method

This second method, for which the requirements on the instrumentation are not as stringent as for the direct method, is based on Kolmogorov's hypothesis (e.g., Ten-

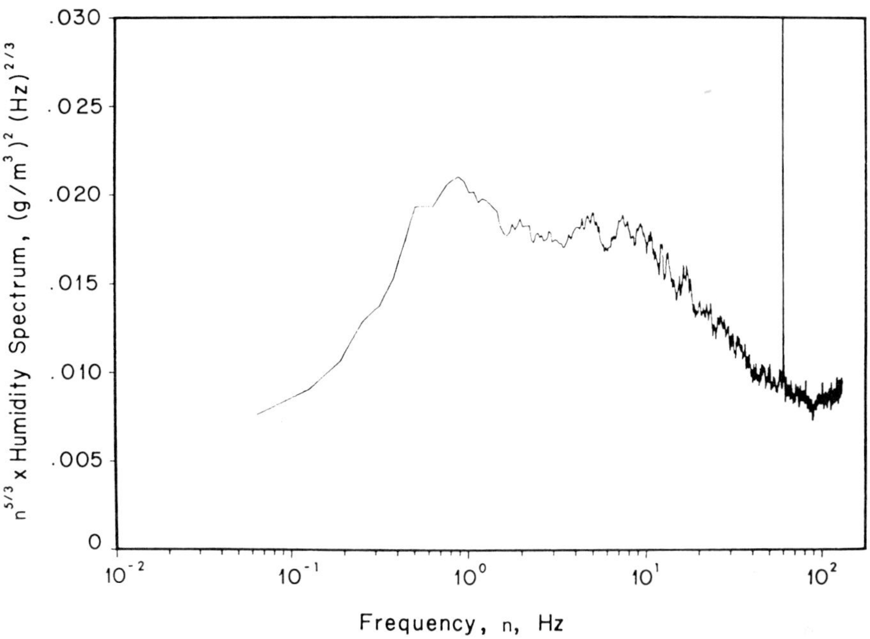

Fig. 8.1. Semi-logarithmic plot of the (5/3) moment of the humidity spectrum measured above plowed farmland in Minnesota (September, 1973). The inertial subrange is manifested by the horizontal plateau between 1 and 8 Hz. The roll-off for higher frequencies is due to the response characteristics of the Lyman-alpha humidiometer; the spike at 60 Hz was caused by line noise (from Champagne *et al.*, 1977).

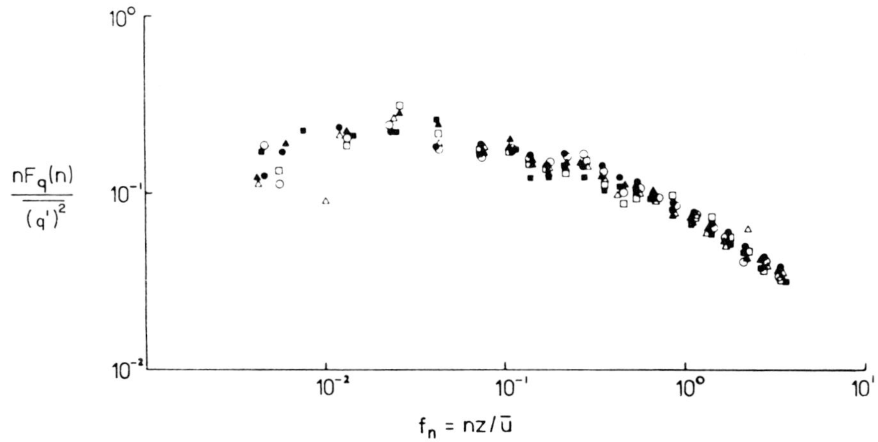

Fig. 8.2. Illustration of the normalized spectrum of specific humidity meaured with an infrared hygrometer. The inertial subrange is manifested by the segment with the slope of $(-2/3)$. The different points represent stability classes over 6 intervals between $-\zeta = 0.07$ to 1 (from Raupach, 1978).

nekes and Lumley, 1972), that over a range of wave numbers κ, referred to as the inertial subrange, the (one-dimensional) spectra can be described as follows

$$F_q(\kappa) = \beta_q \varepsilon^{-1/3} \varepsilon_q \kappa^{-5/3}, \tag{8.14}$$

$$F_{uu}(\kappa) = \alpha_u \varepsilon^{2/3} \kappa^{-5/3}, \tag{8.15}$$

$$F_\theta(\kappa) = \beta_\theta \varepsilon^{-1/3} \varepsilon_\theta \kappa^{-5/3} \tag{8.16}$$

where β_q, α_u and β_θ are the Kolmogorov constants; κ is the wave number, usually in radians per unit length, which may be related to the cyclic frequency, n, by means of Taylor's (1938) assumption so that $\kappa = (2\pi n/\bar{u})$ or, to the reduced frequency, $f_n = (nz/\bar{u})$ by $\kappa = (2\pi f_n/z)$; f_n may be interpreted as the ratio of height over the wave length. The spectra $F(\kappa)$ are defined such that their integrals over the whole range of κ yield the variances of q' (i.e., $2\bar{m}$ in (3.46)), u', and θ' respectively. Examples of specific humidity spectra are shown in Figures 8.1 and 8.2.

Numerous determinations have been made of the Kolmogorov constants, but their values are still somewhat uncertain. The constant α_u is often taken to be typically around 0.55. For example, McBean *et al.* (1971) found 0.54 ± 0.09, Paquin and Pond (1971) 0.57 ± 0.10, and Hicks and Dyer (1972) 0.54 ± 0.03. There are some indications, however, that 0.55 is too large and that α_u lies closer to 0.50 (e.g., Frenzen, 1977; Champagne *et al.*, 1977). The consensus on β_q and β_θ is that both lie around 0.80. For example, in the case of β_θ, Paquin and Pond (1971) found 0.83 ± 0.13, Wyngaard and Coté (1971) 0.79 ± 0.10, Hicks and Dyer (1972) 0.71 ± 0.04 and Champagne *et al.* (1977) 0.82 ± 0.04. In the case of β_q, Paquin and Pond (1971) found 0.80 ± 0.17, Smedman-Högström (1973) 0.58 ± 0.2, Leavitt (1975) 0.81 ± 0.31 and Raupach (1978) 0.88 ± 0.26.

Thus the dissipations ε, ε_θ and ε_q can be obtained by applying (8.14) through (8.16) for one or more values of κ with the measured spectral densities $F(\kappa)$ in the inertial subrange, for example, 1 Hz in Figure 8.1. A perhaps more reliable technique, since it gives an average, consists of integrating the spectra over a range of wave numbers, say $\kappa_2 - \kappa_1$, in the inertial subrange, to obtain the variance contribution over that range; this variance contribution may be calculated from the measurements. A different, but equivalent procedure to determine the dissipation terms makes use of the structure functions in the inertial subrange, involving spatial separation; Paquin and Pond (1971) used this approach inversely to determine the Kolmogorov constants.

The inertial dissipation method was first suggested by Deacon (1959) and Taylor (1961) appears to have first shown its feasibility.

In summary, it is seen that presently there still is no consensus on how the dissipation method – both in its direct and in its inertial version – should be applied to obtain accurate results. Undoubtedly, more research is needed before the method will be fully operational.

CHAPTER 9

Methods Based on Measurements of Mean Profiles

Over a uniform surface with an adequate fetch, these methods are based directly on the similarity theories for the atmospheric boundary layer treated in Chapter 4. In the present chapter, first, a brief account is given of the application of profile expressions; second, a summary is given of different forms and applications of some related bulk transfer coefficients.

9.1. MEAN PROFILE METHOD WITH SIMILARITY FORMULATIONS

The available flux-profile functions for the boundary layer given in Chapter 4 allow the calculation of surface fluxes from measurements of mean concentration at two or more levels. The word 'mean', which appears also in the title of this chapter, refers to the fact that the $\bar{q}$, $\bar{u}$ and $\bar{\theta}$ data used are obtained by averaging over a certain time period. In the light of the findings discussed in Section 8.1b, averaging times of about 30–60 min are appropriate. Although the gradient profiles can be used, it is preferable to use the integral profiles. The functional form of the profile functions depends on the level above the surface, i.e., the sublayer, where the measurements are made. Figure 3.1 shows the different sublayers of the ABL.

a. Measurements in the Surface Sublayer

The surface sublayer is the fully turbulent layer, which is well above the surface roughness elements – say at least three times their height h_0 – but lower than an elevation of the order of 50–100 m. The profiles in this sublayer are given by (4.33) through (4.35), or if the lowest measurement is at the surface, by (4.33′) through (4.35′). The Ψ-functions appearing in these equations are specified in Section 4.2b. Note that the subscripts 1 and 2 in (4.33) through (4.35) refer to a lower and upper level, at which the respective measurements of $\bar{q}$, $\bar{u}$ and $\bar{\theta}$ are made; these elevations need not be the same in all three equations. The same holds, of course, for (4.33′) through (4.35′). Note also that, in general, the surface flux F of any other scalar admixture (beside water vapor or sensible heat) can be calculated by a similar equation. Thus, one has

$$F = a_c k u_* \rho(\bar{c}_1 - \bar{c}_2)\left[\ln\left(\frac{z_2 - d_0}{z_1 - d_0}\right) - \Psi_{sc}(\zeta_2) + \Psi_{sc}(\zeta_1)\right]^{-1} \tag{9.1}$$

where a_c and Ψ_{sc} are the analogs of a_v and Ψ_{sv}, respectively, and c is the mean concentration; it is probably a good approximation to put $a_c = a_v$ and $\Psi_{sc} = \Psi_{sv}$. Obviously, when the lowest measurement is at the surface, the flux is

$$F = a_c k u_* \rho (\bar{c}_s - \bar{c}) \left[\left(\frac{z - d_0}{z_{0c}} \right) - \Psi_{sc}(\zeta) \right]^{-1} \tag{9.1'}$$

where $\bar{c}_s$ and $\bar{c}$ are the concentrations at the surface and at the level z, respectively, and z_{0c} is the scalar roughness for the admixture (cf. Chapter 5).

The flux of any admixture, be it E, u_* or H, cannot be calculated simply from measurements of the corresponding concentration only, because each of the equations (4.33) through (4.35) contains also the momentum flux u_*, and the Obukhov length L, defined in (4.25) which, in turn, contains the three fluxes u_*, H and E. In practice, there are two alternative methods of closing a flux determination problem.

Calculation With Known Mean-Profile Data

The first consists of the simultaneous solution of (4.33) through (4.35) [or (4.33') through (4.35')] for the three unknowns u_*, H, E with known measurements at least at two levels of mean specific humidity, mean wind speed and mean temperature. This numerical problem may be solved in different ways. One simple way is by iteration as follows; initially it is assumed that $L = \infty$ and first estimates of E, u_* and H are obtained with omission of the Ψ_s-terms in (4.33) through (4.35); this allows a first estimate of L with (4.25), which is then used in (4.33) through (4.35) to obtain second estimates of E, u_* and H; these permit a second estimate of L and so on; the iteration is stopped when successive estimates cease to change appreciably. When measurements of $\bar{q}$, $\bar{u}$ and $\bar{T}$ are available at more than two elevations, at each iteration E, u_* and H can be obtained as the slopes from (4.33) through (4.35) by least squares regression through the origin.

A somewhat different procedure for simultaneous solution of (4.33) through (4.35) involves the use of a bulk Richardson number. If $\bar{q}$, $\bar{u}$ and $\bar{T}$ are measured at the same two levels z_1 and z_2, this can be defined as [cf., (4.39)]

$$Ri_B = -\frac{g(z_2 - z_1)[(\bar{\theta}_1 - \bar{\theta}_2) + 0.61 T_a(\bar{q}_1 - \bar{q}_2)]}{T_a(\bar{u}_2 - \bar{u}_1)^2}. \tag{9.2}$$

If it is further assumed that $a_v = a_h$ and $\Psi_{sv} = \Psi_{sh}$, (9.2) can be written for the surface sublayer as

$$Ri_{Bs} = \frac{(\zeta_2 - \zeta_1)[\ln(\zeta_2/\zeta_1) - \Psi_{sh}(\zeta_2) + \Psi_{sh}(\zeta_1)]}{a_h[\ln(\zeta_2/\zeta_1) - \Psi_{sm}(\zeta_2) + \Psi_{sm}(\zeta_1)]^2}. \tag{9.3}$$

For a given experimental set-up, z_2, z_1 and d_0 are known and fixed, so that the right hand side of (9.3) is a function of L only. This function can be calculated by means of the relationships presented in Section 4.2b. Inversion of this function produces $L = L(Ri_{Bs})$, that is L in terms of the profile measurements $\bar{q}_1$, $\bar{q}_2$, $\bar{u}_1$, $\bar{u}_2$, $\bar{T}_1$ and $\bar{T}_2$. The inversion of (9.3) can be carried out graphically or by fitting the obtained curve to a suitable mathematical expression. Once L is known from the inverted (9.3), u_* can be calculated directly by means of (4.34), after which (4.33) and (4.35) produce E and H. When the lowest measurements are taken at the surface, the procedure is essentially the same with (4.33') through (4.35') but the bulk Richardson number can be written as

$$Ri_{Bs} = \frac{\zeta\{\ln[(z - d_0)/z_{0h}] - \Psi_{sh}(\zeta)\}}{a_h\{\ln[(z - d_0)/z_{0m}] - \Psi_{sm}(\zeta)\}^2} \tag{9.4}$$

in which it is assumed that $a_v = a_h$, $\Psi_{sv} = \Psi_{sh}$ and $z_{0v} = z_{0h}$. Note that z_{0h} which is a function of u_* is not known, *a priori*, in (9.4). However, (9.4) is only used to calculate a stability parameter, so that (for this purpose only) a rough estimate of z_{0h} is probably sufficiently accurate; this may be obtained, for example, by using a rough estimate of u_* obtained by means of the log-profile (4.6) or simpler yet (see Chapter 5) by assuming that z_{0h} is some fraction of z_{0m}.

When the flux of another admixture must be determined, its profile equation, (9.1) or (9.1′), is added as a fourth equation with a fourth unknown.

The profile equations are sometimes cast in a form involving bulk transfer coefficients defined in (4.114), (4.115) and (4.117). In the surface sublayer the coefficient for water vapor is given by (4.119) and sensible heat by an analogous expression. Clearly, however, this is only a matter of form, and for a non-neutral surface sublayer (4.114), (4.115) and (4.117) contain all three fluxes E, u_*, H in the coefficients, so that they must be solved simultaneously just like (4.33) through (4.35).

Calculation With Known Mean Profiles and Flux of Another Admixture

The second method consists of using the known mean profile and the surface flux of another but similar scalar, in addition to the mean profile of the scalar under consideration. The requirement of similarity refers in this context to the equality of the Ψ_s-functions in the profile Equations (4.33) through (4.35).

Probably the oldest application of this principle is the Bowen ratio (Bowen, 1926), defined in (1.3). This ratio, which is used mostly in the energy budget method (see Section 10.1a), can be written as follows

$$\mathrm{Bo} = \frac{c_p(\bar{\theta}_1 - \bar{\theta}_2)}{L_e(\bar{q}_1 - \bar{q}_2)} \tag{9.5}$$

under the assumption that in the surface sublayer $a_v = a_h$ and $\Psi_{sh} = \Psi_{sv}$ in (4.33) and (4.35). Over water in (9.5), usually the surface values $\bar{\theta}_s$ and $\bar{q}_s$ are used instead of $\bar{\theta}_1$ and $\bar{q}_1$, respectively, like in (4.33′) and (4.35′). Since z_{0v} is, in general, not equal to z_{0h} (see Chapter 5), this procedure is not quite correct. But the molecular diffusivities for heat, water vapor, and most other gases in air are sufficiently close, so that in most practical problems this error is likely to be negligible. The Bowen ratio concept thus provides a simple expression for evaporation in terms of the sensible heat flux and of mean specific humidity and mean temperature measurements in the surface sublayer,

$$E = \frac{H(\bar{q}_1 - \bar{q}_2)}{c_p(\bar{\theta}_1 - \bar{\theta}_2)}. \tag{9.6}$$

Similarly, the surface flux of any other scalar admixture of the air can be expressed in terms of, say, measurements of mean specific humidity and concentration, and a known rate of evaporation, by virtue of (4.33) and (9.1),

$$F = \frac{E(\bar{c}_1 - \bar{c}_2)}{(\bar{q}_1 - \bar{q}_2)} \tag{9.7}$$

b. Measurements in the Dynamic Sublayer

Sufficiently close to the surface in the lower part of the surface sublayer, the profiles of the mean concentrations are logarithmic. Under neutral conditions, the profiles are logarithmic in the whole surface sublayer, that is up to a level of the order of 50–100 m. Nevertheless, even for very unstable, but not stable, conditions Bradley (1972) has observed that the wind profile over grass is logarithmic as high as one meter above the surface. This is probably the case for scalar admixtures also.

The absence of any L-dependency greatly simplifies the formulations in terms of wind speed measurements in this logarithmic zone. For example, the rate of evaporation is obtained explicitly, by combining the logarithmic parts of (4.33) with (4.34),

$$E = \frac{a_v k^2 \rho(\bar{u}_2 - \bar{u}_1)(\bar{q}_1 - \bar{q}_2)}{\{\ln [(z_2 - d_0)/(z_1 - d_0)]\}^2} \tag{9.8}$$

This equation, with $d_0 = 0$, was first derived by Thornthwaite and Holzman (1939), and over the years it has been the subject of numerous studies; in the form of (9.8) it was suggested by Pasquill (1949b) and also used by Rider (1957) to calculate evapotranspiration from surfaces covered by different crops.

When the lower measurements are taken at the surface, the profile equations contain the roughness parameters. For the rate of evaporation one has, instead of (9.8), the following by combining (4.33′) and (4.34′)

$$E = \frac{a_v k^2 \rho\, \bar{u}_1(\bar{q}_s - \bar{q}_2)}{\ln [(z_2 - d_0)/z_{0v}] \ln [(z_1 - d_0)/z_{0m}]} \tag{9.9}$$

and similar expressions for H and F, from (4.35′) and (9.2), respectively; z_2 is the level of measurement of $\bar{q}_2$ and z_1 that of $\bar{u}_1$. An expression similar to (9.9) was first proposed by Sverdrup (1937) but with $d_0 = 0$ and with $z_{0m} = z_{0v}$ proportional to (ν/u_*) which he obtained by assuming some type of viscous sublayer over a smooth surface: later (1946) he also considered a rough sea surface for which he proposed (9.9) with $d_0 = -z_0$ and $z_{0v} = z_{0m} = z_0$.

c. Upper-Air Measurements: The ABL Profile Method

For steady conditions over a homogeneous surface, it may be possible to calculate surface fluxes by making use of upper-air, aerological or 'rawinsonde' data on the basis of the bulk-transfer equations for the ABL, which are given in Section 4.3b. It is shown there that, in principle, with measurements at a level $z = z_1$ in the surface layer and near $z = \delta$, the top of the boundary layer, the surface fluxes can be calculated by means of (4.74), (4.76), (4.77) and (4.78). When the lowest measurements are at the surface itself, the fluxes are given by (4.75), (4.76), (4.77) and (4.79).

As discussed in Section 4.3a, there are several ways of defining or determining δ, the thickness of the ABL. Nonetheless, it has been found that, for practical calculations of the rate of evaporation and the sensible heat flux, the height scale inferred directly from the mean profile measurements is probably the most reliable. Under unstable conditions this δ is the height of the upper boundary of the layer of convective mixing below the inversion. Under stable conditions it is not as easy to define;

but it can be taken as the thickness of the surface inversion, that is the level up to which significant cooling has propagated from the surface, or the height of the lowest maximum in the wind profile.

Just like for the surface sublayer in Section 9.1a, it is impossible to calculate the flux of any admixture, be it E, u_* or H, from measurements of the corresponding concentration only; each of (4.74) through (4.79) contains functions of L, so that each depends implicitly on E, u_* and H. Therefore, all three fluxes must be solved for simultaneously. This simultaneous solution can be carried out by techniques similar to those for the surface sublayer discussed in Section 9.1a. One such method to determine evaporation was proposed by Mawdsley and Brutsaert (1977). However, this study was only partly conclusive, mainly because at that time the functions $D = D(\delta/L)$ had not been determined. In the meantime, this function has been determined by Brutsaert and Chan (1978). Note that because the functions C are probably different from D, it is impossible to use the Bowen ratio technique [cf. [9.6]) by means of upper air data to calculate the evaporation rate.

To apply this boundary-layer profile method, profiles of mean wind speed, temperature and specific humidity are required in the lowest 2 km or so of the atmosphere. These data are published and available for many stations around the world, where usually twice daily, at 0000 and 1200 GMT, (UT), balloon ascents take place to determine the profiles synoptically. In the United States alone there are over 70 rawinsonde stations.

As noted, the similarity formulations of (4.74) through (4.79) require a steady flow over a homogeneous plane surface. These requirements imply several restrictions. The method is not applicable near coastlines or other locations with severe local advection, nor can it be used when changing weather conditions and frontal activity are present. Moreover, at present, the published rawinsonde data are not always as accurate as they should be, and the vertical intervals between the profile measurements are often larger than desirable. Fortunately, improvements are continuously being made in this technology so that this problem will probably be solved in the future. Another factor is the uncertainty which still surrounds the similarity functions A, B, C and D. The scatter in the experimentally-determined values, as shown in Section 4.3b, is still quite large.

However, despite these restrictions, the availability of the necessary data and the sound basis of the similarity approach provide this method with strong appeal. An interesting feature is that it produces regional estimates when applied over statistically uniform surfaces; indeed, as it takes tens of kilometers downwind before a discontinuity in surface conditions propagates fully to the top of the boundary layer, the observed profiles reflect surface conditions over this whole fetch. The method still needs more research before it will reach its full potential in the solution of practical problems. Nevertheless, it should be useful already under certain conditions, as a supplemental tool to check and verify the results obtained by other methods.

9.2. BULK-TRANSFER APPROACH

This approach consists of the application of bulk-transfer equations, such as given in (4.114), (4.115) and (4.117). The coefficients Ce_r, Cd_r and Ch_r can be determined theoretically or empirically. To determine evaporation by means of (4.114) the specific humidity $\bar{q}_s$ must be known; therefore, this approach is used mostly over water where $\bar{q}_s$ can be taken simply as $q^*(T_s)$, the saturation value at the temperature of the water surface. Since $\bar{q}$ in the air and $\bar{u}$ are not always measured at the same level, it is appropriate to generalize (4.114) as follows

$$E = \mathrm{Ce}\ \rho \bar{u}_1(\bar{q}_s - \bar{q}_2) \tag{9.10}$$

where the subscripts 1 and 2 refer to the levels z_1 and z_2 at which the wind speed and the specific humidity are measured. The main advantage of the bulk-transfer approach, usually with a constant-known coefficient, lies in the fact that it can be applied on a routine basis with regular and easily obtainable data of mean wind speed, water surface temperature and humidity of the air.

a. Over a Uniform Surface

The bulk-transfer approach is most commonly applied with measurements taken in the surface sublayer. This can be justified to some extent by the form of the similarity profiles (4.33′) through (4.35′), which yield a theoretical bulk-transfer coefficient given by (5.21) with (5.10). However, this equation shows that any empirical bulk-transfer coefficient for data taken in the surface sublayer can only be constant if the roughness parameters are constant, and either if the atmosphere is neutral, or if the effect of stability as reflected in $[(z-d_0)/L]$ is negligible or constant. For example under neutral conditions, by virtue of (9.9) and (9.10), the water vapor transfer coefficient is simply

$$\mathrm{Ce} = \frac{a_v k^2}{\ln\left[(z_2 - d_0)/z_{0v}\right] \ln\left[(z_1 - d_0)/z_{0m}\right]}\,. \tag{9.11}$$

Within certain ranges of normal wind speeds these conditions are apparently often satisfied over ocean and sea surfaces. This near-constancy of roughness and atmospheric stability explains why bulk transfer coefficients are widely used to parameterize surface momentum, heat and water vapor transport. Some recently-obtained measurements and empirical equations are reviewed in Sections 5.1b and 5.2b. Still, the scatter among all recent determinations of the bulk-transfer coefficients is quite large. This suggests that when accurate results are required the use of some average coefficient may not be adequate, and that it may be necessary to use a profile technique such as those outlined in Section 9.1a, including consideration of the effects of atmospheric stability, of the scalar roughnesses and probably also of the sea state.

The form of (9.9) is, in a sense, also suggestive of evaporation equations of the type proposed by Stelling (1882) as given in (2.5). As can be seen from (3.2), (3.5) and (3.6), the vapor pressure e is closely proportional to the specific humidity $\bar{q}$ and the introduction of the additional constant A_s may be viewed as a means of improving the correlation between the mean wind speed and the rate of evaporation. Although their theoretical justification is marginal, equations like Stelling's (2.5) have been found useful to describe evaporation from water or wet surfaces. Some examples for various problems and surfaces can be found in papers by Penman (1948, 1956), Brutsaert and Yu (1968), Shulyakovskiy (1969), and Neuwirth (1974), among others.

b. Evaporation from Lakes

Lakes and other water bodies of finite extent rarely satisfy the fetch requirement discussed in Section 7.1c. In such situations, the bulk transfer coefficients, derived for a uniform surface, are often not applicable and it may be necessary to give consideration to the effects of local advection. In practice, no general strategy has evolved thus far, but there are several ways of coping with this difficulty.

Lake Coefficient By Calibration

Any given lake or reservoir has its own particular exchange characteristics with the environment, on account of its geometry, and the topography, the land use and the climate of the surroundings. In addition, the meteorological data must often be recorded at some site which may have its own peculiar features. This means that it is very difficult to develop expressions for the water vapor or heat transfer coefficients, which are valid for any lake under any conditions. For this reason, probably the most accurate application of the bulk-transfer approach is by means of a coefficient obtained by calibration for the lake under study. In its simplest form the calibration consists of establishing a relationship between $\bar{u}_r$ and $\bar{E}/(\bar{q}_s - \bar{q}_a)$ or $\bar{E}/(\bar{e}_s - \bar{e}_a)$ as follows

$$\bar{E} = \mathrm{Ce}\ \rho\bar{u}_r(\bar{q}_s - q_a) \tag{9.12}$$

or

$$\bar{E} = N\ \bar{u}_r(\bar{e}_s - \bar{e}_a) \tag{9.13}$$

where $\bar{E}$ is the evaporation averaged over the entire lake surface, q_s and e_s the saturation-specific humidity and vapor pressure, respectively, corresponding to the mean lake surface temperature or to some representative water surface temperature, q_a and e_a the specific humidity and vapor pressure of the air, respectively, measured at a nearby reference location, $\bar{u}_r$ a representative mean wind speed; Ce and N are the mass transfer coefficients for the entire lake. The values of $\bar{E}$ needed in this correlation are obtained by some other independent method such as the energy budget method, more detailed profile measurements, eddy correlation measurements over the lake surface, or a detailed water budget. Clearly, the values of Ce and of N depend on the location where $\bar{u}_r$ and $\bar{q}_a$ or $\bar{e}_a$ are measured, and on other factors such as wind direction, nature and length of fetch, and the stability of the air. However, for long-term calculations, the effect of the latter three is usually neglected. This procedure can also be used with other equations beside (9.12) and (9.13). Many investigators still prefer Stelling's Equation (2.5). Others have attempted to improve on (9.12) and (9.13) by taking $\bar{u}_r$ to some power different from unity; however, these different forms do not appear to change the accuracy of the method very much so that the simple form of (9.12) and (9.13) is usually considered to be adequate. Examples of the application of this method have been published by Harbeck and Meyers (1970), Neuwirth (1974) and Hoy and Stephens (1979); in these studies $\bar{E}$ for the calibration was determined by the energy budget method.

When the fetch of the lake is not larger than a few kilometers, the use of data at only one location may yield satisfactory results. But when the lake is much larger, meso-scale phenomena tend to cause some variation of the conditions over the water surface, and improved methods are desirable. Phillips (1978) has presented an empirical mass transfer method to calculate evaporation from Lake Ontario on the basis of observations from several land stations upwind from the lake. For each day a mean daily water surface temperature for each of 88 grid points of the lake was obtained from airborne radiation thermometer surveys. For each grid point a stability class was chosen depending on the surface water temperature and an upwind overland air temperature. Mean daily humidity and wind speed data were used in one of a series of five regression models depending on the stability class, to generate over lake humidity and wind speed for each of the 88 grid points. At each grid point, a mass transfer equation was then used to calculate the local evaporation. Such methods for larger lakes will become more useful when water-surface temperature from remote sensing by satellite or other imagery become available on an operational basis.

Derived Coefficient for On-Shore Measurements

When the data necessary to calibrate a given lake are not available, it is necessary to use theoretical models or results from previous experimental studies on other lakes, in order to derive a bulk transfer coefficient.

As shown in Section 7.2a, under certain, not very restrictive assumptions it is possible to obtain theoretical solutions for evaporation or heat transfer from a finite-size wet surface, which are applicable to the lake problem. The solution of Sutton's (1934) problem, (7.56) with (7.57), or the empirical formula (7.59) with (7.60) derived by Harbeck (1962), should be quite useful when no other information is available, to determine lake evaporation for periods of a few weeks or a month. The data required are the size of the lake, the wind speed, preferably over the water, an average temperature of the water surface, and the vapor pressure in the air unaffected by the water body, thus preferably at an upwind location.

The numerical results of Weisman and Brutsaert (1973) can be used to obtain a rough idea of the effect of the instability of the atmosphere on lake evaporation. These results can be applied by means of (7.89) with (7.90) and Tables 7.1 and 7.2. The data required are the same as those required for Harbeck's formula, and also those needed to calculate A_* and B_*. For practical purposes, these can be approximated as

$$A_* = -\frac{(\bar{T}_s - \bar{T}_{as})}{\bar{T}_{as}} \frac{kg\hat{z}_0}{\bar{u}_r^2 \mathrm{Cd}_r} \tag{9.14}$$

and

$$B_* = -0.61(\bar{q}_s - \bar{q}_{as}) \frac{kg\hat{z}_0}{\bar{u}_r^2 \mathrm{Cd}_r}$$

where $\mathrm{Cd}_r = u_*^2/\bar{u}_r^2$ is a drag coefficient representative of the lake and its surroundings.

Derived Coefficient for Over-Water Measurements

In the case of medium size lakes, i.e., with fetches of the order of 1 to 10 km, when the vapor pressure e_a is measured over the center of the lake surface, the mass transfer coefficient is quite insensitive to the fetch. This can readily be shown by inspection of the analytical solution (7.51), but it is also evident from comparison of experimental coefficients from different lakes. This insensitivity to fetch is not surprising: when the shore line is sufficiently remote, an internal boundary layer with a constant flux layer has developed to the elevation where e_a is measured.

Thus under such conditions, it should be possible to use a profile approach, as described in Section 9.1a, or for long-term calculations, a bulk transfer equation of the type (4.114) with a typical value of the coefficient of say (cf., Table 5.3) $Ce_{10} = Ch_{10} = 1.4\ 10^{-3}$, obtained from ocean measurements.

In engineering practice, lake evaporation is more commonly calculated by means of the bulk transfer Equation (9.13) in terms of vapor pressure. For mid-lake observations of $\bar{u}_r$ and e_a an often quoted value of N is that obtained at Lake Hefner (Marciano and Harbeck, 1954)

$$N = 1.215\ 10^{-2} \tag{9.15}$$

when E is in mm hr^{-3}, $\bar{u}_r$ and e_a are measured at 8 m above the water in m s^{-1} and mb, respectively. By virtue of (3.2), (3.5) and (3.6) the relationship between the coefficients in (9.12) and (9.13) is approximately

$$N = 0.622\ \rho p^{-1}\ \mathrm{Ce}. \tag{9.16}$$

Hence, for an air density $\rho = 1.2\ 10^{-3}$ g cm^{-3} and a pressure of 1013.25 mb, (9.15) is equivalent to $Ce_8 = 1.527\ 10^{-3}$. Under neutral conditions, for $d_0 = 0$ and assuming an effective roughness $\hat{z}_0 = z_{0m} = z_{0v}$ in (9.11) one obtains $\hat{z}_0 = 0.0287$ cm; thus, (9.15) is equivalent with $Ce_{10} = 1.463\ 10^{-3}$, approximately, which is well within the range of values observed over sea surfaces (see Table 5.3).

The necessary over-water data for the application of this technique can be obtained from a stable platform, from a raft or from a buoy in the center of the lake. For very large lakes and for lakes with very irregular shapes, not one but several measuring stations should be used to obtain the data. An example of a type of buoy which has been used for this purpose is shown in Figure 9.1.

Design Approximations for Heated Water

The main objective in the predictive modeling of cooling ponds, reservoirs and streams, used for the disposal of a heat load, is usually the description of the internal heat transport processes in the water. The spatial variability in temperature and velocity in such water bodies is considerably larger than in the air flow above. Accordingly, in the presently-available (e.g., Jirka *et al.*, 1978) models the local advection in the atmosphere is neglected. The heat and water vapor transfer at the water surface is calculated with a local bulk transfer equation; the conditions at the water surface are taken as local values of e_s and T_s, which vary spatially as a result of the transport processes in the water, whereas the conditions in the air are assumed to be constant over the entire water body.

In some engineering models (e.g., Edinger *et al.*, 1968; Yotsukura *et al.*, 1973; Jobson, 1973), empirical bulk-transfer equations like (9.12), (9.13) and (2.5) were

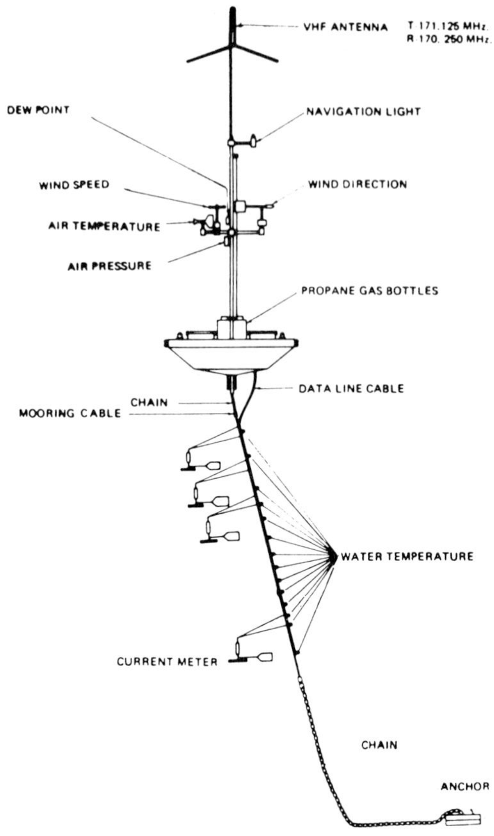

Fig. 9.1. Example of a buoy configuration which was used during the International Field Year for the Great Lakes (IFYGL) in 1972–1973 on Lake Ontario. The height of the sensors was approximately 3 m above the water surface; the propane served as a power supply (adapted from Foreman, 1976).

used. However, such equations, with coefficients developed for water bodies under neutral or long-term average conditions, are often inadequate to describe evaporation and heat transfer from heated water.

Over warm water the air is unstable and the transfer is likely to be enhanced by convective turbulence. When local advection effects are not too severe, the effect of atmospheric stability can be treated quite accurately by the profile method as described in Section 9.1a (e.g., see also Hicks *et al.*, 1977). A different more heuristic approach was suggested by Shulyakovskiy (1969); the underlying idea of his suggestion is that the effects of forced and free convection can be combined by simply adding empirical equations for evaporation under these two special conditions. Evaporation under conditions of forced convection is well understood, and it can be described by any one of the available bulk-transfer equations such as (9.12), (9.13) or (2.5) for a neutral atmosphere. Evaporation under conditions of free convection has not received as much attention, but the assumption commonly made is that it is similar to convective heat transfer. Yamamoto and Miura (1950) made this assumption but their study was restricted to laminar flow. Shulyakovskiy (1969) also applied this assumption, and was thus able to propose for combined free and forced convection the following superposition

$$E = [A + B\bar{u}_2 + C(T_{Vs} - T_{V2})^{1/3}](e_s^* - \bar{e}_2) \tag{9.17}$$

in which the forced convection coefficients [cf., (2.5)] were taken after B. D. Zaykov as $A = 0.15$, $B = 0.112$ and the free convection coefficient after U. Grigull's equation for convective heat transfer as $C = 0.094$; the units are E in mm day^{-1}, $\bar{u}_2$ the wind at 2 m in m s^{-1}, $\bar{e}_2$ the vapor pressure at 2 m in mb and T_{Vs} and T_{V2}, the virtual temperature at the water surface and in the air at 2 m, respectively in K. Ryan *et al.* (1974), who applied Shulyakovskiy's (9.17), found that his proposed coefficients overestimated evaporative heat loss from a cooling pond; therefore, they proposed for forced convection $A = 0$ and $B = 3.2$ W m^{-2} (mb)$^{-1}$ (ms^{-1})$^{-1}$ which is the same value as in Zaykov's but in different units, and for free convection $C = 2.7$ W m^{-2} (mb)$^{-1}$ K$^{-1/3}$, after an equation for free convective heat transfer given by M. Fishenden and D. A. Saunders. Ryan *et al.* (1974) obtained good agreement with available experimental data. Weisman (1975) has compared the results obtainable by means of the Shulyakovskiy approach with those of the turbulent-diffusion approach of Weisman and Brutsaert (1973) given in (7.90) Since the latter results are fetch-dependent, Weisman applied (9.17) with $A = 0$ and with the C-value of Ryan *et al.* (1974), but with $B = N$, as given by (7.60) of Harbeck (1962). He was able to obtain agreement between the results of the eddy diffusion model and those of the modified Shulyakovskiy equation, but only by using an effective roughness $\hat{z}_0 = 0.09$ cm in the diffusion model. This is equivalent to a vapor transfer coefficient in neutral air of about $Ce_{10} = 1.84\ 10^{-3}$, which is somewhat larger than the values reported for open water shown in Table 5.3. Because the eddy-diffusion models of Chapter 7 were found to agree with experimental results with $\hat{z}_0 = 0.02$–0.04 cm, this means that Shulyakovskiy's superposition approach probably overestimates E. The results of Jirka *et al.* (1978) who applied (9.17) with the coefficients of Ryan *et al.* (1974) to several strongly-loaded cooling ponds and found an overprediction in the order of 15 to 20 percent, also point in the same direction, Clearly, the concept of superposition will require further investigation.

9.3. SAMPLING TIMES

Ideally, the data which are used for the mean-profile methods, based on similarity theory, and for the related bulk transfer methods, should be turbulent mean quantities. As discussed in Section 8.1b, in consideration of the turbulence structure near the earth's surface, these mean values should probably be obtained by integrating over an averaging period, which may range between approximately 20 minutes to 1 hour. Thus, in the practical application of the methods treated in this chapter, a reasonable approach may consist of carrying out flux-computations with half-hourly or hourly means of the observed variables. The fluxes for longer periods can then be calculated from these flux values.

Meteorological data are rarely available for hourly and even daily periods; therefore, it is important to know what errors may be expected, when longer-term average values of temperature, specific humidity, wind speed, etc. are used in the calculations.

As regards the bulk transfer approach for water surfaces, several studies have shown that the use of daily mean values usually gives acceptable results, but that monthly mean values of $\bar{q}$, $\bar{T}$ and $\bar{u}$ may produce considerable error. Jobson (1972) analyzed data from Lake Hefner; he found that the value of the bulk transfer coefficient N of (9.13) is independent of the time interval over which the meteorological data have been averaged, provided this interval is not longer than one day. Averaging for one month resulted in a systematic error. The variance in the error distribution increased by a factor of more than 6 as the averaging time increased from 3 hours to 1 day. From an analysis of weather ship data, Kondo (1972b) concluded that time-averaged values of E and H for periods not longer than one day can be approximately calculated by using time-averaged of $\bar{u}$, $(\bar{T}_s - \bar{T}_a)$ and $(\bar{q}_s - \bar{q}_a)$ values for the same period. The bulk-transfer coefficient for periods of 3 months was found to be 1.3 times, and the coefficient for weekly periods still about 1.2 times that for use with short-period observations. In these two studies, the transfer coefficient was taken as independent of atmospheric stability.

The sensitivity of the profile method (with inclusion of atmospheric stability) to the averaging period of the data is not as well known. However, unpublished calculations by the author have suggested that for a water surface the use of daily mean values for the $\bar{q}$, $\bar{T}$ and $\bar{u}$ profiles in the surface sublayer produces calculated E and H values, which are quite close to the means calculated from the hourly profiles. Although the matter requires further study, these results for water surfaces are not unexpected, since the diurnal evaporation cycle over a water body is usually weak.

In contrast over land, the rate of evaporation typically exhibits a pronounced diurnal cycle; this is illustrated in Figures 1.5, 6.1, and 6.2. Therefore, in general, there is probably no good substitute for hourly or half-hourly mean values.

CHAPTER 10

Energy Budget and Related Methods

These methods involve either the direct application, or some approximation of the equation for the energy budget. One form of this equation is given in (6.1) but in many practical situations it can be simplified considerably. A common characteristic of most energy budget methods is that they require the determination of the net radiation, R_n. In general, the energy budget method allows the determination of one of the terms of (6.1), or a simplified form of it, when all the remaining terms can be determined by some independent method.

Since the main objective here is the determination of the rate of evaporation E (or the sensible heat flux into the air H) it is convenient to rewrite (6.1) as

$$L_e E + H = Q_n \tag{10.1}$$

where Q_n is defined as the available energy flux density

$$Q_n = R_n + L_p F_p - G + A_h - \partial W/\partial t \tag{10.2}$$

whose terms are given in Chapter 6. In hydrological applications it is common practice to express the specific energy fluxes as equivalent rates of evaporation; Equation (10.1) is then written as

$$E + H_e = Q_{ne} \tag{10.3}$$

where $H_e = H/L_e$ and $Q_{ne} = Q_n/L_e$. Also, note that in many applications, especially over land surfaces, the terms F_p, A_h and $\partial W/\partial t$ are of little consequence, so that it is sufficiently accurate to put $Q_n = R_n - G$.

As noted in Chapter 2, one of the first to apply an energy-budget approach was probably Homén (1897). Schmidt (1915) applied the method to determine E from the ocean, but he had no practical procedures to determine all the terms in the energy budget equation on a short-term basis. Some ten years later, Bowen (1926) suggested the energy budget Equation (10.1) to determine lake evaporation by using the ratio (9.5) now named after him (cf., (10.4)). This suggestion was concluded to be correct by Cummings and Richardson (1927), when they applied Bowen's ratio in studying the terms in the energy budget equation to estimate lake evaporation. Sverdrup (1935) used the energy budget method to determine snowmelt; H and E were determined from temperature and specific humidity gradients in the air with power-law eddy diffusivities obtained from wind-speed measurements on the basis of the Reynolds analogy. In another paper published the same year (see Albrecht, 1937), he also proposed the use of the Bowen ratio, essentially in the form of (9.5) to determine the rate of evaporation by means of (10.4).

10.1. STANDARD APPLICATION

When Q_n and either H or E can be determined, (10.1) provides directly the remaining unknown flux. Usually, however, both H and E are unknown, and an indirect method must be used. From the methodological point of view, these indirect energy budget methods are analogous to the mean profile methods of Section 9.1. In both, essentially three equations are used which contain E, u_* and H implicitly. In the profile methods these are the equations for $\bar{q}$, $\bar{u}$ and $\bar{\theta}$. In the energy budget methods, (10.1) is used either with equations for $\bar{q}$ and $\bar{\theta}$ (Section a) or with equations for $\bar{u}$ and $\bar{\theta}$ or $\bar{q}$ (Section b).

a. With Bowen Ratio (EBBR)

When Q_n is known, the combination of the energy budget (10.3) with the Bowen ratio Bo defined in (1.3) produces

$$E = \frac{Q_{ne}}{1 + \text{Bo}}. \tag{10.4}$$

Similarly, for the sensible heat flux one has

$$H = \frac{\text{Bo}\, Q_n}{1 + \text{Bo}}. \tag{10.5}$$

Bo can be determined, as shown in (9.5), from profile data of specific humidity and temperature in the atmospheric surface sublayer. In accordance with the findings discussed in Section 8.1b, these data should be taken as averages over 30 min to 1 hour, approximately. Equation (10.4) shows that the energy budget with Bowen ratio (EBBR) method for E is most accurate when Bo is small. Both (10.4) and (10.5) produce a singularity when $\text{Bo} = -1$; but, as pointed out by Tanner (1960), over an active vegetation this is not a problem, as this situation usually occurs only when H is low, around sunrise, sunset and occasionally at night. This situation does occur more often over cold water, and it may be necessary to use an alternative method when $-1 < \text{Bo} < -0.5$ to avoid the problem of a very small denominator in (10.4) and (10.5). Tanner (1960) suggested the use of a bulk transfer method for these special conditions. Another way consists of using mean values of Bo corrected by means of wind measurements, as outlined by Webb (1960, 1964); this method is especially useful when some terms in the available energy Q_n are only known for daily periods or longer.

The EBBR method has the advantage that no similarity functions for the atmospheric turbulence appear explicitly in the formulation. No measurements of turbulence or of the mean wind speed are required, and the formulation, as written in (10.4) with (9.5), is independent of atmospheric stability. In addition, when Bo is small, the EBBR method may be less susceptible, albeit not immune (e.g., Figures 7.11 and

7.16) to imperfect fetch conditions than mean profile methods, in which such effects are more directly apparent.

The validity of the EBBR method depends critically on the similarity of the temperature and humidity profile; for the surface sublayer this requires the equality of the terms in the square brackets in (4.33) and (4.35). It has usually been assumed that $\Psi_{sv} = \Psi_{sh}$; however, recent experimental studies have suggested that this may not always be the case. Verma *et al.* (1978) found that, under conditions of regional (not local) advection, the EBBR method consistently underestimated the evaporation from a well-watered vegetation, in relation to that obtained with a lysimeter. This was interpreted to signify that ϕ_{sv} and ϕ_{sh} are not the same under stable conditions. However, this finding will have to be confirmed by further independent studies (see also Brost, 1979; Hicks and Everett, 1979). A second point is that different sink or source distributions for heat and water vapor, as they may occur in tree canopies, are likely to be reflected in a dissimilarity of the respective profiles above the vegetation as well. Any such effects probably disappear only at some elevation above which the profiles given in (4.33) and (4.35) are valid. Data analyzed by Garratt (1978a) indicate that this minimal elevation is of the order of three to five times the height of the roughness obstacles. Hence, above a forest canopy of the order of 10 m and more this height requirement for similarity may be prohibitive. As pointed out by C.B. Tanner (1976, personal communication), Bo-measurements taken too closely to the top of the canopy may cause serious error in the application of the EBBR method over forests. To avoid this possible source of error, it may be useful to test the similarity and the constancy of Bo with elevation, by measuring $\bar{q}$ and $\bar{\theta}$ at more than two levels. Fortunately, over most other surfaces, such as water, soil and low vegetation, this matter should not pose any difficulties. Figure 6.2 is an example of the accuracy obtainable with the EBBR method with reference to lysimeter measurements.

Other aspects of the EBBR method and its accuracy have been treated by Fuchs and Tanner (1970), Sinclair *et al.* (1975) and Revfeim and Jordan (1976) among others. Examples of its application to lakes have been presented by Anderson (1954), Mahringer (1970), Keijman (1974), Hoy and Stephens (1979); to short vegetation and crops by Fritschen (1966), Lourence and Pruitt (1971), Perrier *et al.* (1976), Verma *et al.* (1978); to forest by McNaughton and Black (1973) and Thom *et al.* (1975). Over snow McKay and Thurtell (1978) found the method difficult to apply because it was generally difficult to determine the heat storage in the snow pack independently. Also Mahringer (1970) encountered difficulties with this method in determining evaporation from a lake covered with ice and snow.

The Bowen ratio concept can be extended to determine the flux of other admixtures as well [cf., (9.6) and (9.7)]. Sinclair *et al.* (1975) applied it to determine the flux of CO_2. For this purpose, beside the Bowen ratio (1.3), a second ratio can be defined, to link the sensible heat flux with the photosynthetic energy flux

$$\mathrm{Bo}_P = \frac{H}{-L_p F_p}. \tag{10.6}$$

Making use, again, of the assumptions that $a_h = a_s$ and $\Psi_{sh} = \Psi_{sc}$ in (4.35) and (9.1) applied to CO_2 with $F = F_p$, one obtains from (10.6)

$$\mathrm{Bo}_p = \frac{c_p(\bar{\theta}_1 - \bar{\theta}_2)}{L_p(\bar{c}_2 - \bar{c}_1)}. \tag{10.7}$$

Substitution of (9.5) and (10.7) in the energy budget Equation (6.1) produces, if A_h and $(\partial W/\partial t)$ are negligible, the photosynthetic energy flux

$$-L_p F_p = \frac{R_n - G}{1 + \mathrm{Bo}_p + (\mathrm{Bo}_p/\mathrm{Bo})}. \tag{10.8}$$

An example of the results obtainable with this approach is shown in Figure 6.8.

b. With Profiles of Mean Wind and of One Scalar (EBWSP)

In case profile data of the mean temperature or of the mean specific humidity are lacking, it is possible to use instead the mean wind-speed profile in the application of the energy budget method. In fact, this procedure is potentially more powerful than the Bowen ratio method, since it not only yields E and H, but also u_*.

As an illustration of this wind and scalar profile-energy budget (EBWSP) method, suppose that specific humidity measurements are not available. One can then use (10.1) together with the profile equations (4.34) (or (4.34′)) and (4.35) (or (4.35′)), as a system of three equations with three implicit unknowns E, u_* and H. This system can be solved on the basis of measurements of Q_n, $\bar{\theta}_1 - \bar{\theta}_2$ (or $\bar{\theta}_s - \bar{\theta}$) and $\bar{u}_2 - \bar{u}_1$ (or $\bar{u}$ and z_0). The method is not very sensitive to the accuracy of d_0, so that this parameter can be estimated, if it cannot be measured. The same procedure can be used if temperature measurements are not available; in this case, measurements of Q_n and of the profiles of mean wind speed and mean specific humidity are used in (10.1), (4.33) (or 4.33′) and (4.34) (or (4.34′); this system of three equations can be solved for E, u_* and H.

Already, in 1938, Albrecht (1950) applied the general idea of this method to the climatological calculation of evaporation. E was obtained by means of (10.1) in which R_n was calculated by means of empirical formulae, G was calculated from soil-temperature profiles, and H was calculated by means of a bulk heat transfer equation of the type $H = (T_s - T_a)\,f_h(\bar{u})$; the wind function $f_h(\bar{u})$ was apparently taken as a constant for a given location. The effect of stability was probably first considered in the study by Fuchs *et al.* (1969). The rate of evaporation from a drying soil was calculated by means of (10.1) with (4.35′) in which u_* was given by (4.34′); however, it was assumed that $z_{0h} = z_0$ and $\Psi_{sh} = \Psi_{sm}$ and the Ψ functions were taken to depend on a bulk Richardson number.

More recently, this energy budget method has been applied by Stricker and Brutsaert (1978) to determine evapotranspiration from grass; use was made of measurements over one hour of net radiation, the mean temperature at 0.10, 1.50 and 3.0 m and the mean wind speed at 2.0 m above the surface. The values of E, u_* and H, which were calculated by an iteration procedure with (10.1), (4.35) and (4.34′), were then averaged over daily periods; the ground heat flux G was neglected. The calculated values of E are shown in Figure 10.1. It was found that the calculated results were quite insensitive to the exact formulation of the profile functions Ψ_{sm} and Ψ_{sh}, but that the effect of the atmospheric stability was not neglibible. Also, the effect of

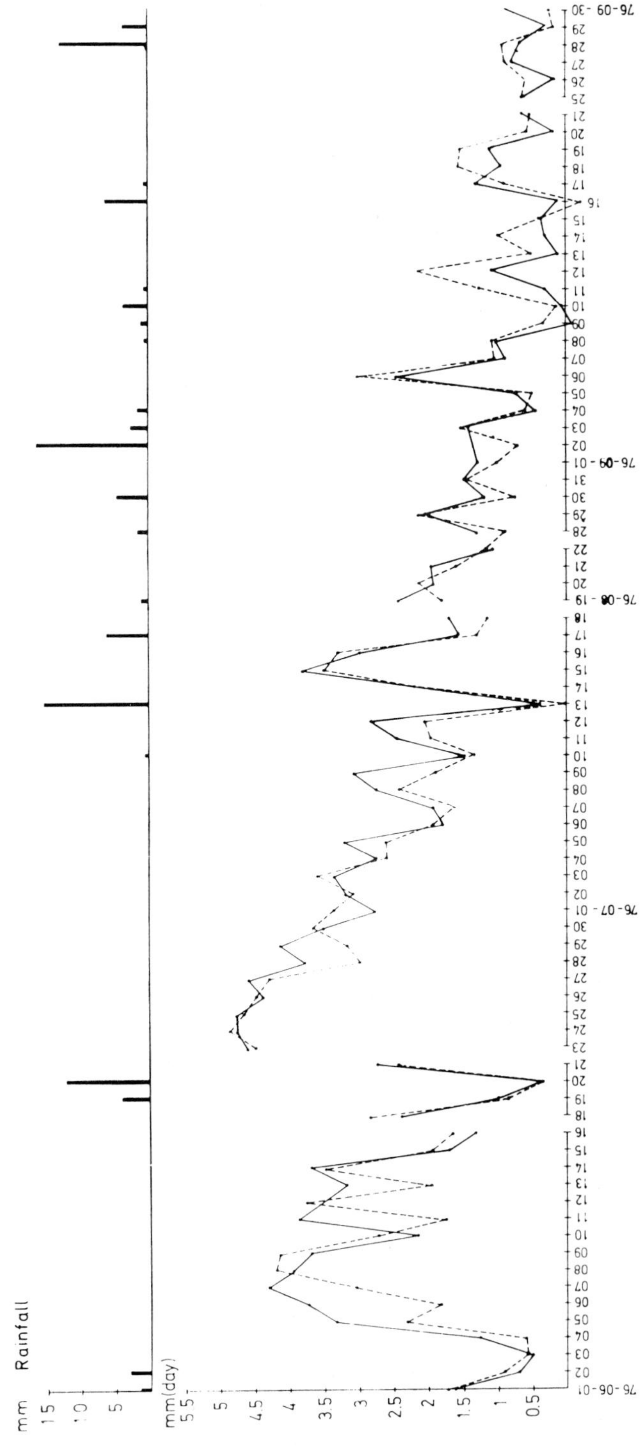

Fig. 10.1 Time series of daily values of E (- - - -) calculated by means of the energy budget with profiles of mean wind and temperature (EBWSP) for relatively dry period in Gelderland; also shown are the rainfall and E calculated with 10.43 (———) (from Brutsaert and Stricker, 1979).

buoyancy, due to water vapor stratification on the overall stability of the atmosphere, was quite noticeable; this effect appears as 0.61 E in (4.25).

Unlike the EBBR method, the EBWSP method is not restricted to the surface sublayer of the atmosphere, since it does not require similarity of temperature and humidity profiles. In principle, the method can also be used with upper air meteorological data. For example, if the specific humidity data are not available or unreliable, (10.1) may be combined with (4.76), (4.77) and (4.78) (or (4.79)) to calculate E, u_* and H from available data of the temperature difference $(\bar{\theta}_1 - \bar{\theta}_\delta)$ (or $(\bar{\theta}_s - \bar{\theta}_\delta)$) and the mean wind speed $(\bar{u}_\delta^2 + \bar{v}_\delta^2)^{1/2}$ at $z = \delta$. This EBWSP method and its simpler derivatives (see Section 10.2b) are sometimes referred to as combination methods on the grounds that both the energy budget and hydrodynamic aspects of evaporation are considered. But this is somewhat misleading, since the Bowen ratio is no less dependent on the validity of the hydrodynamics underlying (say) (4.33) through (4.35) than the formulation of the mean wind speed profile.

10.2. SIMPLIFIED METHODS FOR WET SURFACES

a. Some Comments on Potential Evaporation

Because several of the simple energy budget type methods treated in this Section are often used as measures or indices of potential evapotranspiration, a few comments on this concept are in order. The term potential evapotranspiration appears to have been introduced by Thornthwaite (1948) in the context of the classification of climate. It is now generally understood to refer to the maximum rate of evapotranspiration from a large area covered completely and uniformly by an actively growing vegetation with adequate moisture at all times. The area is specified as large to avoid the possible effects of local advection. Although this concept is widely used, it has also caused confusion, because it does not encompass all possible conditions and it involves several ambiguities. In other words, the term potential evapotranspiration requires closer specification if it is to serve as an unequivocal parameter.

For example, transpiration even at the potential rate, involves such biological effects, as stomatal impedance to the diffusion of water vapor, and the stage in the growth cycle of the vegetation. For this reason, the term potential evaporation is probably preferable. It refers to the evaporation from any large uniform surface which is sufficiently moist or wet, so that the air in contact with it is fully saturated. Such conditions prevail usually only after the occurrence of precipitation and dew. This distinction between potential evapotranspiration and potential evaporation has been found to be quite large over tall vegetation (e.g., McNaughton and Black, 1973; Stewart and Thom, 1973). Over short non-wet vegetation the potential evapotranspiration is often very similar to the evaporation from open water under the same conditions. A possible explanation for this is that the stomatal impedance to water vapor diffusion may be compensated by the larger roughness values, resulting in larger transfer coefficients, of the vegetational surface.

Another point of ambiguity is that potential evaporation is often calculated by means of meteorological data observed under nonpotential conditions. Clearly, this is not the same rate as that which would be calculated (or observed) if the surface had been adequately supplied with water. Indeed, the partition of the available energy at the surface is related to the availability of water for evaporation, and this partition affects the temperature, the humidity and other state variables of the atmosphere. This matter should be kept in mind when the concept is used.

b. The EBWSP Method With Measurements at One Level

When the surface is wet, the surface specific humidity may be assumed to be the saturation value at the surface temperature, i.e., $q_s = q^*(T_s)$. This allows an approximation, first introduced by Penman (1948); the main advantage of this approximation is that it eliminates the need for measurements of $\bar{q}$, $\bar{u}$ and $\bar{\theta}$ at two levels, as in the profile methods (Chapter 9) and standard energy budget methods (Section 10.1), and that measurements at one level suffice.

Penman Approach

The equation derived by Penman (1948) was intended for an open-water surface. In what follows, a somewhat more general derivation is given, which is applicable to any wet surface, but which retains the essential features.

Because the Clausius-Clapeyron equation [see (3.21) and (3.24)] is used, it is preferable to start with the Bowen ratio (9.5) in terms of the vapor pressure; with the lower measurement at the surface, where $e_s = e^*(T_s)$, the Bowen ratio is

$$\mathrm{Bo} = \gamma \frac{(\bar{T}_s - \bar{T}_a)}{(\bar{e}_s - \bar{e}_a)} \tag{10.9}$$

where e_a and T_a are the vapor pressure and temperature in the air, respectively, at some reference level, and where by virtue of (3.2), (3.5) and (3.6),

$$\gamma = \frac{c_p p}{0.622\, L_e} \tag{10.10}$$

which is the psychrometric constant; at 20° C and $p = 1013.25$ mb it is $\gamma = 0.67$ mb K^{-1}. Note that θ is replaced by T, since in the surface sublayer they are practically the same.

A crucial step in Penman's analysis is the assumption

$$\frac{e_s^* - e_a^*}{\bar{T}_s - \bar{T}_a} = \Delta \tag{10.11}$$

where $\Delta = (de^*/dT)$ is the slope of the saturation water vapor pressure curve $e^* = e^*(T)$, at the air temperature T_a, $e_a^* = e^*(T_a)$ the corresponding saturation vapor pressure, and $e^* = e_s^*(T_s)$ the vapor pressure at the wet surface. Since $e_s = e_s^*$ for a saturated surface, the Bowen ratio (10.9) is thus, approximately

$$\mathrm{Bo} = \frac{\gamma}{\Delta}\left[1 - \frac{(e_a^* - \bar{e}_a)}{(\bar{e}_s - \bar{e}_a)}\right]. \tag{10.12}$$

Values of (γ/Δ) for different temperatures at $p = 1000$ mb are presented in Table 10.1 and Figure 10.2; they were obtained by means of (10.10) and values of Δ and L_e listed in Table 3.4.

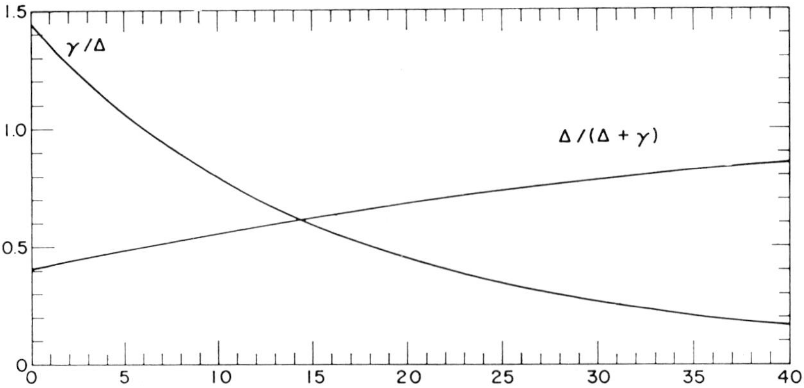

Fig. 10.2. Temperature dependence of (γ/Δ) and $\Delta/(\Delta + \gamma)$, at 1000 mb; γ is defined by (10.10) and Δ can be obtained from Table 3.4 or from (3.24b).

TABLE 10.1
Values of (γ/Δ) at 1000 mb (γ is defined by (10.10) and Δ can be obtained from (3.24b))

Air Temperature T_a(°C)	(γ/Δ)
−20	5.864
−10	2.829
0	1.456
5	1.067
10	0.7934
15	0.5967
20	0.4549
25	0.3505
30	0.2731
35	0.2149
40	0.1707

Substitution of (10.12) in (10.4) yields

$$Q_{ne} = \left(1 + \frac{\gamma}{\Delta}\right) E - \frac{\gamma}{\Delta}\left(\frac{e_a^* - \bar{e}_a}{\bar{e}_s - \bar{e}_a}\right) E. \tag{10.13}$$

In the second term of the right of (10.13), a bulk mass transfer equation can be used, viz.

$$E = f_e(\bar{u}_r)(\bar{e}_s - \bar{e}_a) \tag{10.14}$$

which serves as definition for the wind function $f_e(\bar{u}_r)$. This yields Penman's (1948) equation

$$E = \frac{\Delta}{\Delta + \gamma} Q_{ne} + \frac{\gamma}{\Delta + \gamma} E_A \tag{10.15}$$

where E_A, a drying power of the air, is defined by

$$E_A = f_e(\bar{u}_r)(e_a^* - \bar{e}_a). \tag{10.16}$$

Note that in Penman's (1948) derivation it was assumed that $Q_{ne} = R_n/L_e$ and that all the other terms in (10.2) are negligible. Equation (10.15) has been the subject of numerous theoretical and experimental studies (e.g., Penman, 1956; Tanner and Pelton, 1960; Monteith, 1965, 1973; Van Bavel, 1966; Thom and Oliver, 1977).

As mentioned, from the practical point of view, the main feature of (10.15) is that it requires measurements of mean specific humidity, wind speed and temperature at one level only. For this reason, it is very useful when measurements at several levels, needed for profile methods or standard energy budget methods are unavailable or impractical.

Equation (10.15) has been widely used, but there is still no generally accepted way to formulate $f_e(\bar{u})$, the wind function in E_A. From its definition in (10.14), it is clear that any suitable mass transfer coefficient can be used for this purpose (see Section 9.2). The simplest approach consists of using an empirical wind function. Penman (1948) originally proposed an equation of the Stelling-type (2.5) as follows

$$f_e(\bar{u}_2) = 0.26(1 + 0.54\, \bar{u}_2) \tag{10.17}$$

where $\bar{u}_2$ is the mean wind speed at 2 m above the surface in m s^{-1}, and the constants require that E_A in (10.16) is in mm day^{-1} and the vapor pressure in mb. There are some indications (e.g., Thom and Oliver, 1977) that (10.17) yields reasonable results for natural terrain with small to moderate roughness. Penman (1956) also proposed an intended improvement of (10.17), in which the numerical value 1 between the brackets was replaced by 0.5. Although apparently (Thom and Oliver, 1977) Penman subsequently felt that (10.17) is preferable over this second version, the latter is still widely used in hydrological practice. More recently, on the basis of lysimeter measurements, Doorenbos and Pruitt (1975) have suggested that, for irrigated crops, the constant 0.54 in (10.17) should be replaced by 0.86.

In terms of the bulk water vapor transfer coefficient as defined, for example, in (9.10) one obtains, by virtue of (3.2) (3.5) and (3.6), the wind function,

$$f_e(\bar{u}_1) = 0.622\, \rho p^{-1}\, \mathrm{Ce}\, \bar{u}_1 \tag{10.18}$$

where z_1 is the height of the measurement of $\bar{u}_1$ and z_2 that of e_a. The wind function can also be determined theoretically by means of the similarity profile functions of Chapter 4. Thus, under neutral conditions, by virtue of (9.9) and (10.14), this is

$$f_e(\bar{u}_1) = \frac{0.622 a_v k^2 \bar{u}_1}{R_d T_a \ln[(z_2 - d_0)/z_{0v}] \ln[(z_1 - d_0)/z_{0m}]} \tag{10.19}$$

where, again, z_1 is the level of the wind-speed measurement and z_2 that of the water-vapor pressure; if the vapor pressures e_a^* and e_a are in mb, T_a in K and $\bar{u}_1$ in the same units as E_A, one can put $(0.622/R_d) = 2.167 \times 10^{-4}$, approximately.

When Penman's equation (10.15) is applied to calculate mean values of E over periods of a day or longer, the use of wind functions, such as (10.17)–(10.19), may be adequate. However, when hourly values are required, the effect of atmospheric stability, which varies through the day, may be quite important. A method to include this effect is described next.

The Effect of Atmospheric Stability

Even though (10.15) makes use of measurements at one level only, it is possible to include the effect of stability in the wind function. For this purpose the drying power of the air (10.16) can be written in a form similar to (4.33′)

$$E_A = a_v k u_* \rho (q_a^* - q_a) \left[\ln\left(\frac{z_a - d_0}{z_{0v}}\right) - \Psi_{sv}\left(\frac{z_a - d_0}{L}\right)\right]^{-1} \tag{10.20}$$

where q_a and q_a^* are the specific humidity of the air and the saturation specific humidity at air temperature, respectively. The problem can be solved by an iteration procedure, as follows. An initial value of E is calculated in the usual way by means of (10.15) using an empirical or neutral E_A, say (10.16) with (10.17) or (10.19); it is also possible to use (10.20) with $\Psi_{sv} = 0$, and u_* is calculated by means of (4.34′) with $\Psi_{sm} = 0$. The initial value of E is used to obtain H by means of (10.1). These initial values of E, u_* and H provide a first value of the Obukhov length L by means of (4.25). This value of L now allows the calculation of a second estimate of u_* by means of (4.34′) and a second estimate of E_A by means of (10.20), which produces a second estimate of E by means of (10.15), and so on. The procedure which can easily be programmed on a digital computer, can be halted when successive estimates of E or L are sufficiently close.

c. Advection-Free Evaporation from Wet Surfaces

Equilibrium Evaporation

The two-term structure of (10.15) suggests an interpretation which may serve as an aid in understanding the effect of regional or large-scale advection. When the air has been in contact with a wet surface over a very long fetch, it may tend to become vapor saturated, so that E_A should tend to zero. Accordingly, Slatyer and McIlroy (1961, ch. 3, p.73) reasoned that the first term of (10.15) may be considered to represent a lower limit to evaporation from moist surfaces, which they referred to as equilibrium evaporation. Thus, by definition it is written as

$$E_e = \frac{\Delta\, Q_{ne}}{\Delta + \gamma}. \tag{10.21}$$

The second term of (10.15) can be interpreted as a measure of the departure from equilibrium in the atmosphere. In the absence of cloud condensation or radiative divergence, this departure would stem from large-scale or regional advection effects, involving horizontal variation of surface or atmospheric conditions. Priestley (1959, p. 116) also arrived at (10.21) but on the basis of the EBBR method; he assumed that when the air over a moist surface is vapor-saturated and the ranges of temperature and specific humidity with height z and time t are small, $q^* = q^*(T)$ can be linearized and the Bowen ratio in the surface sublayer is then a constant given by $\mathrm{Bo}_e = c_p/L_e(\mathrm{d}q^*/\mathrm{d}T)$, or (cf. (10.9) and (10.10))

$$\mathrm{Bo}_e = (\gamma/\Delta) \tag{10.22}$$

which is a function of temperature (see Table 10.1 and Figure 10.2). The subscript e

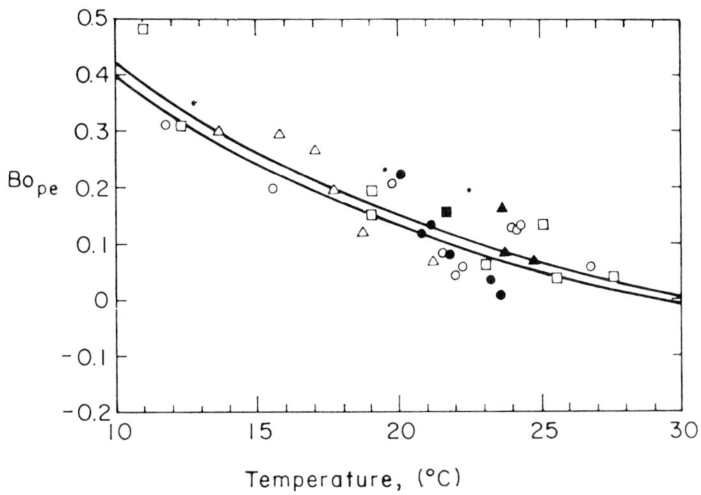

Fig. 10.3. Variation of Bowen ratio Bo_{pe} for moist surfaces with air temperature as given by (10.24); the upper curve represents $\alpha_e = 1.26$ and the lower $\alpha_e = 1.28$. The data points (for daily values) were collected by Davies and Allen (1973) from different sources.

of E_e and Bo_e denotes equilibrium conditions. Substitution of (10.22) in (10.4) yields (10.21).

Average Conditions of Minimal Advection

Over wet surfaces, equilibrium conditions are encountered only rarely, if ever. This is due to the fact that the atmospheric boundary layer is never a truly homogeneous boundary layer, as would be the case in channel flow; rather, it is continually responding to large-scale weather patterns, involving condensation, and unsteady three-dimensional motions, which tend to maintain a humidity deficit even over the ocean. Thus, there is always some degree of advection. Nevertheless, the idea underlying (10.21) has led to further developments. Priestley and Taylor (1972) have taken equilibrium evaporation as the basis for an empirical relationship giving evaporation from a wet surface under conditions of minimal advection, E_{pe}; they analyzed data obtained over ocean and saturated land surfaces in terms of a constant quantity α_e, defined by $E_{pe} = \alpha_e E_e$, that is

$$E_{pe} = \alpha_e \frac{\Delta\, Q_{ne}}{\Delta + \gamma}. \tag{10.23}$$

This is equivalent with a Bowen ratio

$$Bo_{pe} = \alpha_e^{-1}(\gamma/\Delta) + (\alpha_e^{-1} - 1) \tag{10.24}$$

which depends on temperature (see Figure 10.3). For large saturated land and 'advection-free' water surfaces Priestley and Taylor (1972) concluded that the best estimate is $\alpha_e = 1.26$. Also Davies and Allen (1973) for well watered grass, Stewart and Rouse (1976) for shallow lakes and ponds and, later (1977), for saturated sedge

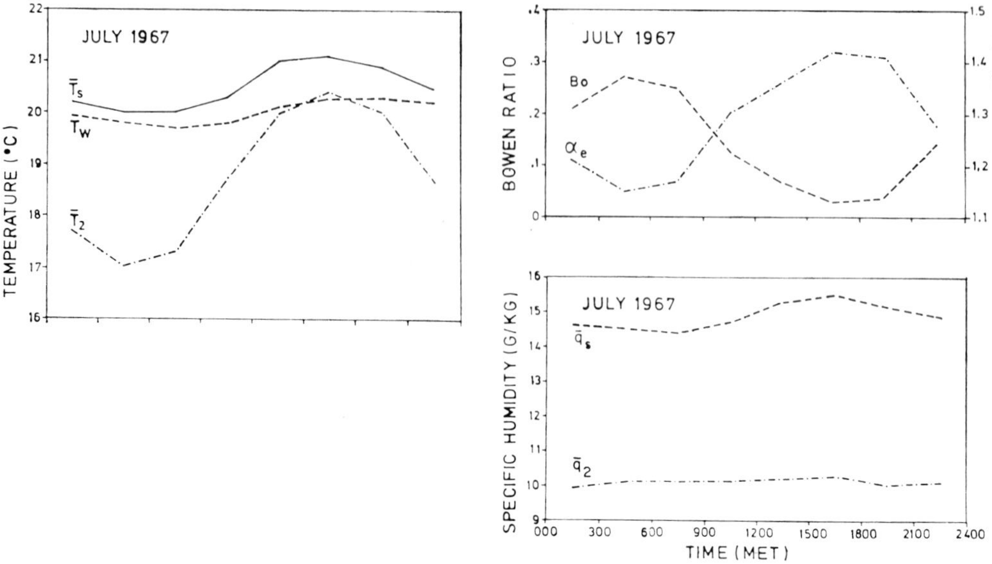

Fig. 10.4. Average diurnal variation of α_e and the Bowen ratio Bo, over a large shallow lake in the Netherlands for July 1967. Also shown are the water surface temperature $\bar{T}_s$, the depth averaged water temperature T_w, the air temperature $\bar{T}_2$, and the specific humidity of the air at the water surface, $\bar{q}_s$ and at 2 m, $\bar{q}_2$ (from DeBruin and Keijman, 1979).

meadow as well, concluded that their data supported the value $\alpha_e = 1.26$. There are some indications however, that α_e may be slightly larger and perhaps closer to 1.28. If one set of data, which was rather discordant ($\alpha_e = 1.08$) with the others were rejected from the sets analyzed by Priestley and Taylor (1972), the mean would be 1.28. The data of Davies and Allen (1973) on perennial ryegrass actually produced $\alpha_e = 1.27 \pm 0.02$, those of Jury and Tanner (1975) on potatoes $\alpha_e = 1.28$ and those of Mukammal *et al.* (1977) on grass $\alpha_e = 1.29$.

The fact that α_e for wet surfaces is of the order of 1.26 to 1.28 shows that 'advection-free' conditions in the sense of Slatyer and McIlroy (1961) hardly ever occur; it shows that over the ocean or over a large saturated land surface the second term of (10.15), that is, large-scale advection, accounts on average for about 21–22 percent of the evaporation rate. That this is indeed only an average and that α_e may assume different values under certain conditions was brought out in the study of DeBruin and Keijman (1979) on a large (460 km^2) shallow (3 m) lake. Comparison of mean daily E_e-values [cf. [10.21)] with EBBR evaporation data showed that $\alpha_e = 1.25 \pm 0.01$ with a correlation coefficient $r = 0.991$ over the summer time and early autumn. However, α_e showed a variation from month to month which is listed in Table 10.2. Furthermore, as illustrated in Figure 10.4 for 3-hourly data in the summer, α_e was also found to exhibit a daily variation between around 1.15 early in the day and 1.42 in the afternoon; still, the daily average for these 3-hourly α_e values for the summer was about 1.29. It seems reasonable at this point to attribute this variation in α_e to changes in the large-scale advection.

It is remarkable that so many land surfaces covered with fairly short vegetation,

TABLE 10.2
Seasonal variation of α_e for a large shallow lake (after DeBruin and Keijman, 1979)

	α_e	Correlation coefficient
April	1.50	0.98
May	1.28	0.98
June	1.25	0.99
July	1.21	0.99
August	1.20	0.99
September	1.25	0.99
October	1.49	0.98

such as grass, which is not actually wet but which has ample water available, yield about the same value of 1.26 to 1.28 in (10.23) as open water surfaces. As mentioned in Section 10.2a, this may be due to a fortuitous compensation of the specific humidity of non-wet leaf surfaces, which is lower than saturation, by a larger effective roughness, and thus transfer coefficient of the vegetative surface. Still, in some studies drastically different values of α_e have been reported. For example, McNaughton and Black (1973) found that $\alpha_e = 1.05$ for a young, 8 m high Douglas fir forest, when it is well supplied with water but not wet; however, on a day after rain had fallen α_e was 1.18, which is closer to the values for open water. Also Barton (1979) who studied evaporation from soil observed that $\alpha_e = 1.05$ under potential conditions. It is clear that the nature of α_e will require further study to explain these discrepancies.

Some attempts have been made to improve on the formulation (10.23) of Priestley and Taylor. DeBruin and Keijman (1979) suggested a linear relation but with a non-zero intercept; however, the intercept constant was only of the order of 10 W m^{-2}, which is quite small compared to the usual values of L_eE, at least in summer. (The slope was found to be 1.17.) Thus, they felt that the difference between the one-parameter (10.23) and a two-parameter model is not significant. Hicks and Hess (1977) related the long-term (mostly 10 to 20 days) average Bowen ratio with the sea surface temperature within a framework suggested by (10.22) and (10.24); they proposed a more general linear relation, as follows

$$\mathrm{Bo}_{pe} = a_e(\gamma/\Delta) + b_e \tag{10.25}$$

where, on the basis of eight data sets over the sea, the constants were found to be $a_e = 0.63$ and $b_e = -0.15$. With (10.4), (10.25) yields the following for evaporation from a wet surface under average conditions of minimal advection

$$E_{pe} = \frac{Q_{ne}}{a_e(\gamma/\Delta) + (1 + b_e)} \tag{10.26}$$

as an alternative to (10.23) of Priestley and Taylor. Both (10.23) and (10.26) produce the same result at about 25°C if $\alpha_e = 1.26$ and at about 20°C if $\alpha_e = 1.28$. Although (10.25) does not allow the determination of α_e as in (10.24), it is of some interest to note that (10.24) with $\alpha_e = 1.28$ yields a median line through the data points of Hicks and Hess (1977).

The energy budget related methods usually require data on the available energy term Q_{ne}, given in (10.2), or some approximation thereof. For situations when such Q_{ne} data are not available DeBruin (1978) has suggested that the calculation of the evaporation from a wet surface under average conditions of minimal advection can be performed by means of

$$E_{pe} = \left(\frac{\alpha_e}{\alpha_e - 1}\right)\left(\frac{\gamma}{\Delta + \gamma}\right) f_e(\bar{u})(e_a^* - \bar{e}_a). \tag{10.27}$$

The main advantage of this equation, which is obtained by combining (10.23) of Priestley and Taylor (1972) with (10.15) of Penman (1948), is that only measurements of $\bar{u}$, $\bar{T}_a$ and $\bar{e}_a$ are required at one level above the surface. Although (10.27) is manifestly quite sensitive to small changes in α_e, DeBruin (1978) obtained fairly good agreement for $\alpha_e = 1.26$ with EBBR data over a large lake; for daily averages the correlation coefficient was $r = 0.85$, but for 10 and 20 day-periods it rose to $r = 0.97$.

Some Related Empirical Formulae

In the literature numerous empirical equations have been proposed to estimate potential evapotranspiration, or so-called consumptive use, of well-watered actively growing crops. Some of these are of interest in the present context since they are related to or derivable from the equilibrium evaporation concept. For example, Makkink (1957) noted that the following

$$E = a \frac{\Delta}{\Delta + \gamma} R_{se} + b \tag{10.28}$$

yielded good results on a monthly basis in the Netherlands. $R_{se} = (R_s/L_e)$ is the global short-wave radiation, as equivalent rate of evaporation, and a and b are constants; their values were found to be $a = 0.61$, $b = -0.12$ (mm day^{-1}) with E-data representing the potential evapotranspiration from a grass-covered lysimeter with the water table maintained at 0.5 m below the soil surface; with a zero-intercept $b = 0$, the only constant was $a = 0.58$. More recently, Stewart and Rouse (1976) pointed out that (10.28) yields good results to determine evaporation from shallow lakes and ponds. From daily means obtained on a small (10^5 m^2) shallow (0.6 m) lake in northern Ontario they derived that $a = 0.9265$ and $b = 1.624$ MJ m^{-2} d^{-1}.

An even simpler equation was proposed by Jensen and Haise (1963) and Stephens and Stewart (1963), viz.

$$E = (aT_a + b)\, R_{se} \tag{10.29}$$

where a and b are constants. Various estimates have been given for their values. From an analysis of about 1000 measurements of consumptive use (but sometimes including seepage losses) from irrigated areas with various crops in the western United States, representing means over periods longer than 5 days, Jensen and Haise (1963) calculated that $a = 0.025$ (°C)$^{-1}$ and $b = 0.078$ with a correlation coefficient $r = 0.86$. For well-watered (but not wet) grass, using mean monthly temperature Stephens and Stewart (1963) obtained $a = 0.015$ (°C)$^{-1}$, $b = 0.072$, $r = 0.81$ in Florida; similarly Stephens (1965) found $a = 0.016$ (°C)$^{-1}$, $b = 0.087$, $r = 0.986$ with data from North Carolina, and $a = 0.019$ (°C)$^{-1}$, $b = 0.12$, $r = 0.967$ with data from California.

Clearly, the form of (10.28) and (10.29) can be derived from the equilibrium concept as given by (10.21) and (10.23): as can be seen in Figure 10.2 $\Delta/(\Delta + \gamma)$ can be fairly well approximated by a linear function of T, and R_s can often be well correlated with R_n, which is the main constituent part of Q_n over daily periods or longer. But since R_s is only one term of the energy budget, a given correlation can only be valid for the type of surface and the location for which it was derived. Equations such as (10.28) and (10.29) may be useful when only short-wave radiation and air temperature are available. They will, however, have to be calibrated by adjusting the constants to suit the local conditions.

10.3. SIMPLIFIED METHODS FOR ACTUAL EVAPOTRANSPIRATION

The methods treated in this section are related to the energy budget method, since they require a knowledge of the available energy Q_n, or as an approximation, the net radiation R_n. When the supply of water is reduced, so that the evaporating surface is no longer effectively wet, the energy budget usually undergoes considerable changes. Although the incoming radiative energy $(R_s(1 - \alpha_s) + \varepsilon_s R_{ld})$ probably does not change much (except for moisture effects on the albedo), the part in (10.1), which is no longer consumed by E, is redistributed primarily among H, R_{lu} and G. Until now, no simple and rigorous methods have been developed to predict this redistribution; in what follows, some approximate methods are presented.

a. Adjustment of Penman's Approach With Bulk Stomatal Resistance

Even when it is well supplied with water near the soil surface or at the roots, an evaporating vegetation cannot, in general, be considered wet, except after rainfall or dew formation. Thus, the specific humidity at the surface of the foliage elements is likely to be somewhat smaller than the saturation value at the corresponding temperature, and equations such as Penman's (10.15) are no longer applicable. To remedy this Penman and Schofield (1951) and later, more formally, Monteith (1963), Thom (1972) and others, have introduced various resistance parameters to characterize the transfer between the supposedly vapor-saturated stomatal cavities and the atmosphere. It should be noted, that when this concept is used in the context of evapotranspiration over dry soil, it may also account for the resistance through the upper layer of the soil.

The bulk stomatal resistance defined in (4.163) can be incorporated in the Penman approach as follows. By virtue of (3.2), (3.5), (3.6) and (10.10), the specific humidity can be written in terms of vapor pressure as

$$q = c_p e/\gamma L_e. \tag{10.30}$$

When the vegetation is not actually wetted, the surface vapor pressure e_s is not equal to e_s^*. However, they can be related by means of (4.168) and (4.169) with (10.30) in the following

$$(\bar{e}_s - \bar{e}_a) = \left(\frac{r_{av}}{r_{st} + r_{av}}\right)(\bar{e}_s^* - \bar{e}_a) \tag{10.31}$$

in which $\bar{e}_a$ is the actual vapor pressure in the air, which corresponds to $\bar{q}_r$ at the reference level $z = z_r$ in the surface sublayer. Thus, instead of (10.12), the Bowen ratio (10.9) becomes with (10.11) and (10.31)

$$\mathrm{Bo} = \frac{\gamma}{\Delta}\left(\frac{r_{st} + r_{av}}{r_{av}}\right)\left[1 - \frac{(e_a^* - \bar{e}_a)}{(e_s^* - \bar{e}_a)}\right]. \tag{10.32}$$

Again, substitution of (10.32) into (10.4) produces, instead of (10.13),

$$Q_{ne} = \left[1 + \frac{\gamma}{\Delta}\left(\frac{r_{st} + r_{av}}{r_{av}}\right)\right]E - \frac{\gamma}{\Delta}\left(\frac{r_{st} + r_{av}}{r_{av}}\right)\left(\frac{e_a^* - \bar{e}_a}{e_s^* - \bar{e}_a}\right)E. \tag{10.33}$$

Replacing E of the second term on the right of (10.33) by (4.169) with (10.30) one obtains the desired result

$$E = \frac{\Delta Q_{ne} + \rho c_p(e_a^* - \bar{e}_a)/(L_e r_{av})}{\Delta + \gamma(1 + r_{st}/r_{av})}. \tag{10.34}$$

Clearly, when $r_{st} = 0$, (10.34) reduces to an expression which is equivalent with Penman's (10.15). Equation (10.34) was probably first given by Thom (1972). Earlier Monteith (1963) had already derived a similar result in terms of r_c and r_a, as defined in (4.170) and (4.171), instead of r_{st} and r_{av}; although the formulation with the canopy resistance r_c is still being used in practice, as noted behind (4.171) the formulation (10.34) with r_{st} is conceptually preferable.

As is the case with Penman's (10.15), a feature which makes (10.34) appealing is that measurements are required at only one height above the surface, rather than profiles at two or more levels. Its main drawback, when it is to be used as a tool to determine E with meteorological data, is that r_{st} is generally unknown. The interactions of such factors as the distribution of heating by radiation and the distribution of vapor sources on different parts of the canopy through the day and through the season, the current state of biological activity and senescence, the moisture stress at the roots, the specific physiological characteristics of the species under consideration, are quite intricate and difficult to quantify. This explains why, until now, there are no general relationships for r_{st} or r_c in terms of some other easily measurable soil, plant or atmospheric parameters. The concept of a stomatal resistance has been found to be quite useful in some simulation models and as a diagnostic index of water stress conditions. However, more research will be needed before it will become a practical device for predictive purposes in hydrology and climatology.

b. Complementary Relationships between Actual and Potential Evaporation

Bouchet's Hypotheses

Bouchet (1963) arrived at the following complementary relationship, shown in Figure 10.5, between the potential evaporation E_p and the actual regional evaporation E

$$E_p + E = 2E_{p0}. \tag{10.35}$$

The actual evaporation rate E is the average value from a large uniform surface of regional size, involving characteristic scale lengths of the order of 1 to 10 km. The potential evaporation E_p is the evaporation which would take place under the prevailing atmospheric conditions if only the available energy were the limiting factor.

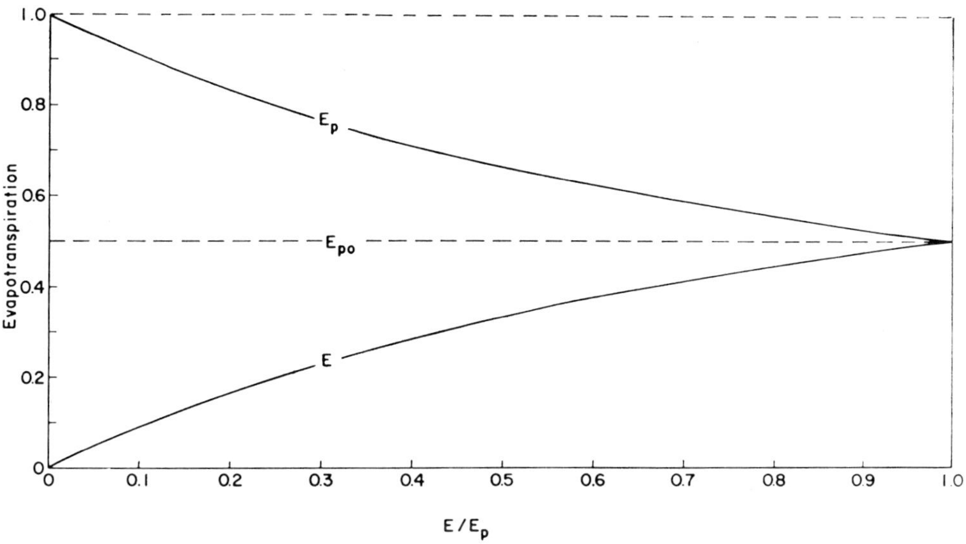

Fig. 10.5. Sketch illustrating Bouchet's (1963a; b) hypothesis; E and E_p are plotted versus E/E_p such that $E + E_p$ = constant as given in (10.35).

Under conditions when E equals E_p, it is denoted by E_{p0}. The derivation of (10.35), which is based on somewhat heuristic arguments, may be given as follows.

If for one or another reason, independent from the available energy, E decreases below E_{p0}, a certain amount of energy becomes available, that is

$$E_{p0} - E = q_1. \tag{10.36}$$

At the scale of the region this decrease of E, with respect to E_{p0}, has probably a relatively small impact on the net radiation, and it affects primarily the temperature, the humidity and the turbulence of the air near the ground. As a result, this available energy flux q_1 causes an increase in E_p. Bouchet's (1963) main hypothesis was, that in the absence of local oasis effects the energy budget remains otherwise unaffected, and that the potential evaporation is increased by exactly that amount, or

$$E_p = E_{p0} + q_1. \tag{10.37}$$

The combination of (10.36) and (10.37) produces the complementary relationship (10.35). Equation (10.35) was the first part of the analysis.

On the basis of additional approximations of the energy budget, a second relationship was obtained

$$2E_{p0} \le (1 - \alpha_s)R_{se} + Q_A \tag{10.38}$$

where $R_{se} = R_s/L_e$ is the global short-wave radiation expressed in units of evaporation rate, α_s the albedo and Q_A a large-scale or regional advection term. By substitution of (10.38) into (10.35), Bouchet obtained his final result

$$E_p + E \le (1 - \alpha_s)R_{se} + Q_A \tag{10.39}$$

which in subsequent applications (cf. Morton, 1969; Fortin and Seguin, 1975) has usually been simplified to the following

$$E_p + E = (1 - \alpha_s) R_{se}. \tag{10.40}$$

Bouchet's (10.39) and (10.40) have not been used very widely, primarily because it is difficult to assess the validity of the assumptions. Also, it is not exactly clear which measures of E_p, E_{p0} and Q_A would yield optimal results in applications. Nevertheless, Bouchet's approach contains worthwhile ideas, and it has led to further developments which are described next.

A Climatological Method

Bouchet's complementary relationship (10.35) was applied by Morton (1975, 1976) to estimate evapotranspiration for climatological purposes. The term E_p in (10.35) was assumed to be given by Penman's Equation (10.15), which was modified by replacing $f_e(\bar{u}_r)$ by an empirical constant f_A and Q_{ne} by $R_{ne}(= R_n/L_e)$. The term E_{p0} in (10.35) was assumed to be the same as E_{pe} of Priestley and Taylor (1972) as given by (10.23) with $\alpha_e = 1.38$. However, this equation was modified by replacing Q_{ne} by $(R_{ne} + M_m)$, where M_m is an empirical advection term; this term which was considered to be zero in spring and summer, and positive in fall and winter when R_n is low or negative, was related to the net long-wave radiation and the incident short-wave radiation as follows

$$M_m = (\theta_m R_{nl} - \phi_m R_s)/L_e \tag{10.41}$$

where R_s and R_{nl} are defined in (6.2) and (6.25) and where θ_m and ϕ_m are empirical constants; in addition, M_m was subjected to the constraint $M_m \geq 0$. Morton's (1976) result can be written as

$$E = \frac{\Delta}{\Delta + \gamma}(1.76 R_{ne} + 2.76 M_m) - \frac{\gamma}{\Delta + \gamma}(e_a^* - e_{da}^*) f_A \tag{10.42}$$

where e_{da}^* is the saturation vapor pressure at the dewpoint temperature of the air, and where M_m is given by (10.41). This equation was calibrated with monthly temperature, humidity, sunshine and precipitation data obtained from climatological stations in arid regions where E equals precipitation and with empirical equations for the radiation terms. The three empirical constants needed for the application of (10.42) were found to be $\theta_m = 1.37$, $\phi_m = 0.394$, $f_A = 47.5$ cal cm^{-2} d^{-1} mb^{-1} for $T \geq 0°$C and $f_A = 54.6$ cal cm^{-2} d^{-1} mb^{-1} for $T < 0°$C. Morton (1976) used this method, which gave good results on an annual basis, to calculate monthly evaporation for large areas and river basins.

The Advection-Aridity Approach

Brutsaert and Stricker (1979) proposed an approach by combining Bouchet's (1963) complementary relationship (10.35) with the ideas on regional advection effects, advanced by Slatyer and McIlroy (1961; ch. 3, p. 73) and outlined behind (10.21). Accordingly, the term E_p in (10.35) was assumed to be obtainable from (10.15) of Penman; but since that equation already contains the effect of large-scale advection

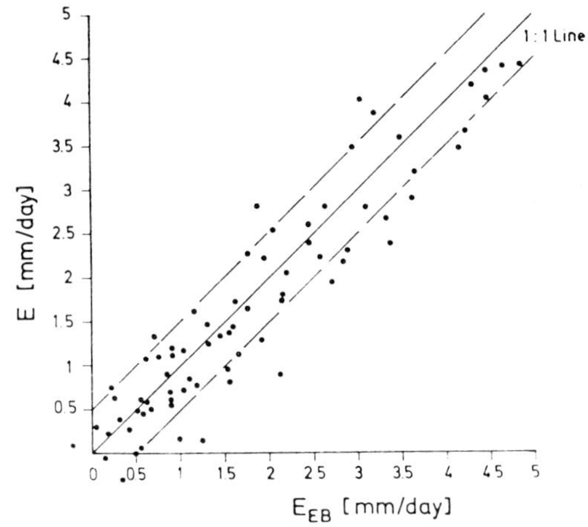

Fig. 10.6. Comparison between daily values of actual evapotranspiration E_{EB} determined by means of the energy budget method (EBWSP), and evapotranspiration E obtained by means of advectionaridity equation (10.43) with $\alpha_e = 1.28$ and $f_e(\bar{u}_2)$ given by (10.17) (from Brutsaert and Stricker, 1979).

in the second term, it was assumed that this obviates the need for additional advection terms. Under conditions of limited water supply at the surface, E_p is not the same as the evaporation that would be observed if the surface were well supplied with water. Thus E_p determined this way may be considered as an apparent potential evaporation. The term E_{p0}, needed in (10.35), was assumed to be given by E_{pe}, the potential evaporation under average conditions of minimal large-scale advection for which several expressions are available (e.g., (10.23), (10.26) and even (10.29.). This approach is referred to as 'advection-aridity' approach because the non-availability of water for evaporation, that is the aridity of the region, is deduced from the large-scale advection of drying power of the air, as implied by the atmospheric conditions. Though developed independently, it is similar to Morton's (1976) approach as given in (10.42); however, it does not require calibration of the constants and it is applied with daily instead of monthly data.

Several alternative formulations are possible, depending on the particular choice of the expressions for E_p and E_{pe}. Substitution of (10.15) for E_p and (10.23) for E_{p0} in (10.35) yields (Brutsaert and Stricker, 1979)

$$E = (2\alpha_e - 1)\frac{\Delta}{\Delta + \gamma} Q_{ne} - \frac{\gamma}{\Delta + \gamma} E_A \tag{10.43}$$

where α_e is of the order of 1.26 to 1.28 on average, and E_A is defined in (10.16). Obviously, an alternative but probably equivalent form could be obtained by using (10.26) or possibly another expression instead of (10.23). Also, beside (10.15), it may be possible to determine E_p by other means, possibly even by a carefully calibrated evaporimeter.

Equation (10.43) was tested by comparing the calculated daily means with corresponding values obtained by means of an EBWSP method (see Section 10.1b) for a dry summer period in Gelderland. As shown in Figures 10.6 and 10.1 the agreement was generally good. It was also found that the method is relatively insensitive to the choice of α_e (viz., 1.26 or 1.28) and the choice of the wind function $f_e(\bar{u}_r)$ in E_A.

In summary this approach requires the same data as Penman's (10.15). The main advantage of all the methods based on Bouchet's complementary relationship is that only meteorological parameters are needed; thus no soil moisture data, no stomatal resistance properties of the vegetation, nor any other measures of the aridity are required, which are used in other methods to reduce a calculated potential evaporation to actual evapotranspiration. The advection-aridity approach, as formulated in (10.43), has the additional appeal that it does not involve any calibration to determine the parameters. So far, this approach has been tested with only one set of data, so that more research is needed to ascertain its applicability.

c. Extensions of Equilibrium Evaporation Concept

Direct Application

True equilibrium conditions, as described in Section 10.2c are probably quite rare. Nevertheless, in several studies it has been observed that under certain conditions the actual rate of evapotranspiration (under non-potential conditions) can be closely approximated by E_e as given by (10.21).

The idea was introduced by Denmead and McIlroy (1970); they suggested that, whereas usually effects of upwind advection tend to keep the potential evaporation above its limiting equilibrium value, the effects of soil moisture deficit act in the opposite direction, reducing the actual evaporation below the potential value E_p. Accordingly, they felt that except in a desert or under conditions of local advection, actual evapotranspiration should never be very different from equilibrium evaporation, and E_e may possibly serve as a simple measure for true crop evapotranspiration. They tested

$$E = \alpha_a E_e \tag{10.44}$$

in which $\alpha_a = 1$, with hourly data above a wheat field near Canberra, Australia. Although the scatter was quite large, the experimental E-data obtained by a lysimeter showed that (10.44) yielded good results up to about 25 mW cm^{-2}, but E was overestimated at higher rates. Also Wilson and Rouse (1972) and Davies and Allen (1973) observed that their experimental data obtained with various field crops under 'moderately dry conditions' in Ontario were not inconsistent with (10.44) with $\alpha_a = 1$, on average.

However, under more arid conditions the matter may not be that simple. Rouse *et al.* (1977), who studied evaporation from subarctic surfaces with lichen heath and lichen woodland, found that for wet soil conditions α_a approaches 1.26 ($= \alpha_e$) but that for dry conditions α_a falls below unity to an average of about 0.95. Brutsaert and Stricker (1979) noted that over the 74-day period of their study, the average actual E was practically the same as E_e; however, the scatter in the comparison of the daily results, as shown in Figure 10.7, was larger than that in the corresponding comparison of the E-values calculated with (10.43) (see Figure 10.6) of the advection-

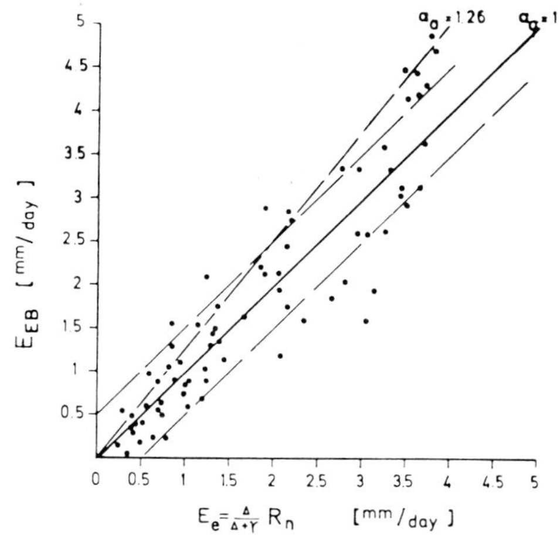

Fig. 10.7. Comparison between daily values of equilibrium evaporation E_e, calculated by means of (10.21), and the corresponding estimates of evapotranspiration by means of an energy budget method E_{EB} in Gelderland during the relatively dry period June–September 1976. The symbol α_a is defined in (10.44) (from Brutsaert and Stricker, 1979).

aridity approach. Williams *et al.* (1978), who studied a seeded rangeland grass surface in British Columbia, confirmed that $\alpha_a = \alpha_e = 1.26$ under moist conditions; however they indicated that (10.44) with $\alpha_a = 1$ seriously overestimated E under dry conditions.

This brief review indicates that (10.44) with $\alpha_a = 1$ may be a useful estimator of average actual evapotranspiration when the water supply is not severely restricted. But the reliability of this approximation is still not well established.

Empirical Modifications Accounting for Drying

As α_a in (10.44) appears to vary, depending on the moisture availability, several attempts have been made to relate α_a to surface moisture or to some other parameter. Noteworthy is the study of Davies and Allen (1973) who fitted their data to

$$\alpha_a = a[1 - \exp(-b\theta/\theta_f)] \qquad (10.45)$$

where θ is the volumetric soil water content in the upper 0.05 m of soil and θ_f the same variable at field capacity; by a least-squares fitting procedure the data (see Figure 10.8) yielded $a = 1.26$ and $b = 10.563$ (for $N = 22$ and $r^2 = 0.98$). Equation (10.45) has been applied by others (e.g., Williams *et al.*, 1978; Barton, 1979); although the trend of the data was generally the same, the parameters came out quite differently. Mukammal and Neumann (1977), who did not fit their data to (10.45), also noted a similar dependency of α_a on θ. The use of surface soil moisture to determine α_a is probably only possible in the case of bare soil surfaces or of vegetation with shallow roots. Also, from the studies reviewed here, it is clear that the relationship between α_a and θ is not likely to be very general and that it must be determined again for any new set of conditions. Nevertheless, a knowledge of $\alpha_a = \alpha_a(\theta)$, (if it exists) is

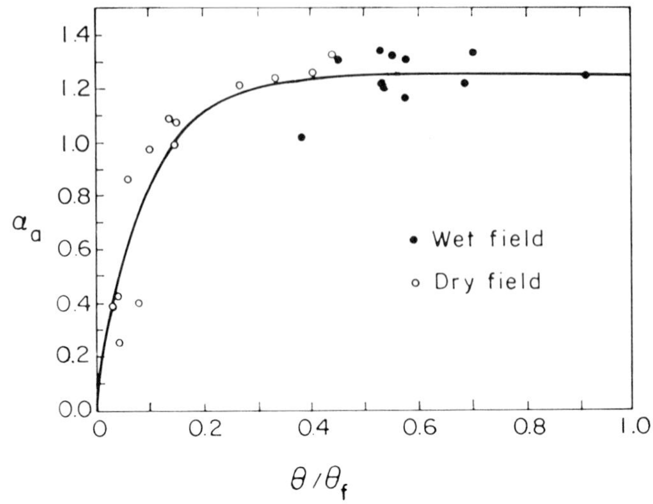

Fig. 10.8. Variation of α_a of (10.44) with surface soil water content θ expressed relative to that at field capacity θ_f for a sandy loam covered with perennial ryegrass in Ontario (adapted from Davies and Allen, 1973).

of interest, since it may provide a way of applying soil moisture data, obtained by remote sensing (e.g., Schmugge, 1978) or by other techniques, to estimate evaporation from land.

Various correlations involving rainfall P have also been considered. Priestley and Taylor (1972) presented α_a of (10.44) as a function of $\int(E - P)\,dt$ as accumulated water deficit in cm of water, for different situations. However, they were unable to generalize their results to determine at what value of this deficit the evaporation rate begins to fall below the potential rate at which $\alpha_a = 1.26$. They concluded that to resolve this issue would require a more detailed study of soil-plant water movement. Shuttleworth and Calder (1979) who compared equilibrium evaporation with long-term (annual) evaporation from a spruce forest in Wales and a Scotspine forest in Norfolk, derived the following correlation

$$E = (0.72 \pm 0.07)E_e + (0.27 \pm 0.08)P. \tag{10.46}$$

Also these authors emphasized that their equation is only applicable for the particular conditions under which it was developed.

CHAPTER 11

Mass Budget Methods

Mass budget methods are based on the principle of conservation of mass applied to some part of the hydrological cycle. Conservation of mass, formulated as a mass budget equation, requires that, in general, for any given control volume, the inflow rate minus the outflow rate equal the rate of change of the water stored. Accordingly, evaporation can be determined as the only unknown rest term in the budget equation if all the other terms can be determined independently. Although, from the conceptual point of view, mass budget methods are by far the simplest, their application is often difficult and impractical. Therefore, they are less commonly used than aerodynamic or energy budget methods. Nevertheless, their conceptual simplicity is an appealing feature and, in certain situations, a mass budget approach can be quite appropriate. In this chapter a brief description is given of several ways in which the mass budget can be applied in practice.

11.1. TERRESTRIAL WATER BUDGET

a. Soil Water Depletion and Seepage

Experimental Determination in the Field

Local evaporation from a land surface, with or without vegetation, can be estimated from the water budget equation for a soil layer. For a control volume consisting of a soil column of thickness h_{s0} and unit horizontal area, and with all the terms taken as mean values over a sampling period, this gives an evaporation rate

$$E = -\frac{1}{h_{s0}}\int_0^{h_{s0}} \frac{\partial \theta}{\partial t}\,dz + (P + q_{ri} + q_{si}) - (q_d + q_{ro} + q_{so}) \tag{11.1}$$

where z is the vertical coordinate pointing down from $z = 0$ at the surface, θ is the specific soil water content as volume fraction, P the rate of precipitation (or irrigation), q_d the rate of downward seepage or drainage through the lower boundary at $z = h_{s0}$, q_{ri} the lateral inflow rate over the soil surface, q_{ro} the corresponding outflow, q_{si} the lateral inflow rate resulting from soil water flow, and q_{so} the corresponding outflow. In most applications, the differences of the lateral flow terms are negligible and the budget Equation (11.1) becomes

$$E = -\frac{1}{h_{s0}}\int_0^{h_{s0}} \frac{\partial \theta}{\partial t}\,dz + p - q_d. \tag{11.2}$$

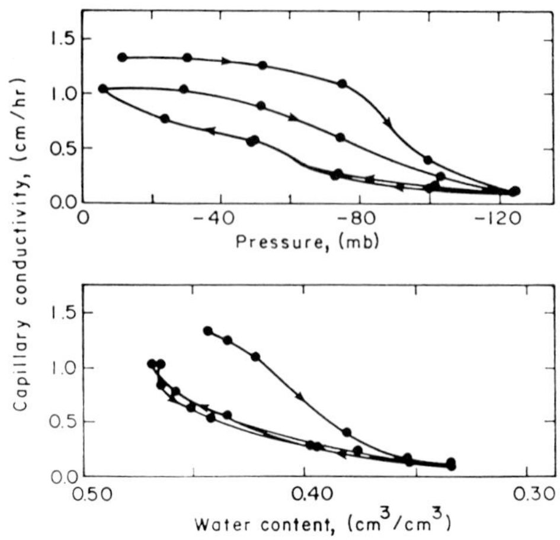

Fig. 11.1. Example of capillary conductivity curves of a silt-loam for cycles of wetting and drying. The $k(p_w)$ relationship exhibits considerable hysteresis but not the $k(\theta)$ relationship. In this experiment, the initial cycle is different from the succeeding ones on account of soil consolidation resulting from the initial application of negative pressure (from Nielsen and Biggar, 1961).

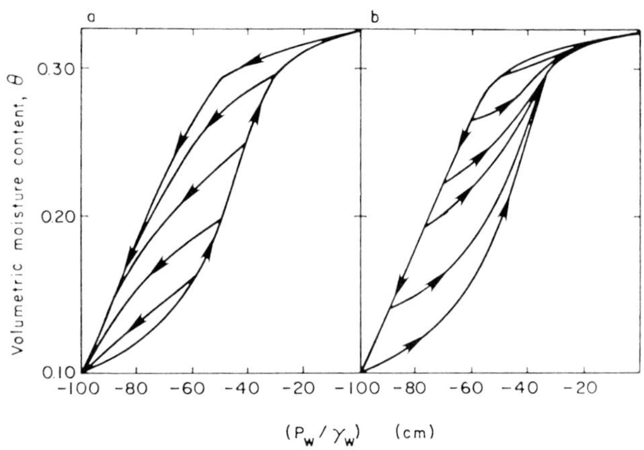

Fig. 11.2. Curves showing the hysteresis in the relationship between the water content and the water pressure (p_w) for Adelaide dune sand; (a) represents the draining scanning curves and (b) the wetting scanning curves; the boundary hysteresis loop is the same in (a) and (b) (adapted from Talsma, 1970).

Mean values of the finite difference form of $(\partial\theta/\partial t)$ over the sampling period as a function of z can be determined by various methods. In the earlier field experiments related to irrigation of agricultural crops (e.g., Israelsen, 1918; Edlefsen and Bodman, 1941), the method consisted of soil sampling and gravimetric analysis before and after drying of the samples in an oven. More recently, the neutron scattering method and other (e.g., Schmugge *et al.*, 1980) techniques have become available which allow *in-situ* soil moisture measurements.

The method is probably most useful in situations where q_d is negligible, so that evaporation is the only depletion mechanism of the moisture content of the soil pro-

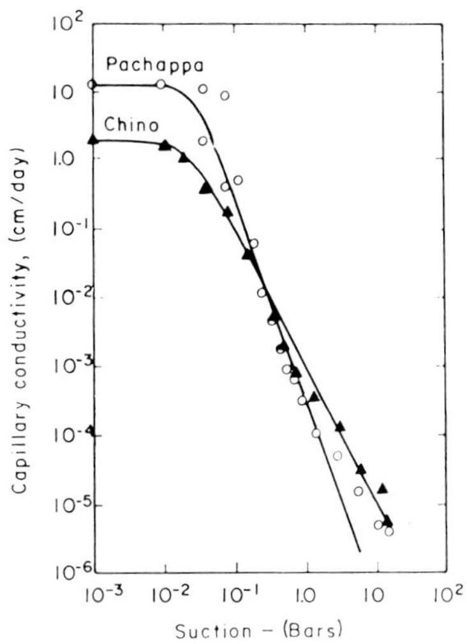

Fig. 11.3. Capillary conductivity k at 25°C for a fine sandy loam (Pachappa) and a clay (Chino) as a function of suction $-p_w$. The curves represent (11.6) (from Gardner and Fireman, 1958).

file. Still, with some additional information it may be possible to obtain reliable estimates of q_d.

The specific flux of water $\mathbf{v}_s$ in an isotropic soil can be described by the extension of Darcy's law to partly saturated soils, suggested by Buckingham (1907) and Richards (1931), viz.

$$\mathbf{v}_s = -k\left(\frac{1}{\gamma_w}\nabla p_w - \nabla z\right) \tag{11.3}$$

where p_w is the soil water pressure (negative pressure is also referred to as suction or tension), $\gamma_w = \rho_w g$ the specific weight of the water, $k = k(\theta)$ is the hydraulic (or capillary) conductivity, and z is pointing down. Since θ is a function of the soil water pressure p_w, k can also be considered as a function of the pressure. In Figure 11.1, an example is shown of $k(\theta)$ and $k(-p_w)$ for a silt loam. Although the $k(-p_w)$ relationship exhibits considerable hysteresis, this is usually not the case for the $k(\theta)$ relationship. In Figure 11.2 an example is given of the relationship between θ and $-p_w$ for a sandy soil; it shows that this relationship is also subject to considerable hysteresis. In other words, both $k = k(-p_w)$ and $\theta = \theta(-p_w)$ depend on the sequence of events of wetting and drying by which the current value of θ is attained. Clearly, when the problem is one of drying only or of wetting only, hysteresis does not have to be considered. Figure 11.3 shows two additional examples of $k(-p_w)$ during a drainage cycle, namely for a fine sandy loam and for a clay.

The vertical component of $\mathbf{v}_s$ can be written as

$$v_{sz} = -k\left(\frac{1}{\gamma_w}\frac{\partial p_w}{\partial z} - 1\right). \tag{11.3'}$$

In the present case (11.3′) can be used to determine the downward drainage rate $q_d = v_{sz}$, provided data are available on $(\partial p_w/\partial z)$ (in finite difference form) and on $k(\theta)$.

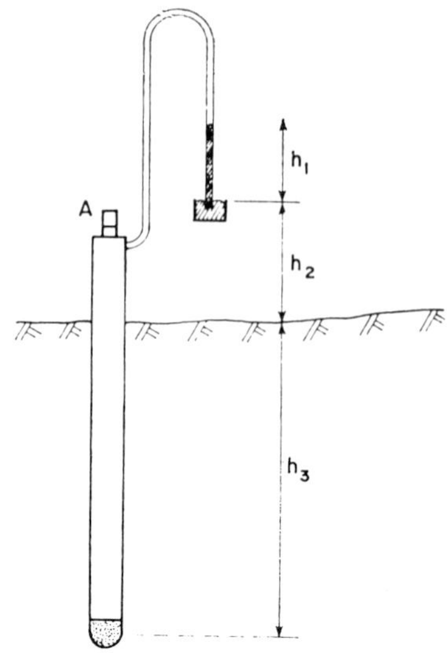

Fig. 11.4. Schematic sketch of a tensiometer installed in the field; a manometer fluid (e.g., mercury) has a column height h_1 above that of the reservoir surface h_2; the porous cup, which is installed at depth h_3, is filled with water in contact with the soil water; at A, the main tube can be opened to fill it with water or to bleed it of air bubbles.

The soil water pressure p_w at a point z in the profile can be measured by means of a tensiometer; this instrument, (see Figure 11.4) which developed through the work of Gardner *et al.* (1922), Kornev (1924), Israelsen (1926), Richards (1949) and others, is a water-filled manometer with a sensing element consisting of a porous cup with sufficiently fine pores to ensure continuous contact (without air leakage) between the water in the soil and that in the manometer. The capillary conductivity k as a function of water content θ can be determined by different methods; for the present purpose field estimates for the undisturbed soil profile are preferable. Various experimental determinations, consisting in general of the inverse application of the finite difference form of (11.2) with (11.3) in the absence of P and by preventing E at the surface, have been carried out by Ogata and Richards (1957), Nielsen *et al.* (1964, 1973), Davidson *et al.* (1969) and Baker *et al.* (1974). The results of such a field determination can also be supplemented by laboratory methods or calculations (e.g., Brutsaert, 1967; Klute, 1972).

In several field studies (e.g., Richards *et al.*, 1956; Nielsen *et al.*, 1973) it has been observed that during the vertical redistribution of soil water at depths of a meter or more, where there is no influence of surface evaporation, the hydraulic gradient is rarely very different from unity. This allows the approximation of (11.3′),

$$q_d = k. \tag{11.4}$$

Thus, in such a case, q_d may be estimated approximately with only a measurement of the soil water content at $z = h_{s0}$, provided, of course, $k = k(\theta)$ is known.

Still other simplications regarding q_d are possible. For example, Tanner and Jury (1976) represented q_d as an experimentally determined exponential function of water content. In many situations, however, especially during the second stage of drying [see (11.12)], the downward drainage rate may simply be neglected; but this needs to be checked in each particular case.

Measurements of soil water content and water pressure at several levels in the profile are not easy and they require many precautions. The soil water depletion method is probably only useful for special experimental situations under favorable conditions, and it is clearly not generally applicable on a routine basis. It is usually hard, if not impossible, to apply when one of the following conditions are present: a water table close to the surface, frequent and large precipitation, non-negligible or unknown net lateral inflows, a large drainage rate q_d, and considerable variability in the soil properties. Thus, the accuracy obtainable with this method depends largely on the local conditions. Examples of the determination of evaporation from measurements in the soil profile can be found in studies by Jensen (1967), Davidson *et al.* (1969) and Scholl and Hibbert (1973).

Some Theoretical Calculations for Bare Soil

The water which evaporates at a bare soil surface is transported to the surface through the underlying layers of the soil profile. The exact formulation of this transport is rather complicated (e.g., Philip, 1957; De Vries, 1958), since the water transport takes place both in the liquid and vapor phase, involving not only pressure gradients and gravity but often also temperature gradients with a soil heat flux, and salt concentration gradients. Nevertheless, it has been found that, in many situations of hydrological interest, the main features of the evaporation at the soil surface can be obtained on the basis of the isothermal flow equation, viz. Darcy's law (11.3). In particular, two such situations are of interest here, namely evaporation in the presence of a water table, and unsteady evaporation from a soil profile without water table.

(*i*) *Steady evaporation from a water table.* The water flows from the water table, through the soil profile to the soil surface, where it is taken away by evaporation.

For a vertical coordinate system pointing upward with $z = 0$ at the water table, where $p_w = 0$, (11.3′) yields, since $E = v_{sz}$,

$$z = \frac{-1}{\gamma_w} \int_0^{x-p_w} \frac{\mathrm{d}x}{1 + [E/k(x)]}. \tag{11.5}$$

This can be integrated readily for a uniform soil profile, provided the capillary conductivity $k(p_w)$ is known as a function of the soil water pressure. Observe that in (11.3) the capillary conductivity was defined as $k(\theta)$; however, since the water content θ is a function of the soil water (capillary) pressure, p_w, k is also a function of p_w. Several $k(p_w)$ have been proposed in the past, but Gardner (1958) concluded that for most soils the following empirical equation can be used to fit the data,

$$k = \frac{a}{(-p_w/\gamma_w)^n + b} \tag{11.6}$$

where a, b and n are constants; this function is shown for two soils in Figure 11.3. Note that (a/b) is the hydraulic conductivity at satiation k_0; b is the value of $(-p_w/\gamma_w)^n$

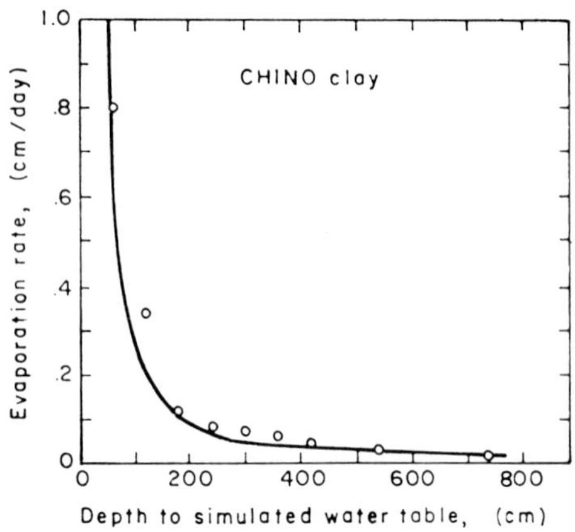

Fig. 11.5. Comparison between experimental rates of steady evaporation from a column of clay soil and the curve calculated by means of the integral of (11.3′) with (11.6) in the form $k = 1100/(p_w^2 + 565)$ cm day^{-1}, where p_w is in millibars (adapted from Gardner and Fireman, 1958).

when $k = k_0/2$, and the range of n lies between about 2 for clayey soils and 4 or more for sandy soils. Gardner (1958) presented the solutions of (11.5) with (11.6) for $n = 1$, 3/2, 2, 3 and 4.

Equation (11.5) produces the vertical pressure distribution of the soil water for any given rate of evaporation E. For relatively low values of E or for a soil profile with a water table at a shallow depth below the surface, the value of $(-p_w)$, that is the soil-water suction at the soil surface, is relatively small, and the soil surface is close to saturated. Hence, in such a case, the rate of evaporation is governed by the prevailing atmospheric conditions, and not by the ability of the soil profile to transmit water. However, as the drying power of the air, or the depth of the water table is increased, so that also the suction $(-p_w)$ at the soil surface increases, the rate, at which water moves upward and evaporates, increases. But eventually a limit is approached beyond which E cannot increase; in the limit E is totally controlled by the ability of the profile to transmit water, regardless of the drying power of the air, that is the potential evaporation. For most practical purposes, it is probably sufficiently accurate to assume that the actual evaporation at any time is the lesser of the potential evaporation and of the limiting evaporation E_{lim}.

A satisfactory approximation of this limiting value E_{lim} can be obtained by assuming that the soil surface at $z = d_w$ is nearly dry or at field capacity, so that $(-p_w) \to \infty$ and $k \to 0$. Integration of (11.5) with (11.6) produces then in general (e.g., Cisler, 1969) the following relationship between the limiting rate of evaporation and the depth of the water table,

$$d_w = \frac{\pi}{n \sin(\pi/n)} \left(\frac{a}{a + b\,E_{\text{lim}}} \right) \left(\frac{a + b\,E_{\text{lim}}}{E_{\text{lim}}} \right)^{1/n}. \qquad (11.7)$$

Since, in many cases, $a \gg bE_{\text{lim}}$, this is to a good approximation

$$E_{\text{lim}} = a \left[\frac{\pi}{n \sin(\pi/n)} \right]^n d_w^{-n}. \qquad (11.7')$$

Equation (11.7′) suggests that $E_{\lim}$ is proportional to d_w^{-n}. As shown in Figure 11.5, experimental results tend to confirm this. Note, however, that although the theoretical curve in Figure 11.5 is similar to (11.7′), it is not quite the same. In the experiments of Gardner and Fireman (1958) the column was only 1 m long and the depth of the water table was simulated by imposing a negative pressure at the bottom of the column; this negative pressure must be taken as the lower limit of integration of (11.3′), rather than zero as in (11.5). Nevertheless, most of the flow resistance occurs near the top of the soil column, where the soil water suction is the largest, so that little is lost by not extending the soil column down to the water table.

The model on which (11.7′) is based is clearly an oversimplification. Especially near the soil surface water vapor transport may be important so that the limiting evaporation rate is probably larger than the predicted value. However, Gardner (1958) has estimated that the increase is not likely to exceed 20 percent. At any rate, the result shown in Figure 11.5 illustrates the adequacy of the isothermal flow model for steady evaporation in the presence of a water table.

Equation (11.5) was used by Willis (1960) to study the steady-state flow from a water table in the case of a soil profile consisting of two layers of different texture. He concluded that for many practical purposes the presence of inhomogeneities may have little effect on E when d_w is relatively large; the effect of stratification was pronounced for a system with the coarse-textured soil overlying the fine-textured soil, but not for the reversed condition.

(*ii*) *Unsteady drying of soil profile without water table.* A high water table at a constant depth is not a common occurrence; more often than not the water which evaporates from the soil surface is supplied by a release from storage in the soil profile. To facilitate the solution of this problem, it is convenient to consider two stages in this drying process of the soil profile.

In the first stage, which prevails as long as the soil is still sufficiently moist, the evaporation rate is primarily controlled by the atmospheric conditions; therefore, it is sometimes referred to as the energy-limiting rate. Obviously, for constant atmospheric conditions, the rate of drying is constant. The duration of the first stage depends on the rate of evaporation and the ability of the soil profile to supply this rate. The rate of evaporation during this stage can thus probably best be calculated on the basis of measurements in the atmosphere.

As the soil near the surface dries out, the water supply to the surface eventually falls below that required by the atmospheric conditions. In this second or falling-rate stage, the rate of evaporation is limited by the conditions and properties of the soil profile. The transition from the first to the second stage may be quite abrupt at a given point on the surface, but on a field-wide scale it is usually more gradual. It was noted by Jackson *et al.* (1976) that the transition from the first to the second stage can be characterized sometimes by a change in albedo.

In the second stage of drying, water moves also through the profile by diffusion of water vapor. And especially after the soil has become quite dry, the water transport in the profile is sensitive to the temperature gradients in the soil. However, when the profile has become dry, the rate of evaporation is usually so small that it is of little significance hydrologically. Thus, at least initially in the falling rate stage, the water moves primarily as a liquid. Although the matter is more complicated (e.g., Philip, 1957; Cary, 1967), just like for the steady case, the available evidence shows that some of the more important features of the falling-rate stage of drying can be obtained by means of the simple isothermal flow model.

The governing equation of this approach is obtained by combining the equation of continuity of soil water, with Darcy's law (11.3). For an incompressible isotropic soil and an incompressible fluid this is the Richards (1931) equation for vertical flow,

$$\frac{\partial \theta}{\partial t} = \frac{\partial}{\partial z}\left(k \frac{\partial (p_w/\gamma_w)}{\partial z} - k\right). \tag{11.8}$$

For mathematical convenience (11.8) is often rewritten as follows

$$\frac{\partial \theta}{\partial t} = \frac{\partial}{\partial z}\left(D \frac{\partial \theta}{\partial z}\right) - \frac{\partial k}{\partial z} \tag{11.9}$$

where, by definition (e.g., Klute, 1952), $D = k[\mathrm{d}(p_w/\gamma_w)/\mathrm{d}\theta]$ is the soil water diffusivity. The solution of (11.9) is not easy, mainly because $D = D(\theta)$ and $k = k(\theta)$ are highly nonlinear, and also because $D(\theta)$ exhibits hysteresis (e.g., Staple, 1976) under conditions of alternate wetting and drying.

A simplified problem formulation, whose solution still has considerable practical relevance, can be obtained by considering the second stage of drying as a problem of desorption; this formulation, which was first used by Gardner (1959) involves the following assumptions. First, it is assumed that the effect of gravity is negligible, so that the second term on the right of (11.9) can be omitted. In other words, it is assumed that the drying of a vertical soil column is the same as that of a horizontal column. As a result, (11.9) becomes

$$\frac{\partial \theta}{\partial t} = \frac{\partial}{\partial z}\left(D \frac{\partial \theta}{\partial z}\right). \tag{11.10}$$

Second, the boundary conditions are taken as

$$\begin{aligned} \theta &= \theta_i, & z &\geq 0, & t &= 0, \\ \theta &= \theta_0, & z &= 0, & t &> 0, \end{aligned} \tag{11.11}$$

where θ_i is the initial water content of the soil and θ_0 the water content at the presumably dry soil surface. In the first of (11.11) it is assumed that the initial water content is uniform, and in the second that the water content at the surface is always very low. These conditions are equivalent with the assumption that the energy-limiting drying rate, i.e., the potential evaporation, is so large, that the duration of the first stage of drying is negligibly short.

To date, no general exact solution has been obtained for (11.10) with (11.11), but only approximate solutions or for certain types of diffusivity functions. Gardner (1959) made use of two solutions. One was the linearized solution obtained by means of a weighted-mean diffusivity calculated by Crank's method. The other solution, which was presented graphically, was obtained by iteration for the exponential-type diffusivity; this diffusivity is quite suitable for most soils, at least for wetting (Brutsaert, 1979b). A detailed discussion of methods of solution is beyond the scope of this

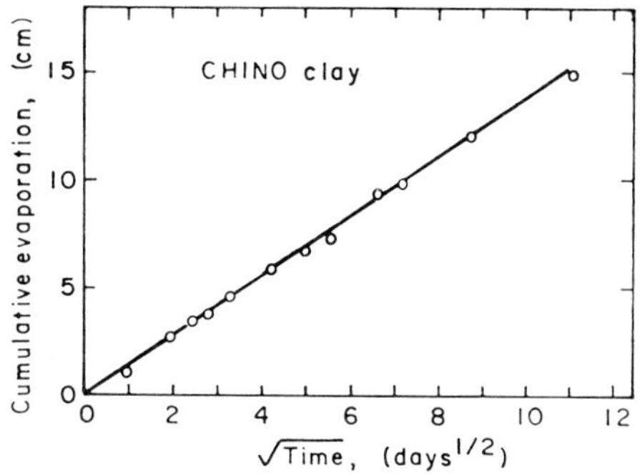

Fig. 11.6. Cumulative evaporation from a bare soil surface as a function of the square root of time obtained in the laboratory for a 1.0 m column of Chino clay (from Gardner, 1959).

book. In the present context, however, the most interesting feature of any solution of (11.10) with (11.11), regardless of the solution method and regardless of the assumed diffusivity function $D(\theta)$, is that the total water volume lost from the soil profile is proportional to the square root of time; this can readily be seen from the Boltzmann transform $\phi = zt^{-1/2}$, which is used to reduce (11.10) to an ordinary differential equation. Thus, the rate of evaporation is given by

$$E = \tfrac{1}{2}De\ t^{-1/2} \tag{11.12}$$

where De, which is a constant for a given soil and given values of θ_i and θ_0, is commonly referred to as desorptivity.

Good agreement was obtained by Gardner (1959) between (11.12) and the evaporation rate from a 100 cm-long, initially uniformly moist column of clay soil, subjected in the laboratory to a large potential evaporation rate of about 4 cm day^{-1}. These results are shown in Figure 11.6. Apparently, the column was long enough to be effectively semi-infinite for about 100 days. Similar field data on daily mean evaporation from a bare sand surface are shown in Figure 11.7. These data, which were obtained by Black *et al.* (1969) by means of a weighable lysimeter in Wisconsin, suggest a value of the desorptivity around $De = 0.496$ cm day$^{-1/2}$. Black *et al.* (1969) compared this result with the linearized solution for the desorptivity, viz. $De = 2(\theta_i - \theta_0)(\bar{D}/\pi)^{1/2}$ in which $\bar{D}$ is the weighted-mean diffusivity. Applying Crank's method, viz. $\bar{D} = [1.85/(\theta_i - \theta_0)^{1.85}]\int_{\theta_0}^{\theta_i}(\theta_i - \theta)^{0.85}D(\theta)\,d\theta$, they estimated $\bar{D}$ from soil samples to be 10 cm^2 day^{-1}; thus with $(\theta_i - \theta_0) = 0.12$ the linearized solution produces $De = 0.43$ cm day$^{1/2}$ which is about 13 percent lower. In light of the natural variability of the soil, and also of the likely errors resulting not only from the limitations of the problem formulation, but also from its linearization, the agreement may be termed good. Black *et al.* (1969) suspected that after a rainfall the evaporation would eventually depart from the $t^{-1/2}$ relationship, because of the finite depth of wetting. Still, they were able to model an entire summer of evaporation from the lysimeter by applying (11.12) after each rainfall event. Evidently, in their experiment the duration of the first stage of drying was sufficiently short that it could be neglected. This may have been due to the fact that the potential evaporation was always larger than the relatively low actual evaporation rates from the sand. Under more moderate drying conditions

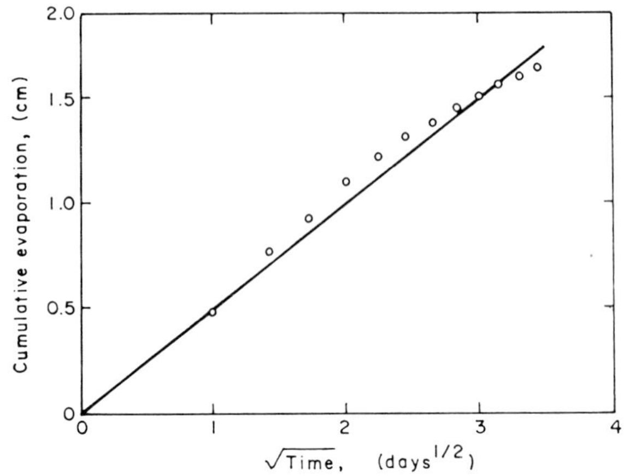

Fig. 11.7. Cumulative evaporation from a bare soil surface as a function of the square root of time, obtained in the field by means of a lysimeter. The straight line represents (11.12) with a desorptivity $De = 0.496$ cm/(day)$^{1/2}$ (adapted from Black *et al.*, 1969).

or for soils of finer texture, the first stage of drying should be included in the analysis.

In (11.11) the initial water content is assumed to be uniform. But the water content distribution at the beginning of the second stage of drying is rarely uniform, and it is clear that the subsequent evaporation rate must depend on this initial θ-distribution; this in turn depends on the rate of evaporation during the first stage and on its duration. None the less, Gardner and Hillel (1962) have observed in the laboratory that this effect is of relatively short duration, so that soon after the end of the first stage the rate of evaporation becomes independent of the initial drying rate, and that it depends only on the water content of the soil. This means that the same drying function, namely the solution (11.12) for high potential evaporation, should yield a good representation of the cumulative evaporation for any potential evaporation, by a proper translation of the time variable. In other words, the evaporation rate, after it drops below the first stage value, can be assumed to follow the same decrease with time as (11.12); as a first approximation, the value of t to be used at the beginning of the second stage of drying can be taken as $t = (\sum E_{p1}/De)^2$ where $\sum E_{p1}$ is the value of the cumulative evaporation at the end of the first stage.

The problem formulation resulting in (11.12) and its experimental verification refer to idealized situations. In most practical field cases, such factors as the stratification of the soil profile, deep percolation or downward seepage, uncertainty concerning the end of the first stage of drying, and others tend to add complications. Despite its shortcomings and possible theoretical objections, however, under certain conditions (11.12) can serve as a simple parametric relationship to predict the daily mean evaporation rate from a bare soil in the second stage of drying. In practice, De is probably best determined from a field experiment during one or two drying episodes when E can be determined independently. When this is impossible, De may be estimated by the solution of (11.10) with (11.11).

b. River Basins and Other Hydrological Catchments

Over extensive land surfaces the mean evaporation rate can be obtained from the following form of the water balance equation

$$E = P + [(Q_{ri} + Q_{gi}) - (Q_{ro} + Q_{go}) - \mathrm{d}S/\mathrm{d}t]/A \tag{11.13}$$

where p is the rate of precipitation (mean value over the sampling period), Q_{ri} and Q_{ro} the surface inflow and outflow rates, Q_{gi} and Q_{go} the ground-water inflow and outflow rates, S the stored water and A the surface area under consideration.

On an Annual Basis

Since it is very difficult to measure the storage and the ground-water flows, (11.13) has been mostly applied in climatological calculations of E. Thus, presumably on an annual basis, $\mathrm{d}S/\mathrm{d}t$ is nearly zero, and over very large areas $(Q_{gi} - Q_{go})$ is usually negligible compared to the other terms. If, moreover, the area is a natural basin, Q_{ri} is zero or, in case of artificial interbasin water exchange, it is usually known exactly. Hence, if $q_r = (Q_{ro} - Q_{ri})/A$ is the mean surface runoff per unit area from the basin, (11.13) can be simplified to

$$E = P - q_r. \tag{11.14}$$

The value of E obtained with (11.14) can be useful to check or to calibrate other methods at least on an average annual basis.

Equation (11.14) has been used to derive some simple heuristic relationships, which are of some interest. Schreiber (1904), upon inspection of earlier work by Ule and by Penck, noted that when P decreases, so does q_r, but that when P increases, q_r tends to the same value without, however, reaching it. Accordingly, he proposed the following interpolation equation for the annual runoff from Central European rivers

$$q_r = P \exp\left(-\frac{a}{P}\right) \tag{11.15}$$

where a is a constant for a given basin; his data showed that a is around 46 to 80 cm for the source areas and for plains and around 80 to 115 cm for the middle reaches. Schreiber estimated that the error to be expected with this approach is of the order of 10 to 15 percent for any year and of the order of 5 percent for means over several years. With (11.14) one obtains for the annual evaporation

$$E = P\left[1 - \exp\left(-\frac{a}{P}\right)\right]. \tag{11.16}$$

According to Budyko (1948; 1974), in 1911 Ol'dekop suggested that a in (11.16) can be taken as the value of the potential evaporation E_p ('the maximum possible value for evaporation in the given conditions'). A similar reasoning then led Ol'dekop to propose

$$E = E_p \tanh\left(\frac{P}{E_p}\right). \tag{11.17}$$

In the same spirit, Budyko (1948; 1974) postulated that the following are valid, in the limit of very dry conditions

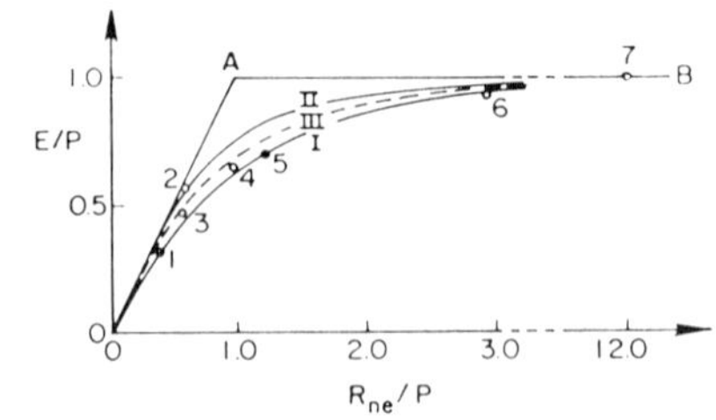

Fig. 11.8. Ratio of annual evaporation to precipitation as a function of the radiative index of dryness. Lines AB and and OA represent (11.18) and (11.19), respectively; curves I, II, III represent (11.20), (11.21) and (11.22), respectively. The numbers are data points 1. Lapland; 2. Central Germany; 3. Java; 4. East Coast U.S.; 5. Irkutsk; 6. Gobi; 7. Egypt (adapted from Budyko, 1948).

$$\frac{q_r}{P} \rightarrow 0 \quad \text{or} \quad \frac{E}{P} \rightarrow 1 \text{ when } \frac{R_{ne}}{P} \rightarrow \infty \tag{11.18}$$

and in the limit of very moist conditions

$$E \rightarrow R_{ne} \quad \text{when} \quad \frac{R_{ne}}{P} \rightarrow 0 \tag{11.19}$$

where R_{ne} is the annual value of net radiation expressed as equivalent height of evaporation. To cover the intermediate range in analytical form, Budyko tried interpolation equations analogous to (11.16) and (11.17), viz

$$E = P\left[1 - \exp\left(-\frac{R_{ne}}{P}\right)\right] \tag{11.20}$$

and

$$E = R_{ne} \tanh\left(\frac{P}{R_{ne}}\right) \tag{11.21}$$

which are shown as curves I and II, respectively, in Figure 11.8. Since the data appeared to lie between these two equations, he then proposed the geometric mean, viz.

$$E = \left\{R_{ne}P \tanh\left(\frac{P}{R_{ne}}\right)\left[1 - \exp\left(-\frac{R_{ne}}{P}\right)\right]\right\}^{1/2} \tag{11.22}$$

which is shown as curve III in Figure 11.8.

Schreiber's reasoning leading to (11.15) and (11.16) is also evident in the proposal of Turc (1954; 1955). From annual rainfall and runoff data in large river basins he concluded that when P increases, E does not increase indefinitely with P, but it does not exceed a maximal value L_T. Thus he tested a heuristic interpolation equation of the general form $(E/L_T) = (P/L_T)/[1 + (P/L_T)^a]^{1/a}$ where a is a constant. Applying this with a large number of data sets in different climatic regions Turc finally proposed for the annual rate of evaporation

$$E = P\left[0.9 + \left(\frac{P}{L_T}\right)^2\right]^{1/2} \tag{11.23}$$

for $(P/L_T) > 0.316$ and $E = P$ for $(P/L_T) < 0.316$. The maximal evaporation rate L_T

was related empirically to the mean annual temperature T_a (in °C) by $L_T = 300 + 25T_a + 0.05T_a^3$. Turc also proposed to extend his method to 10-day periods, by developing additional empirical relationships with lysimeter data. Pike (1964) found that a slightly modified form of (11.23), namely with 1.0 instead of 0.9 and with L_T replaced by E of (10.15) for open water, or adjusted pan data, gave better results in Malawi.

With E Assumed Proportional to Potential Evaporation

Attempts have been made to apply (11.13) for periods shorter than a year by using indirect methods to estimate the term $\mathrm{d}s/\mathrm{d}t$ in the equation

$$E = P - q_r - (\mathrm{d}s/\mathrm{d}t) \tag{11.24}$$

where $s = (S/A)$ is the water in active storage per unit area.

One such method to estimate $\mathrm{d}s/\mathrm{d}t$ is that of Budyko (1974, p. 97) on the basis of meteorological and hydrological data from a network of stations. The main assumption in this method is that the actual rate of evaporation is proportional to potential evaporation E_p, as follows

$$E = E_p \frac{s}{s_0} \tag{11.25}$$

where the storage s is specified as the moisture stored in the upper soil layer of 1 m depth and s_0 a critical value above which E equals E_p. Although the matter is not entirely clear, it appears (Budyko, 1974, p. 335) that this E_p was intended to be the same as that obtainable by means of an approach similar to Penman's (10.15). In other words, E_p is an apparent potential evapotranspiration, since it is calculated with meteorological data observed under the prevailing, i.e., non-potential conditions. As noted in Section 10.2a, these are not the same as those which would prevail if the surface were well supplied with water. It was established that s_0 in (11.25), usually amounts to a layer of 10 to 20 cm water, with seasonal and regional variations; the value of s_0 may be obtained by calibration. At any rate, the storage term s needed to apply (11.25) can not usually be determined for a region. Therefore, it can be eliminated from the problem by means of the additional equation (11.24). If E, P and q_r are monthly means and if s_1 designates the moisture stored in the active soil layer at the beginning of the month, and s_2 that at the end, (11.24) becomes

$$E = P - q_r + s_1 - s_2. \tag{11.26}$$

For these monthly periods (11.25) can be written as

$$\begin{aligned} E &= E_p \frac{s_1 + s_2}{2s_0} \quad \text{for } 0 < \frac{s_1 + s_2}{2} < s_0 \\ E &= E_p \qquad\qquad \text{for } \frac{s_1 + s_2}{2} \geq s_0. \end{aligned} \tag{11.27}$$

The application of this method is straightforward, especially under conditions of extreme insufficiency of moisture, so that runoff ceases, i.e., $q_r = 0$. The rate of evaporation can then be calculated by successive approximation as follows. An initial value of s_1 for the first month is chosen at random; (11.26) yields s_2, which, when substituted in (11.27), produces E for the first month. The same procedure is carried out for the second month, with s_2 of the first serving as s_1 of the second, and so on. The sum of all monthly E values can be compared to the total annual P. The ratio of the two should allow a proportional adjustment of the s_1 value of the first month, and the process can be started over again and continued until the calculated annual E equals the recorded P. In principle, the same method can also be applied when runoff q_r is not negligible and when P is not small compared to E_p. The successive approximations may be halted when the calculated E equals the annual sum of $(P - q_r)$, or, if long term means are calculated, when the calculated s_2 of the last month agrees with the estimate of s_1 of the first month. When q_r is not negligible its determination may present some difficulties, since observational data may be inadequate. For this case Budyko and Zubenok (Budyko, 1974, p. 100) have developed procedures to calculate $\bar{q}_r$ on the basis of empirical runoff coefficients and rainfall data.

The main weakness of any method based on a relationship such as (11.25) or (11.27) is, beside the question of the validity of this proportionality, first the unknown value of the maximum soil moisture parameter s_0, and second the rather ambiguous (see Section 10.2) meaning of the potential evaporation. Of course, the relationship may be calibrated empirically by suitable curve fitting methods, but the physical significance is not clear at present. Budyko's water balance method has been applied extensively for various regions of the U.S.S.R. Similar methods have also been used in numerical climate modeling. For example, Manabe (1969) and Holloway and Manabe (1971) applied (11.25) with $s_0 = 0.75\ s_{FC}$ where s_{FC} is the field capacity of the soil; the latter, which is the upper limit of water that can be stored in the soil, was assumed to be 15 cm everywhere on the land surface of the earth.

On the Basis of the Hydrograph Recession

A second indirect method to estimate (dS/dt) in the water balance is based on the analysis of the streamflow from a watershed in the absence of precipitation. In a natural river system, this streamflow results primarily from the drainage from the ground water aquifers into the river channels of the basin. In the hydrological literature it is variously referred to, among others, as drought flow, recession flow, base flow, low flow, streamflow depletion flow and sustained or fair-weather runoff. Except for the depletion of the water in storage by evapotranspiration, the characteristics of the drought flows from a given basin depend primarily on its geological nature. The recession flow with minimal or no evaporation can thus be called the groundwater recession flow; it has also been called the 'potential' recession.

Several attempts have been described in the literature to relate the observed recession flow to the evapotranspiration from the basin. Tschinkel (1963) computed the evapotranspiration during the dry season for the riparian zone in a 14.5 km^2 forested mountainous watershed in southern California; it was obtained from the difference between the actual streamflow hydrograph and a potential recession, which was assumed to be given by

$$Q_r = Q_{0r} K_r^{\tau} \tag{11.28}$$

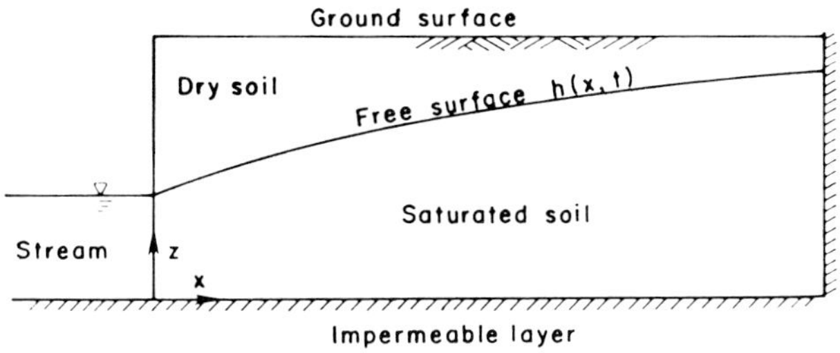

Fig. 11.9. Dupuit-Boussinesq aquifer model on horizontal impermeable bed. The flow rate into the stream is determined under the assumptions that the horizontal flux component is height independent and the hydraulic gradient equals the slope of the 'free surface'.

where Q_{0r} is the rate of flow at any time, Q_r the rate of flow $\tau = (t/\Delta t)$ time units later, Δt is the duration of one time unit and K_r is a constant. Clearly, (11.28) is equivalent to an exponential decay function, so that the groundwater storage is, in fact, assumed to be analogous to a linear reservoir whose outflow rate is proportional to the stored volume. Later, Daniel (1976) followed a different approach by making use of the theoretical rate of outflow from a linearized Dupuit–Boussinesq aquifer model overlying a horizontal impermeable layer (see Figure 11.9) (cf. also Brutsaert and Ibrahim, 1966) with the effect of evapotranspiration included as a constant rate of leakage; thus comparison of the actual hydrograph recession with dimensionless type curves based on the theoretical solution for different values of E, allowed the determination of the basin evapotranspiration. Daniel (1976) applied the method with success on a 23 km^2 basin in Alabama. Again, also in this method the adopted groundwater recession function was obtained by assuming a rather specific conceptual model for the groundwater storage. While the assumption of a linear Dupuit–Boussinesq groundwater aquifer system may be useful to describe certain situations, it cannot be expected to be universally applicable.

A method of describing the ground water recession which is applicable to a wider variety of ground-water aquifer systems, was developed by Brutsaert and Nieber (1977). Although, in that paper, evaporation was not considered, the approach may be extended to include E as follows. The essence of the hydrograph recession method is that during drainage, i.e., in the absence of precipitation, recharge or other input into the basin, there is a unique relationship between the volume of water stored underground in the watershed and the drainage rate into the river channels. A simple assumption is that this is a non-linear function of the type

$$Q_r = aS^b \tag{11.29}$$

where Q_r is the mean rate of flow in the river at the outlet of the watershed, S is the water stored and a and b are constants; many conceptual aquifer models, among which those adopted by Tschinkel (1963) and Daniel (1976), can be considered as special cases of (11.29). In the absence of precipitation P, the water balance equation (11.13) is simply

$$E = -\left(Q_r + \frac{dS}{dt}\right)/A. \tag{11.30}$$

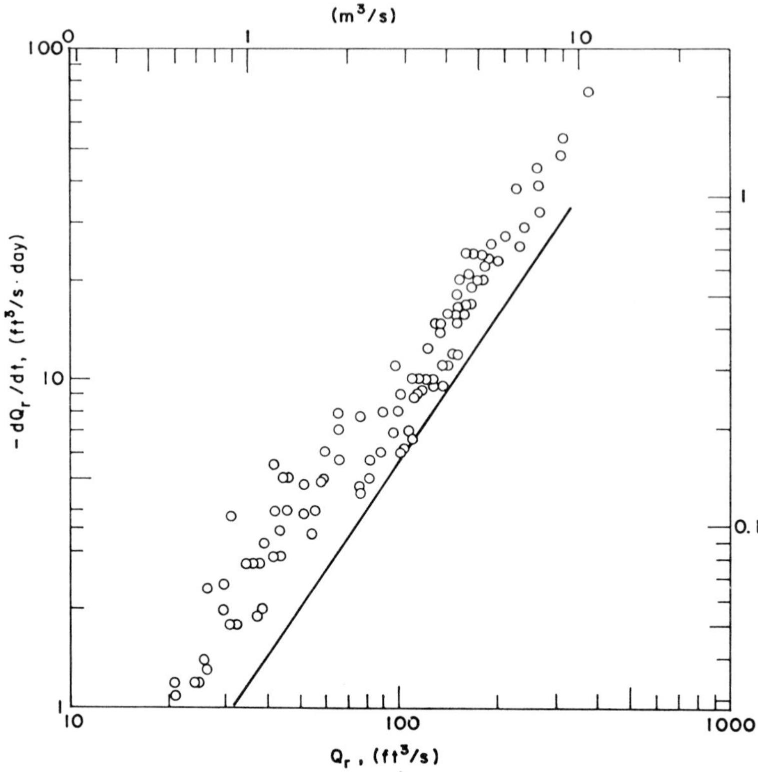

Fig. 11.10. Plot of $-\mathrm{d}Q_r/\mathrm{d}t$ versus Q_r data for low flows in Fall Creek near Ithaca, N.Y., with a drainage area $A = 326$ km². The lower envelope shown has a slope 3/2, which is $(2b-1)/b$ in (11.32), in accordance with the nonlinear aquifer model of Boussinesq (adapted from Brutsaert and Nieber, 1977).

Substitution of (11.29) in (11.30) yields

$$-\frac{\mathrm{d}Q_r}{\mathrm{d}t} = a^{1/b} b Q_r^{(b-1)/b}(Q_r + AE). \tag{11.31}$$

For any given basin the values of the parameters a and b may be determined as follows. The available streamflow record during recession is plotted as log $(-\mathrm{d}Q/\mathrm{d}t)$ versus log (Q); in practice, say with daily data, $-\mathrm{d}Q/\mathrm{d}t$ is taken as $(Q_{i-1} - Q_i)/\Delta t$ and Q as $(Q_i + Q_{i-1})/2$ in which Q_i is the recorded flow on any day and Q_{i-1} is the flow recorded $\Delta t = 24$ hours earlier. The lowest straight envelope of all the data points gives the lowest observed rate of recession corresponding to a minimal or zero value of E in (11.31); this means that the lower envelope represents

$$-\frac{\mathrm{d}Q_r}{\mathrm{d}t} = a^{1/b} b Q_r^{(2b-1)/b} \tag{11.32}$$

and that the parameters can be readily estimated. An example of this type of plot is shown in Figure 11.10. When data are available for a number of stations in the same region, it may be possible to regionalize the parameters by relating them to geomorphological characteristics.

Once the parameters a and b are determined, (11.31) can be reorganized to calculate evaporation from streamflow measurements at successive time periods, viz.

$$E = \frac{1}{A}\left(- \frac{Q_r^{(1-b)/b}}{a^{1/b}b} \frac{dQ_r}{dt} - Q_r \right). \tag{11.33}$$

A weakness of this method is the accuracy of the streamflow measurements Q_r; Equation (11.33) involves the rate of change $-dQ_r/dt$ or $(Q_{i-1} - Q_i)/\Delta t$, which may, on occasion, be of the order of magnitude of the unavoidable error inherent in the measurement. This method, which is merely outlined here, still requires testing to assess its applicability.

c. Lakes and Open-Water Reservoirs

The water balance of a lake can be described by means of (11.13). The relative importance of any of the terms depends on the hydrological and physiographical characteristics of the water reservoir and its surrounding watersheds. The feasibility of determining evaporation by means of (11.13) depends primarily on the relative magnitude of the terms. It is clearly very difficult to obtain a reliable estimate of E whenever it is of the same order of magnitude as the errors inherent in the measurement of any of the terms on the right. Thus, the method is unsuited for a lake with large flows passing through, as surface runoff or groundwater seepage.

Depending on the size of the lake one or more gages are required to estimate precipitation. In most cases, precipitation over the lake must be estimated from gages on the surrounding land. Land and water have different thermal properties, so that the precipitation over a large lake may be considerably different from that on the land. The determination of the distribution of precipitation, to obtain the areal average, is usually quite difficult for short periods of time.

The seepage inflows and outflows are almost impossible to measure. Records of groundwater levels or other piezometric data together with a knowledge of the geological formations are useful but they rarely allow reliable computations. For the purpose of checking its magnitude, the rate of seepage $(Q_{go} - Q_{gi})$ can be determined by means of (11.13), provided all other terms are known and E can be estimated by some independent method during a special testing period. Once this has been accomplished, (11.13) may then be applicable on a routine basis to estimate E.

For the determination of the changes of storage in the lake, waterlevel records and a reliable area-capacity relationship are required. The latter can be obtained from a topographic survey of the lake and its shores. The use of more than one water-level recorder is almost indispensable to avoid errors due to seiches, and wind set-up. When large temperature changes occur the thermal expansion of the water may have to be taken into account by writing (11.13) in terms of mass flows instead of volume flows.

In view of the possible errors, in the case of lakes and reservoirs, the water balance method is not likely to be applicable over periods shorter than a week or, more commonly, a month.

d. Water Budget-Related Instruments; Evaporimeters

Lysimeters

A lysimeter is a container placed in the field and filled with soil, on which a vegetation can be maintained for the purpose of studying various soil-water-plant relationships under natural conditions. The term lysimeter became widely used around the middle of the nineteenth century (e.g., Hoffmann, 1861); although etymologically the word indicates that the main purpose was originally to measure leaching and percolation of solutes through a soil, it was soon also used in the sole hydrological context of the determination of evapotranspiration (e.g., Ebermayer, 1879; Wollny, 1893). In fact however, this approach to estimate components in the hydrology of a land surface on the basis of the water budget of a container isolated hydrologically from the surrounding soil, was attempted a long time before that. As noted in Chapter 2, deLaHire (1703) had already run a rain percolation experiment, but that had ended in failure. Dalton (1802c), in cooperation with his friend Thomas Doyle, conducted an experiment by sinking a metal cylinder, 10 inches in diameter and 3 feet deep, into the ground with one side exposed to allow drainage of surplus water through two lateral tubes into bottles; from three years of observations, at first with a bare surface and later with a grass cover, it was concluded that for an average annual rainfall of about 34 in. and 5 in. of dew, a total of 30 in. are evaporated. Similar work by Dickinson was later reported by Parkes (1845).

In order to produce the same rate of evapotranspiration as the surrounding area a lysimeter should be representative of the conditions of the natural soil profile and of the vegetation around it. In other words, when designing and installing a lysimeter care must be taken to insure the same water flux at the soil surface and the same development of the plant roots in the soil profile. This means that the lysimeter should have its surface flush with the surrounding ground surface, and that it should be at least as deep as the rooting depth of the vegetation; moreover, the profiles of soil structure, soil texture, soil water content and soil temperature in the lysimeter must be made as similar as possible to those on the outside. To maintain the same mechanical properties of the soil, it may be desirable to place it in the container as an undisturbed block or 'monolith' (e.g., Brown *et al.*, 1974); when this is impossible, because of the nature of the soil or the large size of the lysimeter, the soil should be placed in the container layer after layer in the same order and with the same density as in the natural profile. Because the lysimeter has a bottom whereas the surrounding soils usually extend down to much greater depths, it may not always be easy to have the same water content profile even near the surface. In Figure 11.11 the conditions in three types of lysimeters are compared with those in a natural profile. Clearly, in the vicinity of a drain open to the atmosphere, the pressure in the soil water is zero, i.e., atmospheric. Thus, in order to simulate the natural drainage process, a lysimeter with open drainage at the bottom should be sufficiently deep. If the lysimeter is shallow (for example, to maintain adequate sensitivity in weighing) it may be necessary to maintain the suction at the bottom artificially by means of a vacuum supply (e.g., Pruitt and Angus, 1960). The simulation of evapotranspiration also involves the simulation of the surface energy budget. Therefore, the soil temperature profiles inside the lysimeter should not be allowed to differ much from those outside. The

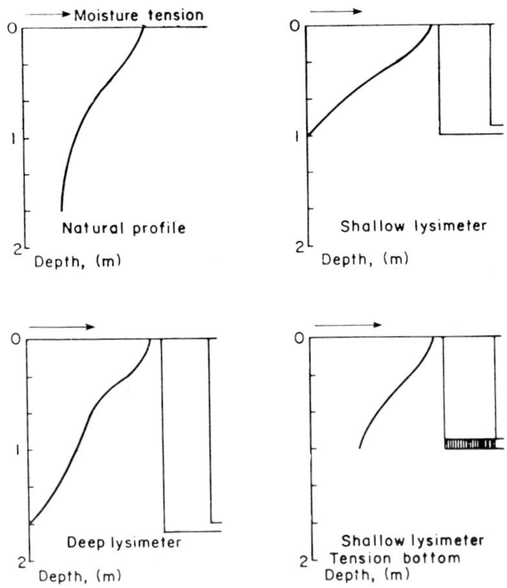

Fig. 11.11. Sketch illustrating how the negative water pressure in the natural soil profile, after precipitation or irrigation, is represented in three types of lysimeters (adapted from Van Bavel, 1961).

difference in thermal regime is probably minimal in the case of a deep lysimeter with a good simulation of the natural water regime and with an abundant vegetational cover. To further reduce thermal discrepancies, it is desirable that the lysimeter container consist of material with a low thermal conductivity and that any gap between container and surrounding retaining walls be sealed off at the surface. The thermal equivalence of the lysimeter to the surroundings can be tested by measuring soil temperature profiles inside and outside the lysimeter. Again, when high accuracy is desired, to obtain thermal similarity, the bottom of the lysimeter may have to be heated or refrigerated artificially at the temperature measured at the same depth in the natural profile outside (e.g., Pruitt and Angus, 1960). Another source of potential error is the effect of the surface discontinuity at the edge of the lysimeter. Thus, the rim of the container should be kept as low as possible above the soil surface; however, if the lysimeter is also used as a precipitation gage, the rim must have a certain height to prevent runoff or spilling. Similarly, the gap between the container wall and the wall, retaining the surrounding profile, should be kept as narrow as possible. The edge effect can be reduced by increasing the surface area of the lysimeter. The size of the lysimeter is also dictated by the scale of the inhomogeneity of the vegetation growing on it. In other words, the number of individual plants on the lysimeter should be sufficiently large to produce an average rate of evapotranspiration which is the same as that of the surrounding land surface. Finally, the placement and maintenance of a lysimeter is subject to fetch requirements just like any other method. Hence, different vegetation on and around the lysimeter, sidewalks, nearby micrometeorological instruments, fences or other obstacles may introduce serious error.

Several types of lysimeter installations have been described in the literature. These

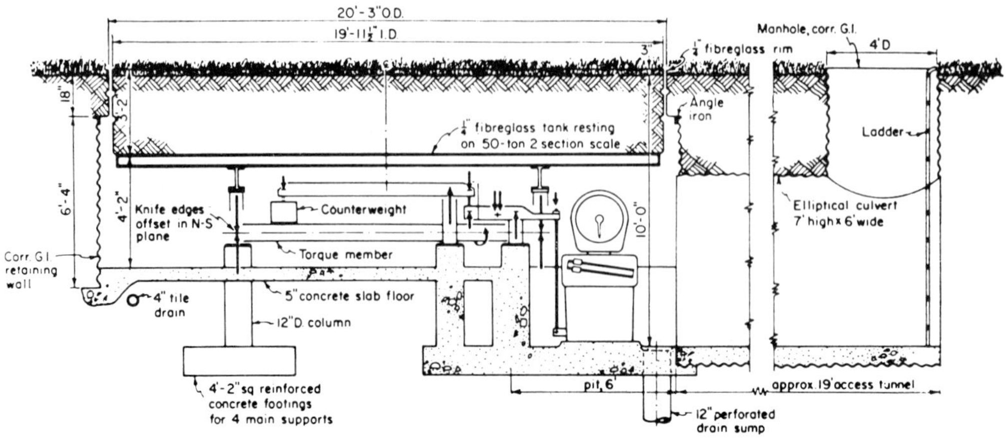

Fig. 11.12. Diagram of the 6.1 m diameter weighable lysimeter at Davis, California, taken as a west-east section. The vegetation is perennial ryegrass (Lolium perenne). The dimensions are given in feet and inches ($1' = 12'' = 30.48$ cm) (from Pruitt and Angus, 1960).

can be classified as non-weighable and weighable lysimeters. Most of the early designs (cf., even Dalton, 1802c; Wollny, 1893) were of the non-weighable type with free drainage at the bottom. This type is easy to install and low in cost. However, unless the moisture content can be measured by some independent technique, percolation-type lysimeters can only be used for long-term measurements, namely between major precipitation events if they only receive natural rainfall, or over weekly or longer periods if they are irrigated. Examples of such installations are described by Harrold and Dreibelbis (1958), and Gilbert and Van Bavel (1954). In areas with high water tables it is necessary to also have a water table in the lysimeter; this usually involves the measurement of the inflow and outflow required to maintain this water table at a given level. However, with water table-type lysimeters it may be necessary to take special precautions to avoid the build-up of salinity (e.g., Williamson, 1963) or to compensate for water-level fluctuations which may conceivably result from other factors such as atmospheric pressure changes (e.g., Van Hylckama, 1968).

Although weighable lysimeters are much more expensive, their appeal is that they produce accurate values of evapotranspiration over daily or shorter periods. Several weighing techniques have been applied in past experiments, namely mechanical scales, load cells with strain gages of variable electrical resistance, and hydraulic weighing systems either with fluid displacements resulting from the changing buoyancy of a floating container, or with fluid pressure changes in hydraulic load cells. The simpler weighable lysimeters of the earlier designs (e.g., Makkink, 1957) are usually lifted out of the ground at regular intervals and weighed by a movable mechanical scale. Examples of lysimeters mounted on a permanently-installed mechanical scale which permits continuous recording, are those described by Harrold and Dreibelbis (1958) and Pruitt and Angus (1960). The latter setup, which has several interesting features, is shown in Figure 11.12; it has a large circular area of 6.1 m diameter and a depth of 0.91 m, and it is equipped with soil temperature control and soil water suction control at the bottom; because of the large area, edge effects are minimal and the gap area between the container wall, consisting of 6.3 mm fiberglass, and the surrounding land surface is smaller than 3 percent of the lysimeter area; the accuracy of the weight readings recorded every 4 min is within 0.03 mm of evaporation. The construction of the weighing and recording system can be simplified considerably by using load cells consisting of electrical resistance devices, utilizing strain gages. Examples of this type of lysimeter are described in papers by Van Bavel and Myers (1962), Ritchie and Burnett (1968), Rosenberg and Brown (1970), and Perrier *et al.* (1974). The hydraulic weighing systems are usually cheaper to build than those with mechanical or strain-gage scales.

In the case of the so-called floating lysimeter, weight changes are determined from the fluid displacements resulting from the changing buoyancy of a floating container. The buoyancy of the container is obtained by attaching buoyancy chambers to the container floating in water (e.g., King *et al.*, 1956) or by floating the lysimeter container in a heavy fluid, such as a zinc-chloride ($Zn\ Cl_2$) solution (e.g., McMillan and Paul, 1961; King *et al.*, 1965; Lourence and Goddard, 1967). The container in the lysimeter described by Lourence and Goddard (1967) has a 6.1 m diameter and is quite similar in construction to that shown in Figure 11.12; later the design of this floating lysimeter was further improved by Goddard (1970) to also permit the direct measurement of the surface shear stress τ_0 or the friction velocity u_*. In the second type of hydraulic weighing systems, weight changes are determined from fluid pressure changes in a hydraulic load cell, by which the lysimeter container is supported. These hydraulic load cells may consist of water-filled bags, bolsters, pillows or tubing made of rubber or other suitable material. Examples of this type of lysimeter have been given by Forsgate *et al.* (1965), Hanks and Shawcroft (1965), Ekern (1967) and Black *et al.* (1968).

Evaporation Pans

Although of uncertain and often dubious applicability as a measure of evaporation in nature, evaporation pans continue to be used widely. It is easy to understand their intuitive appeal, because they model the evaporation from a free water surface in a visible way. Nevertheless, even after many studies have dealt with the pan problem, it is still very difficult, if not impossible, to make a general and practical use of pan data except in special situations. Many types of pans have been tried and used over the years, but some standardization has taken place. To facilitate easy identification when reference is made to them in the literature, a few of the more common pans are now briefly described.

The Colorado Sunken Pan: This is probably one of the oldest standardized pans, since its use goes back to the work of Carpenter (1889; 1891) at Fort Collins. It has a square water surface with 3 ft (91.5 cm) sides, it is usually 1.5 ft (45.7 cm) deep, and it is installed in the ground with its rim approximately 4 in. (10 cm) above the ground surface so that the water surface is approximately maintained at ground level (see also Rohwer, 1934).

The Class A Pan of the U.S. Weather Bureau: This has been the official network instrument in the United States to measure evaporation (e.g., Kadel and Abbe, 1916)

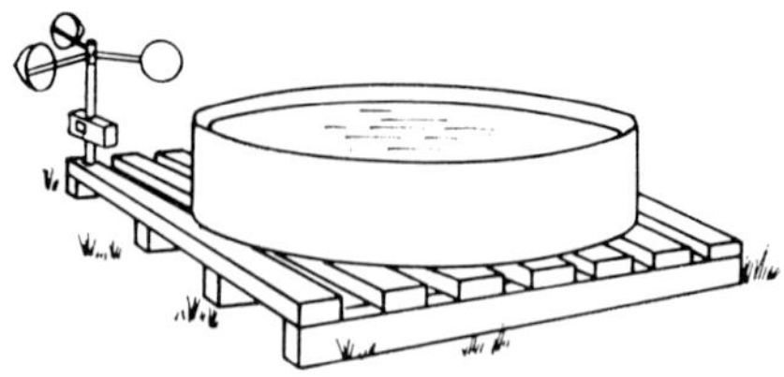

Fig. 11.13. The Class A pan of the U.S. Weather Bureau.

but is used in many other countries as well. It is a cylindrical container, 10 in. (25.4 cm) deep and 4 ft (121.9 cm) in diameter, inside dimensions (Figure 11.13). It is constructed of galvanized iron (22-gage) or some other similar non-rusting metal. It is placed on a grillage of timbers so that its bottom is between 10 to 20 cm off the ground; to anchor this platform, it is sometimes filled with new soil to within approximately 5 cm from the bottom, allowing some ventilation. The water level in the pan should be maintained between 2 to 3 in. (5 to 7.5 cm) from the top of the rim. Usually the water level is measured with a micrometer hook gage in a stilling well. In the standard set-up, a thermometer (preferably max-min) measures the water temperature and a three-cup anometer measures the wind speed at about 15 cm above the rim of the pan.

The Sunken Pan of the Bureau of Plant Industry: Prior to the general acceptance of the Class-A pan, this pan was used at dry land stations of the BPI of the U.S. Department of Agriculture in the western United States (e.g., Horton, 1921). It is 6 ft (182.9 cm) in diameter, and 2 ft (61 cm) deep; it is buried 20 in. (51 cm) in the ground and kept filled with water up to ground level, that is 10 cm below the rim. This pan is usually built of the same type of sheet metal as the Class A pan.

The GGI-3000 Pan: This pan was developed in the U.S.S.R.; it is now widely used as a standard, especially in Eastern Europe (e.g., Gangopadhyaya *et al.*, 1966). It is a cylindrical tank with a conical base; the surface area is 3000 cm^2, with a diameter of 61.8 cm, and a depth of 60 cm at the wall and 68.5 cm at the center. The container is made of galvanized sheet iron and it is buried in the ground, with the rim at about 7.5 cm above ground level.

The 20 m^2 Basin: This type of installation also originated in the U.S.S.R. It is a cylindrical basin with flat base made of 4–5 mm boiler plate sheets or concrete. Its surface area is 20 m^2 with a diameter of 5 m and a depth of 2 m. It is placed in the ground, with the rim at 7.5 cm above ground level and the water level maintained approximately at ground level (e.g., Gangopadhyaya *et al.*, 1966).

There are primarily two types of problems for which considerable efforts have been made to apply pan evaporation data. The first is the determination of evapotranspiration from a well-watered vegetation and the second the determination of lake evaporation.

As a physical phenomenon, the evaporation from any type of pan is quite different from evapotranspiration from a vegetation. Still, field experiments have shown that, over longer periods, pan evaporation is highly correlated with evapotranspiration from the surrounding vegetation under conditions of full cover and good water supply (e.g., Penman, 1948; McIlroy and Angus, 1964; Pruitt, 1966). An example of such a correlation for monthly data at different locations is shown in Figure 11.14.

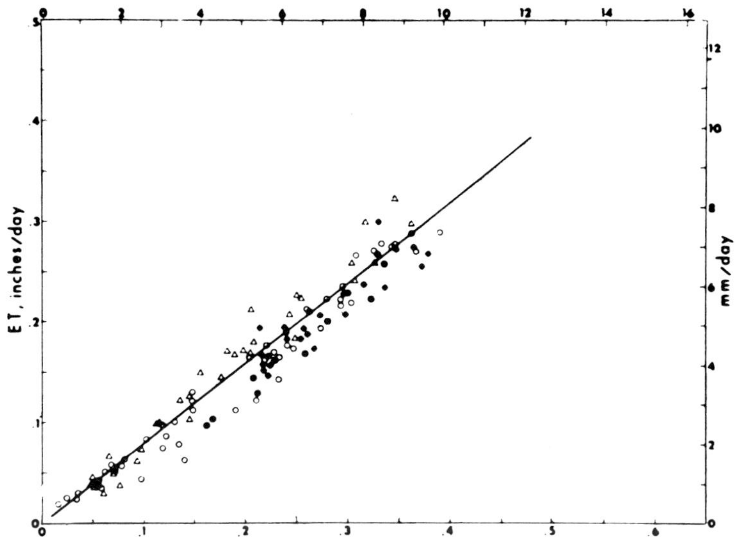

Fig. 11.14. A comparison between monthly mean evapotranspiration *ET* from grass or grass-clover mixture and evaporation from Class A pans at different sites. The slope of the best-fit line is 0.80 (adapted from Pruitt, 1966).

It can be seen that the pan coefficient, defined here as the ratio of evapotranspiration to pan evaporation is approximately 0.8 for grass. Many pan coefficients and regression equations have been reported in the literature. However, these results show considerable variation, depending on the type of vegetation, the pan environment and the climatic conditions. Thus calibration and standardization are essential to render pan data useful in the estimation of ambient potential evaporation.

Many attempts have also been made to relate pan evaporation to lake evaporation. The simplest approach, again, is the use of a pan coefficient, defined as the ratio of lake evaporation to pan evaporation. Typical values (e.g., Rohwer, 1934; Kohler, 1954; Gangopadhyaya *et al.*, 1966; Shnitnikov, 1974) of the pan coefficients on an annual basis are about 0.80 for the Colorado pan, 0.70 for the Class-A pan, 0.92 for the BPI pan, 0.82 for the GGI-3000 and nearly unity for the 20 m^2-basin. With a known coefficient, generally, the larger pans and those that are installed in the ground are the most reliable. Thus, the 20 m^2-basin is preferable, but it is costly to install. However, especially for the smaller types of pans, the coefficient depends not only on the type of pan, but also on its exposure and the climatic conditions. In other words, it varies usually considerably from one lake to another and from month to month at any given location. Thus the pan coefficient approach can be considered useful only to provide a rough estimate of lake evaporation, mostly on an annual basis.

A more satisfactory procedure is the pan conversion method. It consists of taking the ratio of the bulk mass transfer equations [cf., (9.12) or (9.13)] of the lake and the pan to obtain the mean evaporation rate from the lake, as follows,

$$\bar{E} = K_{pa} \frac{(\bar{e}_s - \bar{e}_a)}{(\bar{e}_{pa} - \bar{e}_a)} E_{pa} \tag{11.34}$$

where E_{pa} is the evaporation rate of a pan, e_s and e_{pa} the saturation vapor pressures at the water surface temperatures of the lake and of the pan, respectively, e_a the vapor pressure of the air at a reference station and K_{pa} an empirical constant to be determined for any given situation. The pan conversion method was introduced by Kohler (1954); using daily averages of the vapor pressures and Class-A pan data he obtained $K_{pa} = 0.7$ for Lake Hefner. Webb (1966), who modified the method slightly by using e_{pa} corresponding to the afternoon maximum of the pan temperature and 1200 to 1800 means for e_s and e_a, obtained $K_{pa} = 1.50$ for Lake Hefner. It was felt that using values for the afternoon, when most of the evaporation takes place, should be more reliable than using daily means. The method was tested by Hoy and Stephens (1979). A difficulty, they encountered, was that small values in the denominator of (11.34) may lead sometimes to inordinately large values of $\bar{E}$; this difficulty was resolved by editing the data, that is by accepting only those daily $\bar{E}$ data which satisfied $0 \leq \bar{E} \leq 2$ cm and $\langle \bar{E} \rangle - 3\sigma_E \leq \bar{E} \leq \langle \bar{E} \rangle + 3\sigma_E$ where $\langle \bar{E} \rangle$ and σ_E are the mean and standard deviation over a given period; an alternative method of resolving the problem, they suggested, is to use longer-term means of the variables in (11.34) but then with a different K_{pa} or with a correction term. It was found that Webb's version is more reliable than Kohler's. Hoy and Stephens (1979) obtained a mean $K_{pa} = 1.21$ for the Class-A pan with Webb's procedure, and $K_{pa} = 0.66$ with Kohler's. When the pans were outfitted with a bird guard the means of the coefficients were 1.48 and 0.73, respectively. Although the pan conversion method was found to be less accurate than the bulk mass transfer method (see Section 9.2b), it was considered suitable for operational use.

The pan-conversion method has the advantage that only water-surface temperatures are required on the lake, while the other data can be obtained at a simple on-shore weather station. The constant K_{pa} should be derived by experiment for a given lake and a given measurement set-up. Alternately, taking the ratio of (7.57) with the parameters for the lake and for the pan, respectively, may yield a theoretical value of K_{pa}; the same could be done with the empirical relationships (7.60) and (7.61) or similar ones. However, this has not been tested.

Other Instruments

Beside pans many other types of devices to measure evaporation have been proposed and developed over the past few centuries. A thorough review of them has been made by Livingston (1908; 1909). Most of these are no longer in use today, and only a few have survived. Better known among these are the Piche evaporimeter, the Wild evaporimeter and the porous cup atmometer.

The instrument described by Piche around 1872 in France (Jelinek and Hann, 1873) consists of a glass tube, of 23 to 30 cm length and of 1 cm inside diameter; the tube is closed at the top and provided with a ring from which it can be suspended. The tube is filled with water and at the bottom covered with a disk of moist blotting paper, with a size such that 8 cm^2 are exposed to evaporate. As the water evaporates from the paper, the water level sinks in the tube and indicates the amount of evaporation on etched graduations. At the bottom end of the tube a bent steel wire is mounted as a spring with a small disk to hold the paper disk in place, in case of strong winds.

Fig. 11.15. The Piche evaporimeter (from Abbe, 1905).

Figure 11.15 shows an earlier form of the instrument. In the original version of the instrument a small hole in the paper allowed air to bubble through to replace the evaporated water. In more recent versions the glass tube has a small hole on the side through which the air can enter. Also, in the commercial versions available at present, the total evaporating surface is larger and of the order of 13 cm^2 with a diameter of about 3.2 cm. The instrument is now usually installed with the evaporating surface at about 1.2 m above the ground in a regular meteorological instrument shelter.

On account of its peculiar shape and exposure, it is very difficult to relate the Piche evaporation rate to any other kind of evaporation in nature. Since the instrument is placed in a shelter, it is not exposed to solar radiation, but its evaporation is in response primarily to the humidity deficit in the air, and also, but to a lesser extent, to the wind velocity. This means that the instrument is probably more analogous to a simple leaf placed in the shade, than to a surface of water or wet vegetation exposed to solar radiation. Thus the instrument probably gives a better measure of the drying power E_A, defined in 10.16, rather than of the rate of evaporation E. The latter observation made Stanhill (1962) suggest that it may be possible to estimate the second term in Penman's Equation (10.15) from available Piche evaporation data, E_{pi} as follows,

$$\frac{\gamma}{\Delta + \gamma} E_A = aE_{pi} + b \tag{11.35}$$

where a and b are constants; by calculating E_A by means of the 1956-version of (10.17) in (10.16), Stanhill (1962) derived $a = 0.1469$, $b = 0.1118$ in mm day^{-1} with weekly means for the Negev region. Bouchet (1963b) made the same suggestion with $b = 0$. This idea was worked out further by Brochet and Gerbier (1972) who proposed, as an empirical substitute for (10.15), the following

$$E = aR_s + bE_{pi} \tag{11.36}$$

where R_s is the global short wave radiation. The constants a and b (different from those in (11.35)) can be estimated by relating (11.36) with (11.35) and (10.15); Brochet and Gerbier (1972) presented a procedure to determine the constants for any given

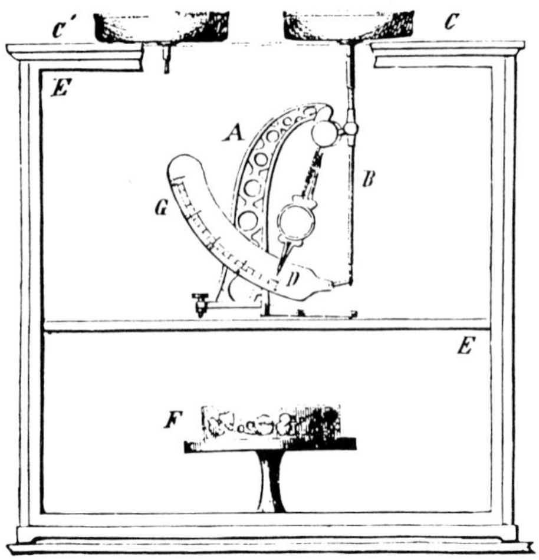

Fig. 11.16. The Wild evaporimeter scale (from Wild, 1874).

latitude and time of the year in France, where E_{pi} records are available for many stations.

The instrument proposed by Wild (1874) in Russia, consists of a shallow cylindrical dish, 2.5 cm deep and with a surface area of 250 cm^2, i.e., 17.84 cm in diameter, filled with water and placed on a balance. The original version of the instrument (see Figure 11.16), still used widely today, is similar to the postal scale: the dish C is supported by the shorter arm of a lever balance and the longer arm, that is the counterweight, is provided with a pointer D indicating the loss of weight due to evaporation on a graduated arc G; however at present, recording balances are also available commercially. The main advantage of this instrument is that it can register evaporation even during winter in colder climates, when the water in the dish is frozen. The balance is commonly placed as it is, in a meteorological shelter at an elevation of approximately 1.2 m. Wild's (1874) original set-up shown in Figure 11.16 is somewhat more elaborate than current practice. The balance was placed inside a glass case E, to protect it from disturbance by the wind during weighing; to also protect it from rusting a dehumidifier (e.g., H_2 SO_4) was placed in F. The cover of the case supported the dish, and it was only withdrawn to weigh the dish at the time of observation. The instrument had a second identical dish C' which was interchanged periodically with C, in order to avoid interruptions of the measurement when the ice mass in C had decreased considerably and before a new quantity of water would freeze to the ambient temperature, and also in order to determine the accretion in case of snow drift. The glass case was placed in the thermometer shelter.

The remarks made about the applicability of the Piche instrument are also valid for the Wild evaporimeter. The instrument is attractive because of its simplicity, but again it is far from clear what the significance of the measurements is.

Porous-bulb atmometers are another type of instrument which is still sometimes used, but they are not as common as the Piche and Wild evaporimeters. The origin of these atmometers can be traced back to Leslie's (1813) instrument, which consisted of a thin ball of porous earthenware, with a diameter of 2 to 3 inches (5 to 7.5 cm), and with a small neck cemented to the lower end of a graduated glass tube. The similar atmometer, which is now still in use, is usually named after Livingston (1935) who propagaged its application in the United States. In its present form, the Livingston atmometer is a hollow porous porcelain sphere, 5 cm in diameter with a wall of about 0.3 cm. When in operation, its moist spherical surface, without any surface film of water, gives a uniform exposure in all directions, except downward, at the narrow cylindrical neck which is glazed and connected to the supply tube; this tube is inserted through the stopper of a bottle with a supply of distilled water. This water is drawn up through the tube by capillarity in the pores of the evaporating surface. A similar instrument was proposed by Bellani in Italy about three years after Leslie's (1813). The Bellani atmometer (cf, Livingston, 1935) consists of a porous ceramic disk, as the upper face of a non-porous hemispheric bulb, which is mounted in the same way as the Livingston atmometer.

As is the case with the other instruments, evaporation data obtained with porous bulb atmometers are difficult to interpret. They have a very peculiar shape and exposure, and it is not obvious how their energy budget, or their aerodynamic properties, can be related to the energy budget or the exchange characteristics of natural surfaces. In addition, these atmometers are easily broken by handling and freezing, and they become easily soiled by dust or other contamination. However, they are more easily installed and maintained than lysimeters or even evaporation pans. Some of the more recent applications of Livingston atmometers have been in connection with the estimation of water needs of irrigated crops. Halkias *et al.* (1955) observed that the difference in evaporation between a black atmometer and a white atmometer has a high correlation with the use of water by irrigated crops; this correlation was considerably higher than that between evaporation from a white atmometer and crop water use. This was confirmed by Shannon (1968) who obtained estimates of monthly evapotranspiration under irrigation with this paired atmometer technique, which were equally good or better than those obtained by means of Class-A pan data. The explanation for all this lies apparently in the fact that the difference between the evaporation from a black and white atmometer should be correlated with the global short-wave radiation, since one absorbs while the other reflects a large part of it; and potential evaporation is well correlated with short wave radiation [e.g., (10.28) and (10.29)]. Thus, this system of paired black and white atmometers is actually used as a substitute short wave radiometer, rather than as an evaporation device. Yu and Brutsaert (1967) made the same observation as regards the difference in evaporation from very shallow (1.58 cm) pans with black and white painted bottoms; the correlation with short wave radiation was quite high.

11.2. ATMOSPHERIC WATER BUDGET

a. Concept and Formulation

This method consists of the determination of evaporation as the, preferably only, unknown term in the water budget equation for a suitably chosen, finite-size control volume in the atmosphere. Just like for the terrestrial water budget, it is possible to derive an equation simply by equating the total inflow minus the outflow of water mass to the time rate of change of stored water in the control volume. However, for a better understanding of this method, it is instructive to take as a starting point (3.44) the equation of conservation of the mean specific humidity. With a source term S_v representing the difference between the local vaporization and condensation that takes place in the air, and with the assumption that the horizontal gradients of the turbulent fluxes and also molecular diffusion are negligible, (3.44) can be written as

$$\frac{\partial \bar{q}}{\partial t} + \bar{u}\frac{\partial \bar{q}}{\partial x} + \bar{v}\frac{\partial \bar{q}}{\partial y} + \bar{w}\frac{\partial \bar{q}}{\partial z} = -\frac{\partial}{\partial z}(\overline{w'q'}) + S_v. \tag{11.37}$$

The addition of the equation of continuity (3.48), multiplied by $\bar{q}$, to (11.37) produces

$$\frac{\partial \bar{q}}{\partial t} + \frac{\partial}{\partial x}(\bar{u}\bar{q}) + \frac{\partial}{\partial y}(\bar{v}\bar{q}) + \frac{\partial}{\partial z}(\bar{w}\bar{q}) = -\frac{\partial}{\partial z}(\overline{w'q'}) + S_v. \tag{11.38}$$

The desired budget equation is obtained by integrating (11.38) over the finite control volume under consideration. Upon multiplication by ρ dz it can first be integrated over the vertical to yield the equation for a column extending from the earth's surface $z = z_s$ to the top of the atmosphere $z = z_t$. Clearly, $(\bar{w}\bar{q})$ equals zero at the upper and lower boundaries of the atmosphere; also, the vertical turbulent vapor flux $\overline{w'q'}$ is zero at the upper boundary and, by virtue of (3.74) equal to (E/ρ) at the lower boundary. Hence, if it is assumed that the net amount of moisture condensed in any air column falls down as precipitation at a rate P, so that

$$-\int_{z_s}^{z_t} S_v \rho \, dz = P \tag{11.39}$$

integration of (11.38) yields

$$\int_{z_s}^{z_t} \frac{\partial \bar{q}}{\partial t}\rho \, dz + \int_{z_s}^{z_t} \nabla\cdot(\mathbf{V}\bar{q})\rho \, dz = E - P. \tag{11.40}$$

The symbol $\mathbf{V} = \mathbf{V}(x, y, z, t)$ denotes the mean (from the turbulence point of view) horizontal velocity, that is $\mathbf{V} = \mathbf{i}\bar{u} + \mathbf{j}\bar{v}$.

Upon multiplication by $(\mathrm{d}A/A)$, (11.40) can now be integrated over the horizontal area A of the control volume. For the integrand of the second term of (11.40) use can be made of the divergence theorem of Gauss, or

$$\frac{1}{A}\iint_A \nabla\cdot(\mathbf{V}\bar{q})\,\mathrm{d}A = \frac{1}{A}\int_C (\mathbf{V}\bar{q})\cdot\mathbf{n}\,\mathrm{d}C \tag{11.41}$$

where $\mathbf{n}$ is the unit vector normal to the periphery C pointing outward, and dC is a differential lineal element of the periphery of the control volume. If $V_n = V_n(x, y, z, t)$ is the wind component normal to the boundary pointing outward, the integral of (11.40) over A can be written as

$$\bar{E} - \bar{P} = \int_{z_s}^{z_t} \frac{\partial \bar{\bar{q}}}{\partial t}\rho \, dz + \frac{1}{A}\int_{z_s}^{z_t}\int_C (\bar{q}V_n)\rho \, \mathrm{d}C \, dz \tag{11.42}$$

where $\bar{E}$, $\bar{P}$ and $\overline{\partial\bar{q}/\partial t}$ are the areal averages of the evaporation rate, precipitation rate and rate of change of $\bar{q}$ at a given level z. Equation (11.42) is commonly written in terms of the pressure p instead of z, as vertical coordinate; since the atmosphere is usually very close to hydrostatic the transformation can be made by means of (3.26). Note that a pressure difference of 100 mb corresponds roughly to a change in elevation of 900 to 1000 m. If $\bar{W}$ denotes the total water vapor content per unit area of the atmospheric column, averaged over the whole control area A, (11.42) can also be written as

$$\bar{E} - \bar{P} = \frac{\partial \bar{W}}{\partial t} + \frac{1}{Ag} \int_{p_t}^{p_s} \int_C (\bar{q} V_n) \, \mathrm{d}C \, \mathrm{d}p \qquad (11.43)$$

where p_s and p_t are the pressures at the surface and at the top of the control volume. Equation (11.43) states that the difference between the average rate of evaporation and precipitation over a given area of the earth's surface equals the rate of increase of water vapor over the area plus the total vapor flux directed away from the region.

b. Application of the Method

At present (11.43) or a form similar to it, is usually solved for $(\bar{E} - \bar{P})$ by means of a set of aerological or so-called rawinsonde observations at the vertices of the polygon enclosing the area A. Hence, the control volume is a prism with base area A and bounded by vertical walls extending from the earth's surface to a level with sufficiently small moisture content where $p = p_t$.

The last term of (11.43) is calculated by summing the product of the normal wind velocity component and the specific humidity over the total side wall area of the prism. When, as is usually the case, the aerological observation stations are not located on the boundary of the area of interest, interpolation (e.g., Cressman, 1959) may allow some adjustment. In some cases, aerological observations can be supplemented by data obtained by aircraft, drop-sonde, or satellite. It should be noted that, if the last term is computed by taking the product of the averages, rather than the average of the products, of V_n and $\bar{q}$, considerable error may result. Also the use of the geostrophic velocity $\mathbf{G}$, instead of the actual profile $\mathbf{V}(z)$, usually causes errors (e.g., Palmén, 1963; Ferguson and Schaefer, 1971). This is due to the fact that, in the vertical, $\bar{q}$ and $\nabla \cdot \mathbf{V}$ are often strongly correlated; if $\nabla \cdot \mathbf{G}$ is used this correlation is neglected [cf., (3.73)].

Previous Studies

There have been two general classes of applications of the atmospheric water budget method. In the first, use was made of the existing network of rawinsonde stations. Among the early attempts to apply it was the analysis of Benton and Estoque (1954) of the water vapor transfer over the entire North American continent. Several other studies have been carried out for large areas at a continental, hemispherical or latitudinal scale; however, except for studies in global circulation, smaller areas are usually of greater interest. Examples of applications over areas smaller than 10^6 km^2 are the studies by Hutchings (1957) for southern England (4 stations with $A = 9 \times 10^4$

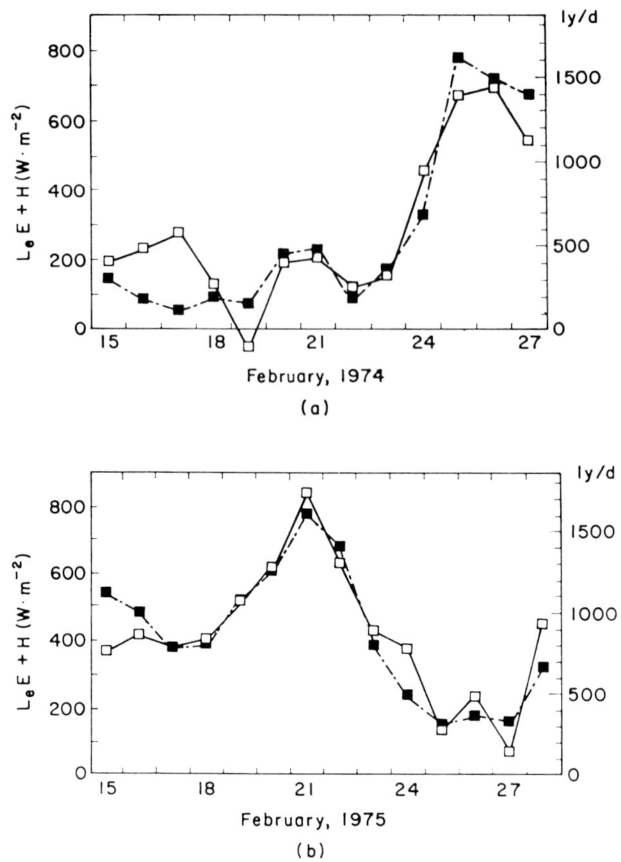

Fig. 11.17. Comparison of the heat flux $(H + L_eE)$ obtained by means of the atmospheric water budget (open squares) by Nitta (1976) and Murty (1976), with the heat flux obtained by a mean profile method by Kondo (1976). These results represent general conditions for the East China Sea. Because the data were not taken at the same locations the areas for which both methods were applied, coincided only approximately. The area enclosed for the atmospheric water budget was of the order of 17 10^4 km^2; its shape was roughly rectangular with center at Okinawa (adapted from Kondo, 1976).

km^2); Palmén (1963) for the Baltic Sea (6 stations with $A = 30.3 \times 10^4$ km^2); Söderman and Wesantera (1966) for Finland (5 stations with $A = 24.7 \times 10^4$ km^2); Rasmusson (1971) by means of the existing network for the Great Lakes ($A = 24.6 \times 10^4$ km^2), the Great Lakes Basin ($A = 48 \times 10^4$ km^2) and the Ohio Basin ($A = 53 \times 10^4$ km^2); Ninomiya (1972) for the East China Sea (8 stations with $A = 63.9 \times 10^4$ km^2); and Shahane *et al.* (1977) and Magyar *et al.* (1978) for the entire United States subdivided in areas with A ranging from 5.18×10^4 to 15.1×10^4 km^2. In the studies just mentioned, the data used were the twice-daily observations of the existing aerological stations, which were available for the standard pressure levels, i.e., 1000, 850, 700, 500, 400 and 300 mb, or for 50 mb increments. It was usually assumed that the water vapor flux divergence and vapor storage changes can be neglected above a level of around 400 to 300 mb, i.e., approximately 7 to 8 km above sea level. In most studies the calculations were made to obtain mean monthly values of evaporation.

In the second class of applications of this method, the data were obtained during short periods of intense measurements, within the framework of large-scale field experiments, as part of the Global Atmospheric Research Program (GARP). In such experiments, the measurements are taken with higher resolution in time and in the vertical and they are also of a higher quality than those of the regular rawinsonde network. A few examples of water budget studies made with such data are those by Augstein *et al.* (1973) for the eastern Atlantic Ocean (observations every 3 hr for 2 weeks, at 3 stations with $A = 25 \times 10^4$ km^2); Holland and Rasmusson (1973) for the western Atlantic Ocean east of Barbados (15 soundings daily for 5 days, at 10 mb intervals at 4 stations with $A = 25 \times 10^4$ km^2); Rasmusson *et al.* (1974) for Lake Ontario (8 soundings daily for 45 days at 10 mb intervals at 6 stations with $A = 1.4 \times 10^4$ km^2); Nitta (1976) and Murty (1976) for the East China Sea (observations every 6 hr for two periods of 14 days, interpolated at 25 mb intervals at 4 stations with $A = 17 \times 10^4$ km^2). In the aforementioned studies mean daily evaporation was calculated and analyzed. The results obtained in these experiments were generally quite satisfactory. For example, as shown in Figure 11.17, the atmospheric budget estimates of Nitta (1976) and Murty (1976) were in good agreement with the daily estimates obtained by Kondo (1976) by means of a mean-profile method (cf. Section 9.1a). An exception among these studies was that on Lake Ontario; Phillips and Rasmusson (1978) reported that, although relative variations of daily evaporation estimates were similar to those obtained from bulk transfer computations, the absolute values were roughly twice those obtained from most other techniques.

Potential and Limitations

The data available for the practical application of the atmospheric water budget method are normally not obtained from a specially designed experiment, but from the operational aerological network. This world-wide grid was designed to observe synoptic-scale features with time scales of a few days and length-scales of the order of 1000 km.

For this reason the operational network, with twice-daily observations, probably cannot provide good resolution for sub-grid changes, and herein lies a serious limitation for budget calculations. This is especially the case in areas with strong diurnal fluctuations and non-homogeneous surface conditions, such as coastal or mountainous regions. Ferguson and Schaefer (1971) noted that severe aliasing errors can result from meso- and micro-fluctuations in wind and moisture near the shore of Lake Ontario; but they were able to lower the errors in the vertically integrated moisture flux by reducing the sampling interval from 12 to 2 hr. Beside the temporal and horizontal spatial resolution there is also the problem of the vertical resolution. The standard levels for which the rawinsonde data are reported, especially the lower ones, viz. 1000, 850, 700 and 500, provide a very poor resolution. Ferguson and Schaefer (1971) calculated that the moisture flux divergence below 700 mb (about 3 km above sea level) averaged 75 percent of the total in the whole 1000–400 mb layer; they concluded that calculations can be reliable only if the profiles are well defined in the lower layers. The lack of adequate vertical resolution in the currently available rawinsonde profiles is, of course, also one of the major difficulties in applying the ABL profile method described in Section 9.1c.

Additional errors may result from the neglect of changes in advection and storage of liquid and solid water in clouds; in (11.39) it is assumed that all net condensation falls out as precipitation. For larger areas the transport and storage of water in clouds is negligible except possibly in cases of cold air outbreaks over relatively warm water.

Finally, there is the problem of the observational accuracy. Rasmusson (1977) has made a detailed analysis of the errors in flux divergence computations resulting from the usual limits of instrumental accuracy in typical rawinsonde observations in networks of different scales. He concluded that with the operational rawinsonde network and current observational schedules, the application of the method to basins with an area smaller than 25×10^4 km^2 is limited, and that the results are probably unreliable. The method can yield good results for areas of the order of 25×10^4 to 10^6 km^2, but it is best suited for areas larger than 10^6 km^2.

The appeal of the method stems mainly from the simplicity of the budget concept and from the availability of the extensive aerological data base for many locations around the world. It is of dubious reliability for surface areas which are usually of practical interest in hydrology. However, it can be very useful and accurate for climatological estimates over larger areas and over monthly or longer periods. Therefore, it can be used to test, compare or complete the results of other methods averaged over larger areas.

Historical References (Prior to 1900)

Achard (1780a), 'Dissertation sur la cause de l'élévation des vapeurs. Observations sur la physique, sur l'histoire naturelle et sur les arts, etc.', *Bureau du Journal de Physique, Paris* **15**, 463–477.

Achard (1780b), 'Mémoire sur le froid produit par l'évaporation. Observations sur la physique, sur l'histoire naturelle et sur les arts, etc.', *Bureau du Journal de Physique, Paris* **16**, 174–186.

Aelfric (1942), *De temporibus anni*, edited by Heinrich Henel, Early English Text Society, Oxford Univ. Press, London, 105 pp.

Aelfric (1961) *De Temporibus Anni*' translation from Anglo-Saxon in T. O. Cockayne (ed.) *Leechdoms, Wortcunning and Starcraft of Early England*, Vol. III, The Holland Press, London, pp. 177–227.

Alexander Neckam (1863), *De naturis rerum*, with the poem of the same author, *De laudibus divinae sapientiae*, edited by T. Wright, Longman, Green, Longman, Roberts and Green, London, 521 pp.

Al-Farabi (1969), *Philosophy of Plato and Aristotle*, translated with an introduction by Mushin Mahdi, Cornell University Press, Ithaca, N.Y., 158 pp.

Ambrose, St. (1961), *Hexameron, Paradise and Cain and Abel*, translated by J.J. Savage, Fathers of the Church, Inc., New York, 449 pp.

Archibald, E. D. (1883), 'The increase in velocity of the wind with altitude', *Nature* **27**, 243–245.

Aristoteles (1574), *Quintum volumen Aristotelis De Coelo, De Generatione & Corruptione, Meteorologicorum, De Plantis*, cum Averrois Cordubensis variis in eosdem commentariis, Venetiis apud Iuntas, 500 pp.

Aristotie (1938a, b), *Problems*, with an English translation by W. S. Hett, Vols. I and II, W. Heinemann, Ltd., London, Harvard Univ. Press, Cambridge, Mass., 461 + 456 pp.

Aristotle (1952), *Meteorologica*, with an English translation by H. D. P. Lee, W. Heinemann, Ltd, London, Harvard Univ. Press, Cambridge, Mass., 433 pp.

Bartholomaeus Anglicus (1601), *De genuinis rerum coelestium, terrestrium et inferarum proprietatibus libri XVIII* [*De rerum proprietatibus*], Francofurti, Apud Wolfgangum Richterum, impensis Nicolai Steinii, Not. & Bibliopolae, 1261 pp. [Facsimile Minerva, G.M.B.H., Frankfurt a.M. 1964].

Basil, St. (1963), *Exegetic Homilies*, translated by A.C. Way, The Fathers of the Church, **46**, The Catholic Univ. of Amer. Press, Washington, D.C., 378 pp.

Beda, Venerabilis (1843) *De natura rerum*, in J. A. Giles, (ed.), *The Miscellaneous Works of Venerable Bede in the Original Latin*, Vol. VI, Whittaker and Co., London, 459 pp.

Black, J. (1803), *Lectures on the Elements of Chemistry*, published from his manuscripts by J. Robison, Longman & Rees, London; W. Creech, Edinburgh, Vol. I., 556 pp.

Boltzmann, L. (1884), 'Ableitung des Stefan'schen Gesetzes, betreffend die Abhängigkeit der Wärmestrahlung von der Temperatur aus der electromagnetischen Lichttheorie', *Ann. Phys. u. Chemie* (*Wiedemann*) **22**, 291–294.

Bouillet, M. (1742), 'Sur l'évaporation des liquides', *Histoire de l'Académie Royale des Sciences, Paris* **55**, 18–21.

Boussinesq, J. (1877), 'Essai sur la théorie des eaux courantes', *Mémoires présentés par div. savants à l'Acad. des Sciences de l'Institut de France* **23**, 1–680.

Bowen, I. S. (1926), 'The ratio of heat losses by conduction and by evaporation from any water surface', *Phys. Rev.* **27**, 779–787.

Burnet, J. (1930), *Early Greek Philosophy*, A. & C. Black, Ltd., London, 375 pp.

Carpenter, L. G. (1889), 'Section of meteorology and irrigation engineering', Second Ann. Rept., Agric. Exp. Station, State Agric. College, Fort Collins, Colo., pp. 49–76.

Carpenter, L. G. (1891), 'Section of meteorology and irrigation engineering', Fourth Ann. Rept., Agric. Exp. Station, State Agric. College, Fort Collins, Colo., pp. 29–97.

Cherniss, H. (1964), *Aristotle's Criticism of Presocratic Philosophy*, Octagon Books, N.Y., 418 pp.

Conrad von Megenberg (1897) *Das Buch der Natur*; *Die erste Naturgeschichte in deutscher Sprache* (In Neu-Hochdeutscher Sprache bearbeitet von Hugo Schulz), Verlag J. Abel, Greifswald, 445 pp.

Coutant, V. and Eichenlaub, V. (1974), 'The *De Ventis* of Theophrastus: its contributions to the theory of winds', *Bull. Amer. Meteorol. Soc.* **55**, 1454–1462.

Dalton, J. (1801), 'New theory of the constitution of mixed aeriform fluids, and particularly of the atmosphere', *J. Nat. Philos., Chemistry and the Arts*, (W. Nicholson) **5**, 241–244.

Dalton, J. (1802a), 'Experimental essays on the constitution of mixed gases; on the force of steam or vapor from water and other liquids in different temperatures, both in a Torricellian vacuum and in air; on evaporation and on the expansion of gases by heat', *Mem. Manchester Lit. and Phil. Soc.* **5**, 535–602.

Dalton, J. (1802b), 'Meteorological observations', *Mem. Manchester Lit. and Phil. Soc.* **5**, 666–674.

Dalton, J. (1802c), 'Experiments and observations to determine whether the quantity of rain and dew is equal to the quantity of water carried off by the rivers and raised by evaporation; with an enquiry into the origin of springs', *Mem. Manchester Lit. and Phil. Soc.* **5**, 346–372.

Daubrée (1847), 'Observations sur la quantité de chaleur annuellement employée à évaporer de l'eau à la surface du globe, . . . etc.', *Comptes Rendus Hebd. Acad. Sc., Paris* **24**, 548–550.

DeLaHire (1703), 'Sur l'eau de pluie, & sur l'origine des fontaines; avec quelques particularitez sur la construction des cisternes', *Histoire de l'Acad. Roy. des Sciences* (Avec les Mémoires de la Mathématique & de Physique pour la même année), *Mem.*, 56–69 (See also Sur l'origine des rivières, *Hist*, 1–6).

De Luc, J. A. (1787), *Idées sur la metéorolgie*, (Partie I *De l'évaporation de l'eau*), Tome I, Veuve Duchesne, Paris, 516 pp.

De Luc, J. A. (1792), 'On evaporation', *Philos. Trans. Roy. Soc. London* **82**, 400–424.

Desaguliers, J. T. (1729), 'An attempt to solve the phenomenon of the rise of vapors, formation of clouds and descent of rain', *Philos. Trans. Roy. Soc.* **36**, 6–22.

Desaguliers, J. T. (1744), *A Course of Experimental Philosophy*, Vol. II, W. Innys, M. Senex, and T. Longman, London, 568 pp.

De Saussure, H. -B. (1783), *Essais sur L'Hygrométrie*, 'III. Essai, Théorie de l'évaporation', Samuel Fauche, Neuchatel, pp. 183–258.

Descartes, René (1637), *Discours de la méthode, plus La Dioptrique, Les Météores et La Géométrie*, De l'Imprimerie de Ian Maire, Leyde [Leiden], 413 pp.

Diels, H. (1879), *Doxographi Graeci*, G. Reimer, Berlin, 854 pp.

Diels, H. (1934), *Die Fragmente der Vorsokratiker*, 5. Aufl. herausgegeben von W. Kranz, Weidmannsche Buchhandlung, Berlin, 1. Band, 482 pp.

Diogenes Laertius (1925), *Lives of Eminent Philosophers*, Vol. II, with an English translation by R. D. Hicks, William Heinemann, London; G. P. Putnam's Sons, New York, 704 pp.

Dobson (1777) 'Observations on the annual evaporation at Liverpool in Lancashire; and on evaporation considered as a test of the moisture or dryness of the atmosphere', *Philos. Trans. Roy. Soc.* **67**, 244–259.

Ebermayer, E. (1879), 'Wie kann man den Einfluss der Wälder auf den Quellenreichthum ermitteln?' *Forstwissenschaftliches Centralblatt* **2**, 77–81.

Fick, A. (1855), 'Ueber Diffusion', *Ann. Phys. u. Chemie* (J. C. Poggendorff) **94** (170), 59–86.

Fitzgerald, D. (1886), 'Evaporation', *Trans. Am. Soc. Civ. Eng.* **15**, 581–646.

Fox, R. (1971), *The Caloric Theory of Gases from Lavoisier to Regnault*, Clarendon Press, Oxford, 378 pp.

Franklin, B. (1765), 'Physical and meterological observations, conjectures, and suppositions', Read June 3, 1756, *Philos. Trans. Roy. Soc.* **55**, 182–192.

Franklin, B. (1887), *The Complete Works* (*Letters to J. Lining*) Vol. 2, 498–505 (April 14, 1757); Vol. 3, 22–27 (June 17, 1758), G. P. Putnam's Sons, New York, London.

Freeman, K. (1953), *The Pre-Socratic Philosophers*, Basil Blackwell, Oxford, 486 pp.

Gershon Ben Shlmoh (1953), *The Gate of Heaven* (Shaar ha-Shamayim) translated and edited by F. S. Bodenheimer, Kiryath Sepher Ltd., Jerusalem, 356 pp.

Gilbert, O. (1907), *Die meteorologischen Theorien des Griechischen Altertums*, B. G. Teubner, Leipzig, 746 pp.

Gilson, E. (1920, 1921), 'Météores Cartésiens et météores scolastiques', *Revue Néoscolastique de Philosophie, Louvain* **22**, 358–384; **23**, 73–84.

Grabmann, M. (1916), 'Forschungen über die Lateinischen Aristotelesübersetzungen des 13. Jahrhunderts', *Beiträge zur Geschichte der Philosophie des Mittelalters*, **18** (5–6), Aschendorffschen Verlagsbuchhandlung, Münster i. W., 270 pp.

Halley, E. (1687), 'An estimate of the quantity of vapour raised out of the sea by the warmth of the sun', *Philos. Trans. Roy. Soc.*, No. 189 (**16**), 366–370.

Halley, E. (1691), 'An account of the circulation of the watery vapours of the sea, and of the cause of springs', *Philos. Trans. Roy. Soc.*, No. 192 (**16**) 468–473.

Halley, E. (1694), 'An account of the evaporation of water, as it was experimented in Gresham Colledge in the year 1693. With some observations thereon', *Philos. Trans. Roy. Soc. London*, No. 212, (**18**), 183–190.

Hamilton, H. (1765), 'A dissertation on the nature of evaporation and several phenomena of air, water, and boiling liquors', *Philos. Trans. Roy. Soc.* **55**, 146–181.

Heller, E. (1800), 'Ueber den Einfluss des Sonnenlichts auf die Verdünstung des Wassers', *Annalen d. Physik* **4**, 210–221.

Heninger, S. K., Jr. (1960), *A Handbook of Renaissance Meteorology*, Duke Univ. Press, Durham, North Carol., 269 pp.

Hesiode (1928), *Théogonie*; *les travaux et les jours*; *le bouclier*, Texte établi et traduit par P. Mazon, Société d'édition Les belles lettres, Paris, 158 pp.

Hesiode (1978), *Works and Days*, edited by M. L. West, Clarendon Press, Oxford, 399 pp.

Hippocrates (1923), *Airs, Waters, Places*, Vol. I, with an English translation by W. H. S. Jones, William Heinemann Ltd., London; Harvard Univ. Press, Cambridge, Mass., pp. 71–137.

Hoffmann, R. (1861), 'Versuche mit Lysimetern', *Jahresber. über die Fortschritte der Argikulturchemie* **2**, 9–15.

Homén, Th. (1897), 'Der tägliche Wärmeumsatz im Boden und die Wärmestrahlung zwischen Himmel und Erde', *Acta Societ. Scientiarum Fennicae* **23** (No. 3) 5–147.

Huxley, T. H. (1900), *Physiography*, MacMillan and Co., Ltd., London, D. Appleton and Co., Ltd., N. Y., 384 pp.

Ikhwan al-Safa (1861), *Meteorologie*, IV in "Die Naturanschauung und Naturphilosophie der Araber im zehnten Jahrhundert, aus den Schriften der lautern Brüder" übersetzt von Fr. Dieterici, Nicolai, Sort. Buchhandlung, Berlin, 216 pp.

Isidore de Séville (1960) *De natura rerum – Traité de la nature*, édité par J. Fontaine, Féret et fils, Eds., Bordeaux, 466 pp.

Isidorus Hispalensis, E. (1911), *Etymologiarum sive originum libri XX* ("Etymologiae"), Oxonii e Typographeo Clarendoniano (Oxford Univ. Press).

Jansen-Sieben, R. (1968) "*De Natuurkunde van het Geheelal*", *een 13–de eeuws middelnederlands Leerdicht*, Académie Royale de Belgique, Brussel, 742 pp.

Jelinek, C. and Hann, J. (1873), 'Der Verdunstungsmesser Piche', *Zeitsch. d. Oesterreich. Gesellsch. Meteorologie* **8**, 270–271.

Job of Edessa (1935), *Book of Treasures*, Syriac text edited and translated by A. Mingana, W. Heffer & Sons Ltd., Cambridge, 470 pp.

Jourdain, A. (1960), *Recherches critiques sur l'âge et l'origine des traductions latines d'Aristote*, (edn. of 1843), Burt Franklin, New York, 472 pp.

Kaufmann, A. (1899), *Thomas von Chantimpré*, Kommissions-Verlag u. Druck von J. P. Bachem, Köln, 137 pp.

Lavoisier (1777), 'De la combinaison de la matière du feu avec les fluides évaporables, et de la formation des fluides élastiques aëriformes', *Mémoires de mathématique et de physique tirés des registres de l'Académie Royale des Sciences* **90**, 420–432.

Lear, F. S. (1936), 'St. Isidore and mediaeval science', *Rice Institute Pamphlet* **23**, 75–105.

Le Roy (1751), 'Mémoire sur l'élévation et la suspension de l'eau dans l'air et sur la rosée', *Mémoires de mathématique et de physique tirés des registres de l'Académie Royale des Sciences*, **64**, 481–518.

Leslie, J. (1813), *Short Account of Experiments and Instruments Depending on the Relations of Air to Heat and Moisture*, Edinburgh: W. Blackwood and J. Ballantyne & Co; Longman, Hurst, Rees, Orme & Brown; and J. Murray, London, 178 pp.

Livingston, G. J. (1908), 'An annotated bibliography on evaporation', *Monthly Weath. Rev.* **36**, 181–186; 301–306; 375–381.

Livingston, G. J. (1909), 'An annotated bibliography on evaporation', *Monthly Weath. Rev.* **37**, 68–72; 103–109; 157–160; 193–199; 248–252.

Lucretius (1924), *De rerum natura*, with an English translation by W. H. D. Rouse, W. Heinemann Ltd., London, Harvard Univ. Press, Cambridge, Mass., 538 pp.

Maury, M. F. (1861), *The Physical Geography of the Sea and its Meteorology*, 8th edn., edited by J. Leighly, 1963, Harvard Univ. Press, Cambridge, Mass., 432 pp.

McKie, D. and de V. Heathcote, N. H. (1935), *The Discovery of Specific and Latent Heats*, E. Arnold & Co., London, 155 pp.

Mieli, A. (1966), *La science Arabe*, E. J. Brill, Leiden, 467 pp.

Monge (1790), 'Mémoire sur la cause des principaux phénomènes de la météorologie', *Annales de Chimie* **5**, 1–71.

Needham, J. (1959), *Science and Civilization in China*, Vol. 3: 'Mathematics and the sciences of the heavens and the earth' (with the collaboration of Wang Ling), Cambridge Univ. Press, 977 pp.

Oxford Study Edition (1976), *The New English Bible, with the Apocrypha*, Oxford Univ. Press, N. Y., 1725 pp.

Parkes, J. (1845), 'On the quantity of rain compared with the quantity of water evaporated from or filtered through soil; with some remarks on drainage', *J. Roy. Agric. Soc. England* **5**, 146–158.

Parrot (1804), 'Bemerkungen über Dalton's Versuche über die Expansivkräfte ...', *Ann. d. Physik* (Gilbert) **17**, 82–101.

Pernoud, R. (1977), *Pour en finir avec la moyen-âge*, Editions du Seuil, Paris, 160 pp.

Perrault, P. (1674), *De l'origine des fontaines*, Pierre Le Petit, Imprimeur & Libraire, Paris, 353 pp.

Perrault, P. (1733), 'Expérience sur le froid', 1670, *Histoire de l'Académie Royale des Sciences*, **1**, 115–116.

Peters, F. E. (1968), *Aristotle and the Arabs*, New York Univ. Press, N.Y., 303 pp.

Pliny (1938), *Natural History*, with an English translation by H. Rackham, Vol. I (Praefatio, Libri I, II), W. Heinemann, Ltd., London, Harvard Univ. Press, Cambridge, Mass., 378 pp.

Pliny (1945), *Natural History*, with an English translation by H. Rackham, Vol. IV (Libri XII-XVI), W. Heinemann, Ltd., London; Harvard Univ. Press, Cambridge, Mass., 551 pp.

Pouillet (1838), 'Mémoire sur la chaleur solaire, sur les pouvoirs rayonnants et absorbants de l'air atmosphérique, et sur la température de l'espace', *Compt. Rendus Hebd. des Séances de l'Acad. Sci.*, **7**, 24–65.

Pouillet (1847), *Eléments de physique expérimentale et de météorologie*, 5me edn., Tome second, Bechet Jeune, Paris, 836 pp.

Rabanus Maurus (1852), 'De universo [De rerum naturis]', Vol. 111, pp. 10–614, in *Patrologiae cursus completus*, J.-P. Migne, Paris.

Reynolds, O. (1874), 'On the extent and action of the heating surface for steam boilers', *Proc. Manchester Liter. Phil. Soc.* **14**, 7–12.

Schmid, E. E. (1860), *Lehrbuch der Meteorologie*, L. Voss, Leipzig, 1009 pp.

Schmidt, W. (1915), 'Strahlung und Verdunstung an freien Wasserflächen; ein, Beitrag zum Wärmehaushalt des Weltmeers und zum Wasserhaushalt der Erde', *Ann. d. Hydrogr. u. Mar. Meteorol.* **43**, 111–124.

Schmidt, W. (1917), 'Der Massenaustausch bei der ungeordneten Strömung in freier Luft und seine Folgen', *Sitzber, Kais. Akad. Wissen. Wien* [2a], **126**, 757–804.

Sedileau (1730a), 'Observations de la quantité de l'eau de pluye tombée à Paris durant près de trois années & de la quantité de l'évaporation, 29 fév. 1692', *Mémoires de l'Académie Royale des Sciences* **10**, 29–36.

Sedileau (1730b), 'De l'origine des rivières & de la quantité de l'eau qui entre dans la mer & qui en sort, 31 mai 1693', *Mémoires de l'Académie Royale des Sciences* **10**, 325–339.

Sedileau (1733a), 'Sur la quantitè d'eau de pluye tombée à Paris', 1692, *Histoire de l'Académie Royale des Sciences* **2**, 133–135 (abstract).

Sedileau (1733b), 'Sur la quantité d'eau tombée à l'Observatoire pendant les quatre dernières années, & sur l'origine des rivières', 1693, *Histoire de l'Académie Royale des Sciences* **2**, 164–167 (abstract).

Seneca (1971; 1972), *Naturales quaestiones*, with an English translation by T. H. Corcoran, W. Heinemann Ltd., London, Harvard Univ. Press, Cambridge, Mass. (Vol. I, 1971, 297 pp.; Vol. II, 1972, 312 pp.).

's Gravesande, G. J. (1742), *Physices elementa mathematica, experimentis confirmata*, Editio tertia, J. A. Langerak, J. and H. Verbeek, Leiden, 1073 pp.

's Gravesande, W. J. (1747), *Mathematical Elements of Natural Philosophy, Confirmed by Experiments*, Vol. II, translated from the Latin by J. T. Desaguliers, 6th edn., W. Innys, T. Longman, T. Shewell, C. Hitch, and M. Senex, London, 389 pp.

Soldner (1804), 'Ueber das allgemeine Gesetz für die Expansivkraft des Wasserdampfes durch Wärme, nach Dalton's Versuchen; Anwendung dieses Gesetzes auf das Verdunsten der Flüssigkeiten', *Annalen d. Phys.* (Gilbert) **17**, 44–81.

Soldner (1807), 'Nachtrag zu der Abhandlung über das allgemeine Gesetz der Expansivkraft der Wasserdämpfe', *Ann. d. Phys.* (Gilbert) **25**, 411–439.

Stefan, J. (1879), 'Ueber die Beziehung zwischen der Wärmestrahlung under der Temperatur', Sitzungsberichte der Math-Naturw. Classe d. Kaiserlichen Akademie d. Wissenschaften, Wien, Vol. **79** (2), 391–428.

Stelling, Ed. (1822), 'Ueber die Abhängigkeit der Verdunstung des Wassers von seiner Temperatur und von der Feuchtigkeit und Bewegung der Luft (vorgelegt 1881)', *Repertorium für Meteorologie, Kaiserliche Akademie der Wissenschaften* (*Meteorologicheskii Sbornik, Imperatorskoj Akademii Nauk*) *St. Petersburg*, **8**, No. 3, 1–49 (Also *Z. Oesterr. Ges. Meteor.*, **17**, (1882), 372–373.)

Stevenson, T. (1880), 'Report on simultaneous observations of the force of wind at different heights above the ground', *J. Scot. Meteor. Soc.* **5**, (103), 348–351.

Tate, T. (1862), 'Experimental researches on the laws of evaporation and absorption, with a description of a new evaporameter and absorbometer', *Philos. Magazine and J. Science*, **23** (IV), 126–135.

Theophrastus (1975), *De Ventis*, edited and translated by V. Coutant and V. L. Eichenlaub, Univ. of Notre Dame Press, Notre Dame, Indiana, 105 pp.

Thomas Cantimpratensis (1973), *Liber de natura rerum* (Vorwort H. Boese), Walter De Gruyter, Berlin, New York, 431 pp.

Van Maerlant, Jacob (1878), *Naturen Bloeme*, uitgegeven door Eelco Verwijs, J. B. Wolters, Groningen, 251 pp.

Van Musschenbroek P. (1732), 'Ephemerides meteorologicae, barometricae, thermometricae, epidemicae, magneticae, ultrajectinae', *Philos. Trans. Roy. Soc.* **37**, 357–384.

Van Musschenbroek, P. (1739), *Essai de physique*, Tome 2, traduit du Hollandais par P. Massuet, Samuel Luchtmans, Leyden.

Van Musschenbroek, P. (1769), *Cours de physique expérimentale et mathématique*, Tome 3, traduit par S. de la Fond, Guillyn Librairie, Paris, 503 pp.

Van Steenberghen, F. (1955), *Aristotle in the West, The Origins of Latin Aristotelianism*, E. Nauwelaerts, Publ., Louvain, 244 pp.

Vincentius Bellovacensis (1624), *Speculum naturale*, Balthazar Bellerus, Duaci (Douai), 1240 pp. (Facsimile of Benedictine Edn., Akademische Druck u. Verlagsanstalt, Graz, 1964).

Vuilhelmus (1567), *Dialogus de substantiis physicis*: ante annos ducentos confectus, à Vuilhelmo Aneponymo philosopho, pp. 1–312, Argentorati, excudebat Iosias Rihelius, (1967 Facsimile edn., Minerva, GMBH, Frankfurt/Main). (Note that the last part, pp. 313–363, is by a different author, who is very familiar with Aristotle's two-exhalation theory; the date of this part is at least a century later than the main part.)

Weilenmann, A. (1877a), 'Die Verdunstung des Wassers', *Schweizerische Meteorologische Beobachtungen, Zürich*, **12**, 7–37.

Weilenmann, A. (1877b), 'Berechnung der Grösse der Verdunstung aus den meteorologischen Factoren, *Zeitsch. der Oesterreichischen Gesellschaft für Meteorologie, Wien.* (*Meteor. Zeitsch.*) **12**, 268–271.

Wild, H. (1874), 'Ueber einen einfachen Verdunstungsmesser für Sommer und Winter', *Bull. de l'-Académie Impériale des Sciences, St. Petersbourg* **19** 440–445.

Woeikoff, (1887), *Die Klimate der Erde*, H. Costenoble, Jena, 396 + 422 pp.

Wollny, E. (1877), *Der Einfluss der Pflanzendecke und Beschattung auf die physikalischen Eigenschaften und die Fruchtbarkeit des Bodens*, Wiegandt, Hempel & Parey Verlag, Berlin, 197 pp.

Wollny, E. (1893), 'Untersuchungen über den Einfluss der Pflanzendecke und der Beschattung auf die physikalischen Eigenschaften des Bodens', *Forsch. Geb. Agrikulturphysik* **18**, 486–516.

References

Aase, J. K. and Idso, S. B. (1978), 'A comparison of two formula types for calculating long wave radiation from the atmosphere', *Water Resour. Res.* **14**, 623–625.

Abbe, C. (1905), 'The Piche evaporometer', *Monthly Weath. Rev.* **33**, 253–255.

Abramowitz, M. and Stegun, I. A. (eds.) (1964), *Handbook of Mathematical Functions*, National Bureau of Standards, Applied Math. Ser. No. 55, Washington, D.C., 1046 pp.

Albrecht, F. (1937), 'Messgeräte des Wärmehaushaltes an der Erdoberfläche als Mittel der bioklimatologischen Forschung', *Meteorol. Z.* **54**, 471–475.

Albrecht, F. (1950), 'Die Methoden zur Bestimmung der Verdunstung der natürlichen Erdoberfläche', *Arch. Met. Geoph. Biokl.* **B2**, 1–38.

Anderson, E. R. (1954), 'Energy-budget studies', *Water loss investigations: Lake Hefner Studies, Tech. Report. Prof. Paper* 269, Geol. Survey, U.S. Dept. Interior, pp. 71–119.

Ångström, A. (1924), 'Solar and terrestrial radiation', *Quart. J. Roy. Meteorol. Soc.* **50**, 121–125.

Arya, S. P. S. (1975), 'Geostrophic drag and heat transfer relations for the atmospheric boundary layer', *Quart. J. Roy. Meteorol. Soc.* **101**, 147–161.

Arya, S. P. S. (1977), 'Suggested revisions to certain boundary layer parameterization schemes used in atmospheric circulation models', *Monthly Weath. Rev.* **105**, 215–227.

Arya, S. P. S. and Wyngaard, J. C. (1975), 'Effect of baroclinicity on wind profiles and the geostrophic drag law for the convective planetary boundary layer', *J. Atmos. Sci.* **32**, 767–778.

Assaf, G. and Kessler, J. (1976), 'Climate and energy exchange in the Gulf of Aqaba (Eilat)', *Monthly Weath. Rev.* **104**, 381–385.

Augstein, E., Riehl, H., Ostapoff, F., and Wagner, V. (1973), 'Mass and energy transports in an undisturbed Atlantic trade-wind flow', *Monthly. Weath. Rev.* **101**, 101–111.

Baker, D. G. and Haines, D. A. (1969), 'Solar radiation and sunshine duration relationships in the North-Central Region and Alaska', Tech. Bull. No. 262, Univ. Minnesota Ag. Exp. Station, 372 pp.

Baker, F. G., Veneman, P. L. M., and Bouma, J. (1974), 'Limitations of the instantaneous profile method for field measurement of unsaturated hydraulic conductivity', *Soil Sci. Soc. Am. Proc.* **38**, 885–888.

Barton, I. J. (1979), 'A parameterization of the evaporation from nonsaturated surface', *J. Appl. Meteorol.* **18**, 43–47.

Baumgartner, A. and Reichel, E. (1975), *The World Water Balance*, Elsevier Sci. Publ. Com., Amsterdam and N. Y., 179 pp, 31 maps.

Benton, G. S. and Estoque, M. A. (1954), 'Water-vapor transfer over the North American continent', *J. Meteorol.* **11**, 462–477.

Black, J. N. (1956), 'The distribution of solar radiation over the earth's surface', *Arch. Meteorol. Geophys. Bioklim.* **7**, 165–189.

Black, T. A., Gardner, W. R., and Thurtell, G. W. (1969), 'The prediction of evaporation, drainage, and soil water storage for a bare soil', *Soil Sci. Soc. Am. Proc.* **33**, 655–660.

Black, T. A., Thurtell, G. W., and Tanner, C. B. (1968), 'Hydraulic load-cell lysimeter, construction, calibration, and tests', *Soil Sci. Soc. Am. Proc.* **32**, 623–629.

Blackadar, A. K. (1962), 'The vertical distribution of wind and turbulent exchange in a neutral atmosphere', *J. Geophys. Res.* **67**, 3095–3102.

Blackadar, A. K. and Tennekes, H. (1968), 'Asymptotic similarity in neutral barotropic planetary boundary layers', *J. Atmos. Sci.* **25**, 1015–1020.

Blom, J. and Wartena, L. (1969), 'The influence of changes in surface roughness on the development of the turbulent boundary layer in the lower layers of the atmosphere', *J. Atmos. Sci.* **26**, 255–265.

Bolz, H. M. (1949), 'Die Abhängigkeit der infraroten Gegenstrahlung von der Bewölkung', *Z. Meteorol* **3**, 201–203.

Boston, N. E. J. and Burling, R. W. (1972), 'An investigation of high wave-number temperature and velocity spectra in air', *J. Fluid Mech.* **55**, 473–492.

Bouchet, R. J. (1963a), 'Evapotranspiration réelle et potentielle, signification climatique', General Assembly Berkeley, Intern. Assoc. Sci. Hydrol., Publ. No. 62, Gentbrugge, Belgium, pp. 134–142.

Bouchet, R. J. (1963b), 'Evapotranspiration réelle, évapotranspiration potentielle, et production agricole', *Ann. Agron.* **14**, 743–824.

Bowen, I. S. (1926), 'The ratio of heat losses by conduction and by evaporation from any water surface', *Phys. Rev.* **27**, 779–787.

Bradley, E. F. (1968), 'A micrometeorological study of velocity profiles and surface drag in the region modified by a change in surface roughness', *Quart. J. Roy. Meteorol. Soc.* **94**, 361–379.

Bradley, E. F. (1972), 'The influence of thermal stability on a drag coefficient close to the ground', *Agric. Meteorol.* **9**, 183–190.

Bradshaw, P., Ferris, D. H., and Atwell, N. P. (1967), 'Calculation of boundary-layer development using the turbulent energy equation', *J. Fluid Mech.* **28**, 593–616.

Brakke, T. W., Verma, S. B., and Rosenberg, N. J. (1978), 'Local and regional components of sensible heat advection', *J. Appl. Meteorol.* **17**, 955–963.

Brochet, P. and Gerbier, N. (1972), 'Une méthode pratique de calcul de l'évapotranspiration potentielle', *Ann. Agron.* **23**, 31–49.

Brost, R. A. (1979), 'Comments on "Turbulent exchange coefficients for sensible heat and water vapor under advective conditions"', *J. Appl. Meteorol.* **18**, 378–380.

Brown, G. W. (1969), 'Predicting temperatures of small streams', *Water Resour. Res.* **5**, 68–75.

Brown, K. W., Gerard, C. J., Hipp, B. W., and Ritchie, J. R.T. (1974), 'A procedure for placing large undisturbed monoliths in lysimeters', *Soil Sci. Soc. Am. Proc.* **38**, 981–983.

Brown, K. W. and Covey, W. (1966), 'The energy-budget evaluation of the micro-meteological transfer processes within a cornfield', *Agric. Meteorol.* **3**, 73–96.

Brunt, D. (1932), 'Notes on radiation in the atmosphere: I', *Quart. J. Roy. Meteorol. Soc.* **58**, 389–420.

Brutsaert, W. (1967a), 'Evaporation from a very small water surface at ground level: three-dimensional turbulent diffusion without convection', *J. Geophys. Res.* **72**, 5631–5639.

Brutsaert, W. (1967b), 'Some methods of calculating unsaturated permeability', *Trans. Am. Soc. Agric. Eng.* **10**, 400–404.

Brutsaert, W. (1970), 'On the anisotropy of the eddy diffusivity', *J. Meteorol. Soc. Japan* **48**, 411–416.

Brutsaert, W. (1973), 'Similarity functions for turbulence in neutral air above swell', *J. Phys. Oceanog.* **3**, 479–482.

Brutsaert, W. (1975a), 'A theory for local evaporation (or heat transfer) from rough and smooth surfaces at ground level', *Water Resour. Res.* **11**, 543–550.

Brutsaert, W. (1975b), 'The roughness length for water vapor, sensible heat, and other scalars', *J. Atmos. Sci.* **32**, 2028–2031.

Brutsaert, W. (1975c), 'Comments on surface roughness parameters and the height of dense vegetation', *J. Meteorol. Soc. Japan* **53**, 96–97.

Brutsaert, W. (1975d), 'On a derivable formula for long-wave radiation from clear skies', *Water Resour. Res.* **11**, 742–744.

Brutsaert, W. (1979a), 'Heat and mass transfer to and from surfaces with dense vegetation or similar permeable roughness', *Boundary-Layer* Meteorol. **16**, 365–388.

Brutsaert, W. (1979b), 'Universal constants for scaling the exponential soil water diffusivity?', *Water Resour. Res.* **15**, 481–483.

Brutsaert, W. and Chan, F. K.-F. (1978), 'Similarity functions D for water vapor in the unstable atmospheric boundary layer', *Boundary-Layer Meteorol.* **14**, 441–456.

Brutsaert, W. and Ibrahim, H. A. (1966), 'On the first and second linearization of the Boussinesq equation', *Geophys. J. Roy. Astron. Soc.* **11**, 549–554.

Brutsaert, W. and Mawdsley, J. A. (1976), 'The applicability of planetary boundary layer theory to calculate regional evapotranspiration', *Water Resour. Res.* **12**, 852–858.

Brutsaert, W. and Nieber, J. L. (1977), 'Regionalized drought flow hydrographs from a mature glaciated plateau', *Water Resour. Res.* **13**, 637–643.

Brutsaert, W. and Stricker, H. (1979), 'An advection-aridity approach to estimate actual regional evapotranspiration', *Water Resour. Res.* **15**, 443–450.

Brutsaert, W. and Yeh, G.-T. (1969), 'Evaporation from an extremely narrow wet strip at ground level', *J. Geophys. Res.* **74**, 3431–3433.

Brutsaert, W. and Yeh, G.-.T. (1970a), 'Implications of a type of empirical evaporation formula for lakes and pans', *Water Resour. Res.* **6**, 1202–1208.

Brutsaert, W. and Yeh, G.-T. (1970b), 'A power wind law for turbulent transfer computations', *Water Resour. Res.* **6**, 1387–1391.

Brutsaert, W. and Yu, S. L. (1968), 'Mass transfer aspects of pan evaporation', *J. Appl. Meteor.* **7**, 563–566.

Buck, A. L. (1976), 'The variable-path Lyman-alpha hygrometer and its operating characteristics', *Bull. Am. Meteorol. Soc.* **57**, 1113–1118.

Buckingham, E. (1907), 'Studies on the movement of soil moisture', Bureau of Soils, Bull. No. 38, U.S. Dept. Agr., Washington, 61 pp.

Budyko, M. I. (1948), *Evaporation under Natural Conditions*, GIMIZ, Leningrad, English Translation, Isreal Progr. Sci. Translations, Jerusalem (1963), 130 pp.

Budyko, M. I. (1970), 'The water balance of the oceans', Symposium on world water balance, Proc. Reading Sympos. Vol. I, Internat. Assoc. Sci. Hydrol., Public. No. 92, pp. 24–33.

Budyko, M. I. (1974), *Climate and Life*, Academic Press, N.Y., 508 pp.

Bunker, A. F. and Worthington, L. V. (1976), 'Energy exchange charts of the North Atlantic Ocean', *Bull. Am. Meteorol. Soc.* **57**, 670–678.

Burger, H. C. (1915), 'Das Leitvermögen verdünnter mischkristallfreier Lösungen', *Phys. Zeits.* **20**, 73–76.

Busch, N. E. and Panofsky, H. A. (1968), 'Recent spectra of atmospheric turbulence', *Quart. J. Roy. Meteorol. Soc.* **94**, 132–140.

Businger, J. A. (1966), 'Transfer of momentum and heat in the planetary boundary layer', Proc. Symp. Arctic Heat Budget and Atmos. Circulation, RAND Corp. RM–5233–NSF, pp. 305–332.

Businger, J. A., Wyngaard, J. C., Izumi, Y., and Bradley, E. F. (1971), 'Flux-profile relationships in the atmospheric surface layer', *J. Atmos. Sci.* **28**, 181–189.

Businger, J. A. and Yaglom, A. M. (1971), 'Introduction to Obukhov's paper on "Turbulence in an atmosphere with a non-uniform temperature"', *Boundary-Layer Meteorol.* **2**, 3–6.

Calder, K. L. (1949), 'Eddy diffusion and evaporation in flow over aerodynamically smooth and rough surfaces: A treatment based on laboratory laws of turbulent flow with special reference to conditions in the lower atmosphere', *Quart. J. Mech. Appl. Math.* **2**, 153–176.

Cary, J. W. (1967), 'The drying of soil: thermal regimes and ambient pressure', *Agric. Meteorol.* **4**, 357–365.

Chamberlain, A. C. (1966), 'Transport of gases to and from grass and grass-like surfaces', *Proc. Roy. Soc. London* **A290**, 236–265.

Chamberlain, A. C. (1968), 'Transport of gases to and from surfaces with bluff and wave-like roughness elements', *Quart. J. Roy. Meteorol. Soc.* **94**, 318–332.

Champagne, F. H., Friehe, C. A., LaRue, J. C., and Wyngaard, J. C. (1977), 'Flux measurements, flux estimation techniques and fine-scale turbulence measurements in the unstable surface layer over land', *J. Atmos. Sci.* **34**, 515–530.

Charnock, H. (1955), 'Wind Stress on a water surface', *Quart. J. Roy. Meteor. Soc.* **81**, 639–640.

Charnock, H. (1958), 'A note on empirical wind-wave formulae', *Quart. J. Roy. Meteorol. Soc.* **84**, 443–447.

Chudnovskii, A. F. (1962), *Heat Transfer in the Soil*, translated from Russian, Israel Program for Scientific Translations, Jerusalem, 164 pp.

Cionco, R. M. (1965), 'A mathematical model for air flow in a vegetative canopy', *J. Appl. Meteorol.* **4**, 517–522.

Cionco, R. M. (1972), 'A wind-profile index for canopy flow', *Boundary-Layer Meteorol.* **3**, 255–263.

Cionco, R. M. (1978), 'Analysis of canopy index values for various canopy densities', *Boundary-Layer Meteorol.* **15**, 81–93.

Cisler, J. (1969), 'The solution for maximum velocity of isothermal steady flow of water upward from water table to soil surface', *Soil Sci.* **108**, 148.

Clarke, R. H. (1970), 'Observational studies in the atmospheric boundary layer', *Quart. J. Roy. Meteorol. Soc.* **96**, 91–114.

Clarke, R. H. (1972), 'Discussion of "Observational studies in the atmospheric boundary layer"', *Quart. J. Roy. Meteorol. Soc.* **98**, 234–235.

Clarke, R. H., Dyer, A. J., Brook, R. R., Reid, D. G., and Troup, A. J. (1971), 'The Wangara experiment: boundary layer data', Tech. Paper No. 19, Div. Meteor. Physics., CSIRO, Australia.

Clarke, R. H. and Hess, G. D. (1973), 'On the appropriate scaling for velocity and temperature in the planetary boundary later', *J. Atmos. Sci.* **30**, 1346–1353.

Clarke, R. H. and Hess, G. D. (1974), 'Geostrophic departure and the functions A and B of Rossby-number similarity theory', *Boundary-Layer Meteorol.* **7**, 267–287.

Corino, E. R. and Brodkey, R. S. (1969), 'A visual investigation of the wall region in turbulent flow', *J. Fluid Mech.* **37**, 1–30.

Corrsin, S. (1974), 'Limitations of gradient transport models in random walks and in turbulence', *Adv. Geophys.* **18A**, 25–60.

Coulson, K. L. (1975), *Solar and Terrestrial Radiation*, Academic Press, N.Y., 322 pp.

Cowan, I. R. (1968), 'Mass, heat and momentum exchange between stands of plants and their atmospheric environment', *Quart. J. Roy. Meteorol. Soc.* **94**, 523–544.

Crawford, T. V. (1965), 'Moisture transfer in free and forced convection', *Quart. J. Roy. Meteorol. Soc.* **91**, 18–27.

Crow, F. R. and Hottman, S. D. (1973), 'Network density of temperature profile stations and its influence on the accuracy of lake evaporation calculations', *Water Resour. Res.* **9**, 895–899.

Csanady, G. T. (1967), 'On the resistance law of a turbulent Ekman layer', *J. Atmos. Sci.* **24**, 467–471.

Csanady, G. T. (1974), 'The "roughness" of the sea surface in light winds', *J. Geophys. Res.* **79**, 2747–2751.

Cummings, N. W. and Richardson, B. (1927), 'Evaporation from lakes', *Phys. Rev.* **30**, 527–534.

Daniel, J. F. (1976), 'Estimating groundwater evapotranspiration from streamflow records', *Water Resour. Res.* **12**, 360–364.

Davenport, D. C. and Hudson, J. P. (1967a), 'Changes in evaporation rates along a 17-km transect in the Sudan Gezira', *Agric. Meteorol.* **4**, 339–352.

Davenport, D. C. and Hudson, J. P. (1967b), 'Meteorological observations and Penman estimates along a 17-km transect in the Sudan Gezira', *Agric. Meteorol.* **4**, 405–414.

Davidson, J. M., Stone, L. R., Nielsen, D. R., and LaRue, M. E. (1969), 'Field measurement and use of soil-water properties', *Water Resour. Res.* **5**, 1312–1321.

Davidson, K. L. (1974), 'Observational results on the influence of stability and wind-wave coupling on momentum transfer and turbulent fluctuations over ocean waves', *Boundary-Layer Meteorol.* **6**, 305–331.

Davies, J. A. (1965), 'Estimation of insolation for West Africa', *Quart. J. Roy. Meteorol. Soc.* **91**, 359–363.

Davies, J. A. and Allen, C. D. (1973), 'Equilibrium, potential and actual evaporation from cropped surfaces in southern Ontario', *J. Appl. Meteorol.* **12**, 649–657.

Davies, J. A., Robinson, P. J., and Nunez M. (1971), 'Field determinations of surface emissivity and temperature for Lake Ontario', *J. Appl. Meteorol.* **10**, 811–819.

Dawson, D. A. and Trass O. (1972), 'Mass transfer at rough surfaces', *Int. J. Heat Mass Transfer* **15**, 1317–1336.

Deacon, E. L. (1949), 'Vertical diffusion in the lowest layers of the atmosphere', *Quart. J. Roy. Meteorol. Soc.* **75**, 89–103.

Deacon, E. L. (1950), 'The measurement and recording of the heat flux into the soil', *Quart. J. Roy. Meteorol. Soc.* **76**, 479–483.

Dascon, E. L. (1959), 'The measurement of turbulent transfer in the lower atmosphere', *Adv. Geophys.* **6**, 211–228.

Deacon, E. L. (1970), 'The derivation of Swinbank's long-wave radiation formula', *Quart. J. Roy. Meteorol. Soc.* **96**, 313–319.

Deacon, E. L. (1973), 'Geostrophic drag coefficients', *Boundary-Layer Meteorol.* **5**, 321–340.

Deacon, E. L. and Webb, E. K. (1962), 'Small-scale interactions', in M. N. Hill (ed.), *The Sea*, Vol. 1, Interscience, New York and London, pp. 43–87.

Deardorff, J. W. (1972), 'Numerical investigation of neutral and unstable planetary boundary layers', *J. Atmos. Sci.* **29**, 91–115.

DeBruin, H. A. R. (1978), 'A simple model for shallow lake evaporation', *J. Appl. Meteorol.* **17**, 1132–1134.

DeBruin, H. A. R. and Keijman, J. Q. (1979), 'The Priestley-Taylor evaporation model applied to a large shallow lake in the Netherlands', *J. Appl. Meteorol.* **18**, 898–903.

DeCoster, M. and Schuepp, W. (1957), 'Mesures de rayonnement effectif à Léopoldville', *Koninklijke Academie voor Koloniale Wetensch. (Brussel), Mededel. der Zittingen* **3** (Nieuwe Reeks), 642–651.

Denmead, O. T. (1964), 'Evaporation sources and apparent diffusivities in a forest canopy', *J. Appl. Meteorol.* **3**, 383–389.

Denmead, O. T. (1976), 'Temperate cereals', in J. L. Monteith (ed.), *Vegetation and the Atmosphere*, Vol. 2, Academic Press, London, pp. 1–31.

Denmead, O. T. and McIlroy, I. C. (1970), 'Measurements of non-potential evaporation from wheat', *Agric. Meteorol.* **7**, 285–302.

Denmead, O. T., Nulsen, R., and Thurtell, G. W. (1978), 'Ammonia exchange over a corn crop', *Soil Sci. Soc. Am. J.* **42**, 840–842.

Desjardins, R. L. and Lemon, E. R. (1974), 'Limitations of an eddy-correlation technique for the determination of the carbon dioxide and sensible heat fluxes', *Boundary-Layer Meteorol.* **5**, 475–488.

De Vries, D. A. (1955), 'Solar radiation at Wageningen', *Meded. Landbouwhogeschool, Wageningen* **55**, 277–304.

De Vries, D. A. (1958), 'Simultaneous transfer of heat and moisture in porous media', *Trans. Am. Geophys. Un.* **39**, 909–916.

De Vries, D. A. (1959), 'The influence of irrigation on the energy and the climate near the ground', *J. Meteorol.* **16**, 256–270.

De Vries, D. A. (1963), 'Thermal properties of soils', in W. R. van Wijk, W. R. (ed.), *Physics of Plant Environment*, North-Holland Pub. Co., Amsterdam, pp. 210–235.

De Vries, D. A. and Peck, A. J. (1958a, b), 'On the cylindrical probe method of measuring thermal conductivity with special reference to soils', *Australian J. Phys.* **11**, 225–271; **11**, 409–423.

Dipprey, D. F. and Sabersky, R. H. (1962), 'Heat and momentum transfer in smooth and rough tubes at various Prandtl numbers', Tech. Rep. 32–269, Jet Propul. Lab., Calif. Inst. of Technol., Pasadena, 1962. (Also published in (1963), *Int. J. Heat Mass Transfer* **6**, 329–335.)

Donaldson, C. du P. (1973), 'Construction of a dynamic model of the production of atmospheric turbulence and the dispersal of atmospheric pollutants', in D. A. Haugen, (ed.) *Workshop on Micrometeorology*, Amer. Meteorol. Soc., Boston, MA, pp. 313–392.

Doorenbos, J. and Pruitt, W. O. (1975), 'Crop water requirements', Irrigation and Drainage Paper No. 24, FAO (United Nations) Rome, 179 pp.

Dunckel, M., Hasse, L., Krügermeyer, L., Schriever, D., and Wucknitz, J. (1974), 'Turbulent fluxes of momentum, heat and water vapor in the atmospheric surface layer at sea during ATEX', *Boundary-Layer Meteorol.* **6**, 81–106.

Dyer, A. J. (1961), 'Measurements of evaporation and heat transfer in the lower atmosphere by an automatic eddy-correlation technique', *Quart. J. Roy. Meteorol. Soc.* **87**, 401–412.

Dyer, A. J. (1967), 'The turbulent transport of heat and water vapour in an unstable atmosphere', *Quart. J. Roy. Meteorol. Soc.* **93**, 501–508.

Dyer, A. J. (1974), 'A review of flux-profile relationships', *Boundary-Layer Meteorol.* **7**, 363–372.

Dyer, A. J. and Crawford, T. V. (1965), 'Observations of the modification of the microclimate at a leading edge', *Quart. J. Roy. Meteorol. Soc.* **91**, 345–348.

Dyer, A. J. and Hicks, B. B. (1970), 'Flux-gradient relationships in the constant flux layer', *Quart. J. Roy. Meteorol. Soc.* **96**, 715–721.

Dyer, A. J., Hicks, B. B., and Sitaraman, V. (1970), 'Minimizing the levelling error in Reynolds stress measurement by filtering', *J. Appl. Meteorol.* **9**, 532–534.

Edinger, J. E., Duttweiler, D. W., and Geyer, J. C. (1968), 'The response of water temperatures to meteorological conditions', *Water Resour. Res.* **4**, 1137–1143.

Edlefsen, N. E. and Bodman, G. B. (1941), 'Field measurements of water movement through a silt loam soil', *J. Am. Soc. Agron.* **33**, 713–731.

Ekern, . C. (1967), 'Evapotranspiration of pineapple in Hawaii', *Plant Physiol.* **40**, 736–740.

Elliott, W. P. (1958), 'The growth of the atmospheric internal boundary layer', *Trans. Am. Geophys. Un.* **39**, 1048–1054.

Emmuanuel, C. B. (1975), 'Drag and bulk aerodynamic coefficients over shallow water', *Boundary-Layer Meteorol.* **8**, 465–474.

Ertel, H (1933), 'Beweis der Wilh. Schmidtschen konjugierten Potenzformeln für Austausch und Windgeschwindigkeit in den bodennahen Luftschichten', *Meteorol. Z.* **50**, 386–388.

Estoque, M. A. and Bhumralkar, C. M. (1970), 'A method for solving the planetary boundary layer equations', *Boundary-Layer Meteorol.* **1**, 169–194.

Federer, C. A. (1977), 'Leaf resistance and xylem potential differ among broadleaved species', *Forest Sci.* **23**, 411–419.

Ferguson, H. L. and Schaefer, D. G. (1971), 'Feasibility studies for the IFYGL atmospheric water balance project', Proc. 14-th Conf. Great Lakes Res., Internat. Assoc. Great Lakes Res., pp. 439–453.

Fermi, E. (1956), *Thermodynamics*, Dover Pub. Inc., N.Y. 160 pp.

Fichtl, G. H. and McVehil, G. E. (1970), 'Longitudinal and lateral spectra of turbulence in the atmospheric boundary layer at Kennedy Space Center', *J. Appl. Meteorol.* **9**, 51–63.

Fitzpatrick, E. A. and Stern, W. R. (1965), 'Components of the radiation balance of irrigated plots in a dry monsoonal environment', *J. Appl. Meteorol.* **4**, 649–660.

Foreman, J. (1976), 'IFYGL physical data collection system: description of archived data', NOAA Tech. Report EDS 15, Nat. Oceanic Atmosph. Admin., U.S. Dept. Commerce, Washington, D.C., 175 pp.

Forsgate, J. A., Hosegood, P. H., and McCulloch, J. S. G. (1965), 'Design and installation of semi-enclosed hydraulic lysimeters', *Agric. Meteorol.* **2**, 43–52.

Fortin, J. P. and Seguin, B. (1975), 'Estimation de l'ETR régionale à partir de l'ETP locale: utilisation de la relation de Bouchet à différentes échelles de temps', *Ann. Agron.* **26**, 537–554.

Frenzen, P. (1977), 'A generalization of the Kolmogorov–von Kármán relationship and some further implications on the values of the constants', *Boundary-Layer Meteorol.* **11**, 375–380.

Friehe, C. A. and Schmitt, K. F. (1976), 'Parameterization of air-sea interface fluxes of sensible heat and moisture by the bulk aerodynamic formulas', *J. Phys. Oceanogr.* **6**, 801–809.

Friend, W. L. and Metzner, A. B. (1958), 'Turbulent heat transfer inside tubes and the analogy among heat, mass, and momentum transfer', *AICh E J.* **4**, 393–402.

Fritschen, L. J. (1966), 'Evapotranspiration rates of field crops determined by the Bowen ratio method', *Agron. J.* **58**, 339–342.

Fritton, D. D., Busscher, W. J., and Alpert, J. E. (1974), 'An inexpensive but durable thermal conductivity probe for field use', *Soil Sci. Soc. Am. Proc.* **38**, 854–855.

Frost, R. (1946), 'Turbulence and diffusion in the lower atmosphere', *Proc. Roy. Soc. London* **A186**, 20–35.

Fuchs, M. and Hadas A. (1972), 'The heat flux density in a non-homogeneous bare loessial soil', *Boundary-Layer Meteorol.* **3**, 191–200.

Fuchs, M. and Tanner, C. B. (1968), Calibration and field test of soil heat flux plates', *Soil Sci. Soc. Am. Proc.* **32**, 326–328.

Fuchs, M. and Tanner, C. B. (1970), 'Error analysis of Bowen ratios measured by differential psychrometry', *Agric. Meteorol.* **7**, 329–334.

Fuchs, M., Tanner, C. B., Thurtell, G. W., and Black, T. A. (1969), 'Evaporation from drying surfaces by the combination method', *Agron. J.* **61**, 22–26.

Gangopadhyaya, M., Harbeck, G. E. Jr., Nordenson, T. J., Omar, M. H., and Uryvaev, V. A. (1966), 'Measurement and estimation of evaporation and evapotranspiration', World Met. Organ., Tech. Note No. 83, WMO–No. 201. TP. 105, 121 pp.

Gardner, W., Israelsen, O. W., Edlefsen, N. E., and Clyde, H. (1922), 'The capillary potential function and its relation to irrigation practice', *Phys. Rev. Ser. 2*, **20**, 196.

Gardner, W. R. (1958), 'Some steady-state solutions of the unsaturated moisture flow equation with application to evaporation from a water table', *Soil Sci.* **85**, 228–232.

Gardner, W. R. (1959), 'Soiutions of the flow equation for the drying of soils and other porous media', *Soil Sci. Soc. Am. Proc.* **23**, 183–187.

Gardner, W. R. and Fireman, M. (1958), 'Laboratory studies of evaporation from soil columns in the presence of a water table', *Soil Sci.* **85**, 244–249.

Gardner, W. R. and Hillel, D. I. (1962), 'The relation of external evaporative conditions to the drying of soils', *J. Geophys. Res.* **67**, 4319–4325.

Garratt, J. R. (1977), 'Review of drag coefficients over oceans and continents', *Monthly Weath. Rev.* **105**, 915–929.

Garratt, J. R. (1978a), 'Flux profile relations above tall vegetation', *Quart. J. Roy. Meteorol. Soc.* **104**, 199–211.

Garratt, J. R. (1978b), 'Transfer characteristics for a heterogeneous surface of large aerodynamic roughness', *Quart. J. Roy. Meteorol. Soc.* **104**, 491–502.

Garratt, J. R. and Francey, R. J. (1978), 'Bulk characteristics of heat transfer in the unstable, baroclinic atmospheric boundary layer', *Boundary-Layer Meteorol.* **15**, 399–421.

Garratt, J. R. and Hicks, B. B. (1973), 'Momentum, heat and water vapour transfer to and from natural and artificial surfaces', *Quart. J. Roy. Meteorol. Soc.* **99**, 680–687.

Geiger, R. (1961), *Das Klima der bodennahen Luftschicht*, 4. Aufl., Friedr. Vieweg & Sohn, Braunschweig, 646 pp.

Gilbert, M. J. and Van Bavel, C. H. M. (1954), 'A simple field installation for measuring maximum evapotranspiration', *Trans. Am. Geophys. Un.* **35**, 937–942.

Glover, J. and Mc Culloch, J. S. G. (1958), 'The empirical relation between solar radiation and hours of sunshine', *Quart. J. Roy. Meteorol. Soc.* **84**, 172–175.

Goff, J. A. and Gratch. S. (1946), 'Low-pressure properties of water from–160 to 212°F', *Trans. Am. Heat. Vent. Eng.* **52**, 95–121.

Goddard, W. B. (1970), 'A floating drag-plate lysimeter for atmospheric boundary layer research', *J. App. Meteorol.* **9**, 373–378.

Goltz, S. M., Tanner, C. B., Thurtell, G. W., and Jones, F. E. (1970), 'Evaporation measurements by an eddy correlation method', *Water Resour. Res.* **6**, 440–446.

Goody, R. M. (1964), *Atmospheric Radiation*, Clarendon Press, Oxford, 436 pp.

Goss, J. R. and Brooks, F. A. (1956), 'Constants for empirical expressions for down-coming atmospheric radiation under cloudless sky', *J. Meterorol.* **13**, 482–488.

Grass, A. J. (1971), 'Structural features of turbulent flow over smooth and rough boundaries', *J. Fluid Mech.* **50**, 233–255.

Günneberg, F. (1976), 'Abkühlungsvorgänge in Gewässern', *Deutsche Gewässerk. Mitteil.* **20**, 151–161.

Hadas, A. (1974), 'Problems involved in measuring the soil thermal conductivity and diffusivity in a moist soil', *Agric. Meteorol.* **13**, 105–113.

Halkias, N. A., Veihmeyer, F. J., and Hendrickson, A. H. (1955), 'Determining water needs for crops from climatic data', *Hilgardia*, **24**, 207–233.

Haltiner, G. J. and Martin, F. L. (1957), *Dynamical and Physical Meteorology*, McGraw-Hill Book Co., N.Y., 470 pp.

Hanks, R. J. and Shawcroft, R. W. (1965), 'An economical lysimeter for evapotranspiration studies', *Agron. J.* **57**, 634–636.

Harbeck, G. E., Jr. (1962), 'A practical field technique for measuring reservoir evaporation utilizing mass-transfer theory', *U.S. Geol. Surv. Prof. Paper*, 272-E. pp. 101–105.

Harbeck, G. E., Jr. and Meyers, J. S. (1970), 'Present-day evaporation measurement techniques', *J. Hydraul. Div., Proc. ASCE* **96** (HY7), 1381–1390.

Harrold, L. L. and Dreibelbis, F. R. (1958), 'Evaluation of agricultural hydrology by monolith lysimeters, 1944–55', U.S. Dept. Agr., Tech. Bull. 1179, 166 pp.

Hastenrath, S. and Lamb, P. J. (1978), *Heat Budget Atlas of the Tropical Atlantic and Eastern Pacific Oceans*, Univ. Wisconsin Press, Madison, 104 pp.

Heskestad, G. (1965), 'A generalized Taylor hypothesis with application for high Reynolds number turbulent shear flows', *J. Appl. Mech.* **87**, 735–740.

Hess, G. D. (1973), 'On Rossby-number similarity theory for a baroclinic planetary boundary layer', *J. Atmos. Sci.* **30**, 1722–1723.

Hicks, B. B. (1970), 'The measurement of atmospheric fluxes near the surface: a generalized approach', *J. Appl. Meteorol.* **9**, 386–388.

Hicks, B. B (1972a), 'Some evaluations of drag and bulk transfer coefficients over water bodies of different sizes', *Boundary-Layer Meteorol.* **3**, 201–213.

Hicks, B. B. (1972b), 'Propeller anemometers as sensors of atmospheric turbulence', *Boundary-Layer Meteorol.* **3**, 214–228.

Hicks, B. B. (1976a), 'Reply', *Boundary-Layer Meteorol.* **10**, 237–240.

Hicks, B. B. (1976b), 'Wind profile relationships from the "Wangara" experiment', *Quart. J. Roy. Meteorol. Soc.* **102**, 535–551.

Hicks, B. B., Drinkrow, R. L., and Grauze, G. (1974), 'Drag and bulk transfer coefficients associated with a shallow water surface', *Boundary-Layer Meteorol.* **6**, 287–297.

Hicks, B. B. and Dyer, A. J. (1970), 'Measurements of eddy-fluxes over the sea from an off-shore oil rig', *Quart. J. Roy. Meteorol. Soc.* **96**, 523–528.

Hicks, B. B. and Dyer, A. J. (1972), 'The spectral density technique for the determination of eddy fluxes', *Quart. J. Roy. Meteorol. Soc.* **98**, 838–844.

Hicks, B. B. and Everett, R. G. (1979), 'Comments on "Turbulent exchange coefficients for sensible heat and water vapor under advective conditions"', *J. App. Meteorol.* **18**, 381–382.

Hicks, B. B. and Goodman, H. S. (1971), 'The eddy-correlation technique of evaporation measurement using a sensitized quartz-crystal hygrometer', *J. Appl. Meteorol.* **10**, 221–223.

Hicks, B. B. and Liss, P. S. (1976), 'Transfer of SO_2 and other reactive gases across the air-sea interface', *Tellus* **28**, 348–354.

Hicks, B. B. and Sheih, C. M. (1977), 'Some observations of eddy momentum fluxes within a maize canopy', *Boundary-Layer Meteorol.* **11**, 515–519.

Hicks, B. B., Wesely, M. L., and Sheih, C. M. (1977), 'A study of heat transfer processes above a cooling pond', *Water Resour. Res.* **13**, 901–908.

Hinze, J. O. (1959), *Turbulence*, McGraw-Hill Book Co., N.Y., 586 pp.

Högström, U. (1974), 'A field study of the turbulent fluxes of heat, water vapour and momentum at a "typical" agricultural site', *Quart. J. Roy. Meteorol. Soc.* **100**, 624–639.

Holland, J. Z. and Rasmusson, E. M. (1973), 'Measurements of the atmospheric mass, energy, and momentum budgets over a 500-kilometer square of tropical ocean', *Monthly Weath. Rev.* **101**, 44–55.

Holloway, J. L., Jr. and Manabe, S. (1971), 'Simulation of climate by a global general circulation model', *Monthly Weath. Rev.* **99**, 335–370.

Horton, R. E. (1921), 'Results of evaporation observations', *Monthly Weath. Rev.* **49**, 553–566.

Hoy, R. D. and Stephens, S. K. (1979), 'Field Study of lake evaporation-Analysis of data from phase 2 storages and summary of phase 1 and phase 2', Austral. Water Resour. Council, Dept. of Nation. Development, Tech. Paper No. 41, 177 pp.

Huang, C.-H. and Nickerson, E. C. (1974a), 'Stratified flow over non-uniform surface conditions: mixing-length model', *Boundary-Layer Meteorol.* **5**, 395–417.

Huang, C.-H. and Nickerson, E. C. (1974b), 'Stratified flow over non-uniform surfaces: turbulent energy model', *Boundary-Layer Meteorol.* **7**, 107–123.

Hutchings, J. W. (1957), 'Water-vapour flux and flux-divergence over southern England: summer 1954', *Quart. J. Roy. Meteorol. Soc.* **83**, 30–48.

Hyson, P. and Hicks, B. B. (1975), 'A single-beam infrared hygrometer for evaporation measurement', *J. App. Meteorol.* **14**, 301–307.

Idso, S. B. (1972), 'Calibration of soil heat flux plates by a radiation technique', *Agric. Met.* **10**, 467–471.

Idso, S. B., Aase, J. K., and Jackson, R. D. (1975), 'Net radiation – soil heat flux relations as influenced by soil water content variations', *Boundary-Layer Meteorol.* **9**, 113–122.

Idso, S. B. and Jackson, R. D. (1969), 'Thermal radiation from the atmosphere', *J. Geophys. Res.* **74**, 5397–5403.

Impens, I. (1963), 'Thermodynamische en physiologisch-ecologische aspecten van de potentiele evapotranspiratie uit grasland', *Mededel. Landbouwhogeschool en Opzoekingsstations v.d. Staat, Gent* **28**, 429–485.

Inoue, E. (1963), 'On the turbulent structure of airflow within crop canopies', *J. Meteorol. Soc. Japan* **41**, 317–326.

Inoue, K. and Uchijima, Z. (1979), 'Experimental study of microstructure of wind turbulence in rice and maize canopies', *Bull. Nat. Inst. Agric. Sci. Tokyo, Japan, Ser. A* **26**, 1–88.

Isrealsen, O. W. (1918), 'Studies on capacities of soils for irrigation water, and on a new method of determining volume weight', *J. Agric. Res.* **13**, 1–37.

Israelsen, O. W. (1927), 'The application of hydrodynamics to irrigation and drainage problems', *Hilgardia* **2**, 479–528.

Itier, B. and Perrier, A. (1976), 'Présentation d'une étude analytique de l'advection', *Ann. Agron.* **27**, 111–140.

Jackson, R. D., Idso, S. B., and Reginato, R. J. (1976), 'Calculation of evaporation rates during the transition from energy-limiting to soil-limiting phases using albedo data', *Water Resour. Res.* **12**, 23–26.

Jaeger, J. C. (1945), 'Diffusion in turbulent flow between parallel plates', *Quart. Appl. Math.* **3**, 210–217.

Janse, A. R. P. and Borel, G. (1965), 'Measurement of thermal conductivity *in situ* in mixed materials, e.g., soils', *Netherl. J. Agric. Sci.* **13**, 57–62.

Jensen, M. E. (1967), 'Evaluating irrigation efficiency', *J. Irrig. Drain. Div., Proc. ASCE* **93** (IR1), 83–98.

Jensen, M. E. and Haise, H. R. (1963), 'Estimating evapotranspiration from solar radiation', *J. Irrig. Drain. Div., Proc. ASCE* **89**, (IR4), 15–41.

Jensen, N. O. (1978), 'Change of surface roughness and the planetary boundary layer', *Quart. J. Roy. Meteorol. Soc.* **104**, 351–356.

Jerlov, N. G. (1968), *Optical Oceanography*, Elsevier, Amsterdam, 194 pp.

Jirka, G. H. (1978), 'Discussion of "Bed conduction computation for thermal models"', *J. Hydraul. Div., Proc. ASCE* **104**, 1204–1206.

Jirka, G. H., Watanabe, M., Hurley-Octavio, H., Cerco, C. F., and Harleman, D. R. F. (1978), 'Mathematical predictive models for cooling ponds and lakes; Part A: Model development and design considerations', Ralph M. Parsons Labor. Rept. No. 238, Dept. of Civil Eng., MIT, Cambridge, Mass.

Jobson, H. E. (1972), 'Effect of using averaged data on the computation of evaporation', *Water Resour. Res.* **8**, 513–518.

Jobson, H. E. (1973), 'The dissipation of excess heat from water systems', *J. Power Div., Proc. ASCE* **99** (PO1), 89–103.

Jobson, H. E. (1977), 'Bed conduction computation for thermal models', *J. Hydraul. Div., Proc. ASCE* **103**, (HY10), 1213–1217.

Kadel, B. C. and Abbe, C., Jr. (1916), 'Current evaporation observations by the Weather Bureau',

Monthly Weath. Rev. **44**, 674–677.

Kader, B. A. and Yaglom, A. M. (1972), 'Heat and mass transfer laws for fully turbulent wall flows', *Int. J. Heat. Mass Transfer* **15**, 2329–2351.

Kaimal, J. C. and Haugen, D. A. (1971), 'Comments on "Minimizing the levelling error in Reynolds stress measurement by filtering" by Dyer *et al.* (1970)', *J. App. Meteorol.* **10**, 337–339.

Kaimai, J. C., Wyngaard, J. C., and Haugen, D. C. (1968), 'Deriving power spectra from a three-component sonic anemometer', *J. Appl. Meteorol.* **7**, 827–837.

Kasahara, A. and Washington, W. M. (1971), 'General circulation experiments with a six-layer NCAR model, including orography, cloudiness and surface temperature calculation', *J. Atmos. Sci.* **28**, 657–701.

Kazanski, A. B. and Monin, A. S. (1960), 'A turbulent regime above the ground atmospheric layer', *Bull. (Izv.) Acad. Sci., U.S.S.R., Geophys. Ser.* **1** (Engl. Edn.), 110–112.

Kazanski, A. B. and Monin, A. S. (1961), 'On the dynamic interaction between the atmosphere and the earth's surface', *Bull. (Izv.) Acad. Sci., U.S.S.R., Geophys. Ser.* **5** (Engl. Edn.), 514–515.

Keijman, J. Q. (1974), 'The estimation of the energy balance of a lake from simple weather data', *Boundary-Layer Meteorol.* **7**, 399–407.

Kersten, M. S. (1949), 'Thermal properties of soils', *Bull. Univ. Minnesota Inst. Tech., Eng. Exper. Stat., Bull.* **28**, 227 pp.

Kim, H. T., Kline, S. J., and Reynolds, W. C. (1971), 'The production of turbulence near a smooth wall in a turbulent boundary layer', *J. Fluid Mech.* **50**, 133–160.

Kimball, B. A. and Jackson, R. D. (1975), 'Soil heat flux determination: a null-alignment method', *Agric. Meteorol.* **15**, 1–9.

Kimball, B. A., Jackson, R. D., Reginato, R. J., Nakayama, F. S., and Idso, S. B. (1976a), 'Comparison of field-measured and calculated soil-heat fluxes', *Soil Sci. Soc. Am. J.* **40**, 18–25.

Kimball, B. A., Jackson, R. D., Nakayama, F. S., Idso, S. B., and Reginato, R. J. (1976b), 'Soil-heat flux determination: temperature gradient method with computed thermal conductivities', *Soil Sci. Soc. Am. J.* **40**, 25–28.

Kimball, H. H. (1928), 'Amount of solar radiation that reaches the surface of the earth on the land and on the sea, and methods by which it is measured', *Monthly Weath. Rev.* **56**, 393–399.

King, K. M., Mukammal, E. I., and Turner, V. (1965), 'Errors involved in using zinc chloride solution in floating lysimeters', *Water Resour. Res.* **1**, 207–217.

King, K. M., Tanner, C. B., and Suomi, V. E. (1956), 'A floating lysimeter and its evaporation recorder', *Trans. Am. Geophys. Un.* **37**, 738–742.

Kitaygorodskiy, S. A. (1969), 'Small-scale atmospheric-ocean interactions', *Izv. Acad. Sci., U.S.S.R., Atmos. Ocean. Phys.* **5**, 641–649.

Kitaigorodskiy, S. A., Kuznetsov, O. A., and Panin, G. N. (1973), 'Coefficients of drag, sensible heat and evaporation in the atmosphere over the surface of a sea', *Izv. Acad. Sci., U.S.S.R., Atmos. Oceanic Phys.* **9**, (11) 1135–1141 (English Edn., 644–647).

Kitaigorodskii, S. A. and Volkov, Y. A. (1965), 'Calculation of turbulent heat and humidity fluxes in an atmospheric layer near a water surface', *Izv. Acad. Sci., U.S.S.R., Atmos. Oceanic Phys.* (Engl. Transl. AGU) **1**, 1317–1336.

Klute, A. (1952), 'A numerical method for solving the flow equation for water in unsaturated materials', *Soil Sci.* **73**, 105–116.

Klute, A. (1972), 'The determination of the hydraulic conductivity and diffusivity of unsaturated soils', *Soil Sci.* **113**, 264–276.

Kohler, M. A. (1954), 'Lake and pan evaporation', *Water-loss investigations: Lake Hefner Studies, Tech. Report, Prof. Paper* 269, Geol. Survey, U.S. Dept. Interior, pp. 127–148.

Kondo, J. (1962), 'Observations on wind and temperature profiles near the ground', *Sci. Rep. Tohoku Univ. (Sendai, Japan), Ser. 5, Geophys.* **14**, 41–56.

Kondo, J. (1967), 'Analysis of solar radiation and downward long-wave radiation data in Japan', *Sci. Rep. Tohoku Univ. (Sendai, Japan), Ser. 5, Geophys.* **18**, 91–124.

Kondo, J. (1971), 'Relationship between the roughness coefficient and other aerodynamic parameters', *J. Meteorol. Soc. Japan* **49**, 121–124.

Kondo, J. (1972), 'On a product of mixing length and coefficient of momentum absorption within plant canopies', *J. Meteorol. Soc. Japan* **50**, 487–488.

Kondo, J. (1972b), 'Applicability of micrometeorological transfer coefficient to estimate the long-period means of fluxes in the air-sea interface', *J. Meteorol. Soc. Japan* **50**, 570–576.

Kondo, J. (1975), 'Air-sea bulk transfer coefficients in diabatic conditions', *Boundary-Layer Meteorol.* **9**, 91–112.

Kondo, J. (1976), 'Heat balance of the East China Sea during the Air Mass Transformation Experiment', *J. Meteorol. Soc. Japan* **54**, 382–398.

Kondo, J. (1977), 'Geostrophic drag and the cross-isobar angle of the surface wind in a baroclinic convective boundary layer over the ocean', *J. Meteorol. Soc. Japan* **55**, 301–311.

Kondo, J. and Akashi, S. (1976), 'Numerical studies of the two-dimensional flow in horizontally homogeneous canopy layers', *Boundary-Layer Meteorol.* **10**, 255–272.

Kondo, J. and Fujinawa, Y. (1972), 'Errors in estimation of drag coefficients for sea surface in light winds', *J. Meteorol. Soc. Japan* **50**, 145–149.

Kondo, J., Kanechika, O., and Yasuda N. (1978), 'Heat and momentum transfers under strong stability in the atmospheric surface layer', *J. Atmos. Sci.* **35**, 1012–1021.

Kondo, J., Sasano, Y., and Ishii, T. (1979), 'On wind-driven current and temperature profiles with diurnal period in the ocean planetary boundary layer', *J. Phys. Oceanog.* **9**, 360–372.

Kondratyev, K. Ya (1969), *Radiation in the Atmosphere*, Academic Press, N.Y., 912 pp.

Kornev, B. G. (1924), 'The absorbing power of soils and the principle of automatic self-irrigation of soils', *Soil Sci.* **17**, 428–429 (abstract).

Korzoun, V. I. *et al.* (eds.) (1977), *Atlas of World Water Balance*, U.S.S.R. National Committee for the International Hydrological Decade, U.N.E.S.C.O. Press, Paris.

Korzun, V. I. *et al.* (eds.) (1978), *World water balance and water resources of the earth*, U.S.S.R. National Committee for the International Hydrological Decade, U.N.E.S.C.O. Press, Paris, 663 pp.

Krischer, O. and Rohnalter, H. (1940), 'Wärmeleitung und Dampfdiffusion in feuchten Gütern', *VDI Forschungsheft* **402**.

Krügermeyer, L., Grünewald, M., and Dunckel, M. (1978), 'The influence of sea waves on the wind profile', *Boundary-Layer Meteorol.* **14**, 403–414.

Laikhtman, D. L. (1964), *Physics of the Boundary Layer of the Atmosphere*, Israel Program for Scientif. Transl. Ltd., Jerusalem, 200 pp.

Landau, L. D. and Lifshitz, E. M. (1959), *Fluid Mechanics*, Pergamon Press, London, 536 pp.

Landsberg, J. J. and James, G. B. (1971), 'Wind profiles in plant canopies: studies on an analytical model', *J. Appl. Ecol.* **8**, 729–741.

Lang, A. R. G., Evans, G. N., and Ho, P. Y. (1974), 'The influence of local advection on evapotranspiration from irrigated rice in a semi-arid region', *Agric. Meteorol.* **13**, 5–13.

Launder, B. E. (1975), 'On the effects of a gravitational field on the turbulent transport of heat and momentum', *J. Fluid Mech.* **67**, 569–581.

Leavitt, E. (1975), 'Spectral characteristics of surface-layer turbulence over the tropical ocean', *J. Phys. Oceanog.* **5**, 157–163.

Leavitt, E. and Paulson, C. A. (1975), 'Statistics of surface-layer turbulence over the tropical ocean', *J. Phys. Oceanog.* **5**, 143–156.

Lemon, E. (1965), 'Micrometeorology and the physiology of plants in their natural environment', in F. C. Steward (ed.) *Plant Physiology* Vol. 4A, Academic Press, N.Y., pp. 203–227.

Lemon, E., Allen, L. H., Jr., and Mueller, L. (1970), 'Carbon dioxide exchange of a tropical rain forest, Part II', *Bioscience* **20**, 1054–1059.

Lettau, H. (1959), 'Wind profile, surface stress and geostrophic drag coefficients in the atmospheric surface layer', *Adv. Geophys.* **6**, 241–257.

Lettau, H. (1969), 'Note on aerodynamic roughness-parameter estimation on the basis of roughness element description', *J. Appl. Meteorol.* **8**, 828–832.

Lettau, H. and Davidson, B. (1957), *Exploring the Atmosphere's First Mile*, Vols. 1–2, Pergamon Press, N.Y.

Lettau, H. and Zabransky, J. (1968), 'Interrelated changes of wind profile structure and Richardson number in air flow from land to inland lakes', *J. Atmos. Sci.* **25**, 718–728.

Linacre, E. T. (1967), 'Climate and the evaporation from crops', *J. Irrig. Drain. Div., Proc. ASCE* **93** (IR4), 61–79.

List, R. J. (1971), *Smithsonian Meteorological Tables*, Smithsonian Institution Press, City of Washington, 6th Edn., 5th Reprint, 527 pp.

Liu, C. K., Kline, S. J., and Johnston, J. P. (1966), 'An experimental study of turbulent boundary layer on rough walls', Report MD-15, Dept. of Mech. Eng., Stanford Univ., Calif.

Livingston, B. E. (1935), 'Atmometers of porous porcelain and paper, their use in physiological ecology', *Ecology* **16**, 438–472.

Löf, G. O. G., Duffie, J. A., and Smith, C. O. (1966), 'World distribution of solar radiation', *Sol. Energy* **10**, 27–37.

Logan, E., Jr., and Fichtl, G. H. (1975), 'Rough-to-smooth transition of an equilibrium neutral constant stress layer', *Boundary-Layer Meteorol.* **8**, 525–528.

Lourence, F. J. and Goddard, W. B. (1967), 'A water-level measuring system for determining evapotranspiration rates from a floating lysimeter', *J. Appl. Met.* **6**, 489–492.

Lourence, F. J. and Pruitt, W. O. (1971), 'Energy balance and water use of rice grown in the Central Valley of California', *Agron. J.* **63**, 827–832.

Lowe, P. R. (1977), 'An approximating polynomial for the computation of saturation vapor pressure', *J. Appl. Meteorol.* **16**, 100–103.

Lumley, J. L. (1965), 'Interpretation of time spectra measured in high-intensity shear flows', *Phys. Fluids* **8**, 1056–1062.

Lumley, J. L. (1978), 'Computational modeling of turbulent flows', *Adv. Appl. Mech.* **18**, 124–176.

Lvovitch, M. I. (1970), 'World water balance', Symposium on world water balance, Proc. Reading Sympos., Vol. II, Inter. Assoc. Sci. Hydrol., Public. No. 93, pp. 401–415.

Lvovitch, M. I. (1973), 'The global water balance', *Trans. Am. Geophys. Un.* **54**, 28–42 (U.S.—IHD Bulletin, No. 23).

Magyar, P., Shahane, A. N., Thomas, D. L., and Bock, P. (1978), 'Simulation of the hydrologic cycle using atmospheric water vapor transport data', *J. Hydrol.* **37**, 111–128.

Mahringer, W. (1970), 'Verdunstungsstudien am Neusiedler See', *Arch. Meteorol. Geophys. Biokl.* **B18**, 1–20.

Makkink, G. F. (1957), 'Ekzameno de la formulo de Penman', *Netherl. J. Agric. Sci.* **5**, 290–305.

Manabe, S. (1969), 'Climate and the ocean circulation', *Monthly Weath. Rev.* **97**, 739–774.

Mangarella, P. A., Chambers, A. J., Street, R. L., and Hsu, E. Y. (1971), 'Energy and mass transfer through an air-water interface', Tech. Rept. No. 134, Dept. of Civil Eng., Stanford Univ., Calif.

Marciano, J. J. and Harbeck, G. E. Jr., (1954), 'Mass-transfer studies', *Water-loss investigations: Lake Hefner Studies, Tech. Report, Prof. Paper* 269, Geol. Survey, U.S. Dept. Interior, 46–70.

Martin, H. C. (1971) 'The humidity microstructure: a comparison between refractometer and a Ly-α humidiometer', *Boundary-Layer Meteorol.* **2**, 169–172.

Mawdsley, J. A. and Brutsaert, W. (1977), 'Determination of regional evapotranspiration from upper air meteorological data', *Water Resour. Res.* **13**, 539–548.

McBean, G. A., Stewart, R. W., and Miyake, M. (1971), 'The turbulent energy budget near the surface', *J. Geophys. Res.* **76**, 6540–6549.

McGavin, R. E., Uhlenhopp, P. B., and Bean, B. R. (1971), 'Microwave evapotron', *Water Resour. Res.* **7**, 424–428.

McIlroy, I. C. and Angus, D. E. (1964), 'Grass, water and soil evaporation at Aspendale', *Agric. Meteorol.* **1**, 201–224.

McKay, D. C. and Thurtell, G. W. (1978), 'Measurements of the energy fluxes involved in the energy budget of a snow cover', *J. Appl. Meteorol.* **17**, 339–349.

McMillan, W. B. and Paul, H. A. (1961), 'Floating lysimeter uses heavy liquid for buoyancy', *Agric. Eng.* **42**, 498–499.

Mc Naughton, K. G. and Black, T. A. (1973), 'A study of evapotranspiration from a Douglas fir forest using the energy balance approach', *Water Resour. Res.* **9**, 1579–1590.

McVehil, G. E. (1964), 'Wind and temperature profiles near the ground in stable stratification', *Quart. J. Roy. Meteor. Soc.* **90**, 136–146.

Melgarejo, J. W. and Deardorff, J. W. (1974), 'Stability functions for the boundary-layer resistance laws based upon observed boundary-layer heights', *J. Atmos. Sci.* **31**, 1324–1333.

Melgarejo, J. W. and Deardorff, J. W. (1975), 'Revision to "Stability functions for the boundary-layer resistance laws, based upon observed boundary-layer heights"', *J. Atmos. Sci.* **32**, 837–839.

Mellor, G. L. and Yamada, T. (1974), 'A hierarchy of turbulence closure models for planetary boundary layers', *J. Atmos. Sci.* **31**, 1791–1806.

Merlivat, L. (1978), 'The dependence of bulk evaporation coefficients on air-water interfacial conditions as determined by the isotopic method', *J. Geophys. Res.* (*Oceans and Atmos.*) **83** (C6), 2977–2980.

Merlivat, L. and Coantic, M. (1975), 'Study of mass transfer at the air-water interface by an isotopic method', *J. Geophys. Res.* **80**, 3455–3464.

Mermier, M. and Seguin, B. (1976), 'Comment on "On a derivable formula for long-wave radiation from clear skies" by W. Brutsaert', *Water Resour. Res.* **12**, 1327–1328.

Meroney, R. N. (1970), 'Wind tunnel studies of the air flow and gaseous plume diffusion in the leading edge and downstream regions of a model forest', *Atmos. Environ.* **4**, 597–614.

Millar, B. D. (1964), 'Effect of local advection on evaporation rate and plant water status', *Australian. J. Agric. Res.* **15**, 85–90.

Millikan, C. B. (1938), 'A critical discussion of turbulent flows in channels and circular tubes', *Proc. 5th Internat. Congr. Appl. Mech., Cambridge, MA*, John Wiley & Sons, Inc., N.Y., pp. 386–392.

Mitsuta, Y. and Fujitani, T. (1974), 'Direct measurement of turbulent fluxes on a cruising ship', *Boundary-Layer Meteorol.* **6**, 203–217.

Miyake, M. and McBean, G. (1970), 'On the measurement of vertical humidity transport over land', *Boundary-Layer Meteorol.* **1**, 88–101.

Moench, A. F. and Evans, D. D. (1970), 'Thermal conductivity and diffusivity of soil using a cylindrical heat source', *Soil Sci. Soc. Am. Proc.* **34**, 377–381.

Monin, A. S. (1959), 'Smoke propagation in the surface layer of the atmosphere', *Adv. Geophys.* **6**, 331–343.

Monin, A. S. (1970), 'The atmospheric boundary layer', *Ann. Rev. Fluid Mech.* **2**, 225–250.

Monin, A. S. and Obukhov, A. M. (1954), 'Basic laws of turbulent mixing in the ground layer of the atmosphere', *Tr. Geofiz. Instit. Akad. Nauk, S.S.S.R.*, No. 24 (151), pp. 163–187 (German translation (1958), 'Sammelband zur Statistischen Theorie der Turbulenz', H. Goering, (ed.), Akademie Verlag, Berlin, 228 pp.

Monin, A. S. and Yaglom, A. M. (1971), *Statistical Fluid Mechanics: Mechanics of Turbulence*, Vol. 1, The MIT Press, Cambridge, Mass., 769 pp.

Monteith, J. L. (1965), 'Evaporation and environment', in G. E. Fogg, (ed.) *The State and Movement of Water in Living Organisms*, Sympos. Soc. Exper. Biol., Vol. 19, Academic Press, N.Y., pp. 205–234.

Monteith, J. L. (1973), *Principles of Environmental Physics*, American Elsevier Publ. Co., N.Y., 241 pp.

Montgomery, R. B. (1940), 'Observations of vertical humidity distribution above the ocean surface and their relation to evaporation', *MIT Woods Hole Oceanogr. Instn., Pap. Phys. Oceanog. Meteorol.* **7** (4), 30 pp.

Montgomery, R. B. and Spilhaus, A. F. (1941), 'Examples and outline of certain modifications in upper-air analysis', *J. Aeronaut. Sci.* **8**, 276–283.

Morton, F. I. (1969), 'Potential evaporation as a manifestation of regional evaporation', *Water Resour. Res.* **5**, 1244–1255.

Morton, F. I. (1975), 'Estimating evaporation and transpiration from climatological observations', *J. Appl. Meteorol.* **14**, 488–497.

Morton, F. I. (1976), 'Climatological estimates of evapotranspiration', *J. Hydraul. Div., Proc. ASCE,* **102** (HY3), 275–291.

Mukammal, E. I. and Neumann, H. H. (1977), 'Application of the Priestley–Taylor evaporation model to assess the influence of soil moisture on the evaporation from a large weighing lysimeter and Class A pan', *Boundary Layer Meteorol.* **12**, 243–256.

Mulhearn, P. J. (1977), 'Relations between surface fluxes and mean profiles of velocity, temperature and concentration, downwind of a change in surface roughness', *Quart. J. Roy. Meteorol. Soc.* **103**, 785–802.

Mulhearn, P. J. (1978), 'A wind-tunnel boundary-layer study of the effects of a surface roughness change: rough to smooth', *Boundary-Layer Meteorol.* **15**, 3–30.

Müller-Glewe, J. and Hinzpeter, H. (1974), 'Measurements of the turbulent heat flux over the sea', *Boundary-Layer Meteorol.* **6**, 47–52.

Munk, W. H. (1955), 'Wind stress on water: an hypothesis', *Quart. J. Roy. Meteorol. Soc.* **81**, 320–332.

Munro, D. S. and Oke, T. R. (1975), 'Aerodynamic boundary-layer adjustment over a crop in neutral stability', *Boundary-Layer Meteorol.* **9**, 53–61.

Murty, L. K. (1976), 'Heat and moisture budgets over AMTEX area during AMTEX'75', *J. Meteorol. Soc. Japan* **54**, 370–381.

Nagpal, N. K. and Boersma, L. (1973), 'Air entrapment as a possible source of error in the use of a cylindrical heat probe', *Soil Sci. Soc. Am. Proc.* **37**, 828–832.

Neuwirth, F. (1974) 'Ueber die Brauchbarkeit empirischer Verdunstungsformeln dargestellt am Beispiel des Neusiedler Sees nach Beobachtungen in Seemitte und in Ufernähe', *Arch. Meteorol. Geophys. Biokl., Ser. B* **22**, 233–246.

Nickerson, E. C. (1968), 'Boundary-layer adjustment as an initial value problem', *J. Atmos. Sci.* **25**, 207–213.

Nickerson, E. C. and Smiley, V. E. (1975), 'Surface layer and energy budget parameterizations for mesoscale models', *J. Appl. Met.* **14**, 297–300.

Nielsen, D. R. and Biggar, J. W. (1961), 'Measuring capillary conductivity', *Soil Sci.* **92**, 192–193.

Nielsen, D. R., Biggar, J. W., and Erh, K. T. (1973), 'Spatial variability of field-measured soil-water properties', *Hilgardia* **42**, 215–259.

Nielsen, D. R., Davidson, J. M., Biggar, J. W., and Miller, R. J. (1964), 'Water movement through Panoche clay loam soil', *Hilgardia* **35**, 491–506.

Niiler, P. P. and Kraus, E. B. (1977), 'One-dimensional models of the upper ocean', in E. B. Kraus, (ed.) *Modeling and Prediction of the Upper Layers of the Ocean*, Pergamon Press, New York, pp. 143–172.

Nikuradse, J. (1933), 'Strömungsgesetze in rauhen Rohren', *VDI Forschungsheft 361* (*Beilage Forsch. Geb. Ingenieurw.* **B4**), 22 pp.

Ninomiya, K. (1972), 'Heat and water-vapor budget over the East China Sea in the winter season', *J. Meteorol. Soc. Japan* **50**, 1–17.

Nitta, T. (1976), 'Large-scale heat and moisture budgets during the air mass transformation experiment', *J. Meteorol. Soc. Japan* **54**, 1–14.

Nunner, W. (1956), 'Wärmeübergang und Druckabfall in rauhen Rohren', *VDI-Forschungsh. 455* (*Beilage Forsch. Geb. Ingenieurw.* **B22**), 39 pp.

Obukhov, A. M. (1946), 'Turbulence in an atmosphere with non-uniform temperature', *Trudy Instit. Teoret. Geofiz: AN – S.S.S.R.*, No. 1 (English Translation: (1971), *Boundary-Layer Meteorol.* **2**, 7–29).

Ogata, G. and Richards, L. A. (1957), 'Water content changes following irrigation of bare-field soil that is protected from evaporation', *Soil Sci. Soc. Am. Proc.* **21**, 355–356.

Oliver, H. R. (1975), 'Ventilation in a forest', *Agric. Meteorol.* **14**, 347–355.

Onishi, G. and Estoque, M. A. (1968), 'Numerical study on atmospheric boundary-layer flow over inhomogeneous terrain', *J. Meteorol. Soc. Japan* **46**, 280–286.

Ordway, D. E., Ritter, A., Spence, D. A., and Tan, H. S. (1963), 'Effects of turbulence and photosynthesis on CO_2 profiles in the lower atmosphere', in E. R. Lemon (ed.), *The Energy Budget at the Earth's Surface*, ARS, USDA. Production Res. Rept. No. 72, Washington, DC, pp. 3–6.

Owen, P. R. and Thomson, W. R. (1963), 'Heat transfer across rough surfaces', *J. Fluid Mech.* **15**, 321–334.

Paeschke, W. (1937), 'Experimentelle Untersuchungen zum Rauhigkeits- und Stabilitätsproblem in der bodennahen Luftschicht', *Beiträge z. Phys. d. freien Atmos.* **24**, 163–189.

Palmén, E. (1963), 'Computation of the evaporation over the Baltic Sea from the flux of water vapor in the atmosphere', General Assembly Berkeley, Intern. Assoc. Sci. Hydrol., Publ. No. 62, Gentbrugge, Belgium, pp. 244–252.

Paltridge, G. W. and Platt, C. M. R. (1976), *Radiative Processes in Meteorology and Climatology*, Elsevier Sci. Pub. Co., Amsterdam, 318 pp.

Panchev, S., Donev, E., and Godev, N. (1971), 'Wind profile and vertical motions above an abrupt change in surface roughness and temperature', *Boundary-Layer* Meteorol. **2**, 52–63.

Panofsky, H. A. (1963), 'Determination of stress from wind and temperature measurements', *Quart. J. Roy. Meteorol. Soc.* **89**, 85–93.

Panofsky, H. A. (1973), 'Tower meteorology', in D. A. Haugen, (ed.) *Workshop on Micrometeorology*, Amer. Met. Soc., pp. 151–176.

Panofsky, H. A. and Petersen, E. L. (1972), 'Wind profiles and change of terrain roughness at Risø', *Quart. J. Roy. Meteorol. Soc.* **98**, 845–854.

Panofsky, H. A. and Townsend, A. A. (1964), 'Change of terrain roughness and the wind profile', *Quart. J. Roy. Meteorol. Soc.* **90**, 147–155.

Paquin, J. E. and Pond, S. (1971), 'The determination of the Kolmogoroff constants for velocity, temperature and humidity fluctuations from second- and third-order structure functions', *J. Fluid Mech.* **50**, 257–269.

Pasquill, F. (1949a), 'Eddy diffusion of water vapour and heat near the ground', *Proc. Roy. Soc. London* **A198**, 116–140.

Pasquill, F. (1949b), 'Some estimates of the amount and diurnal variation of evaporation from a clayland pasture in fair spring weather', *Quart. J. Roy. Meteorol. Soc.* **75**, 249–256.

Paulson, C. A. (1970), 'The mathematical representation of wind speed and temperature profiles in the unstable atmospheric surface layer', *J. Appl. Meteorol.* **9**, 857–861.

Paulson, C. A., Leavitt, E., and Fleagle, R. G. (1972), 'Air-sea transfer of momentum, heat and water determined from profile measurements during BOMEX', *J. Phys. Oceanog.* **2**, 487–497.

Payne, R. E. (1972), 'Albedo of the sea surface', *J. Atmos. Sci.* **29**, 959–970.

Penman, H. L. (1948), 'Natural evaporation from open water, bare soil, and grass', *Proc. Roy. Soc. London* **A193**, 120–146.

Penman, H. L. (1956), 'Evaporation: an introductory survey', *Netherl. J. Agric. Sci.* **4**, 9–29.

Penman, H. L. and Schofield, R. K. (1951), 'Some physical aspects of assimilation and transpiration', *Sympos. Soc. Exper. Biol.* **5**, 115–129.

Perrier, A. (1975a), 'Etude physique de l'évapotranspiration dans les conditions naturelles', *Ann. Agron.* **26**, 1–18.

Perrier, A. (1975b), 'Assimilation nette, utilisation de l'eau et microclimat d'un champ de maïs', *Ann. Agron.* **26**, 139–157.

Perrier, A., Archer, P., and Blanco de Pablos, A. (1974), 'Etude de l'évapotranspiration réelle et maximale de diverses cultures: dispositif et mesures', *Ann. Agron.* **25**, 697–731.

Perrier, A. Itier, B., Bertolini, J. M., and Katerji, N. (1976), 'A new device for continuous recording of the energy balance of natural surfaces', *Agric. Meteorol.* **16**, 71–84.

Perry, A. E. and Joubert, P. N. (1963), 'Rough-wall boundary layers in adverse pressure gradients', *J. Fluid Mech.* **17**, 193–211.

Petersen, E. L. and Taylor, P. A. (1973), 'Some comparisons between observed wind profiles at Risø and theoretical predictions for flow over inhomogeneous terrain', *Quart. J. Roy. Meteorol. Soc.* **99**, 329–336.

Peterson, E. W. (1969a), 'Modification of mean flow and turbulent energy by a change in surface roughness under conditions of neutral stability', *Quart. J. Roy. Meteorol. Soc.* **95**, 561–575.

Peterson, E. W. (1969b), 'On the relation between the shear stress and the velocity profile after a change in surface roughness', *J. Atmos. Sci.* **26**, 773–774.

Peterson, E. W. (1971), 'Predictions of the momentum exchange coefficient for flow over heterogeneous terrain', *J. Appl. Meteorol.* **10**, 958–961.

Peterson, E. W. (1972), 'Relative importance of terms in the turbulent-energy and momentum equations as applied to the problem of a surface roughness change', *J. Atmos. Sci.* **29**, 1470–1476.

Petukhov, B. S. and Kirillov, V. V. (1958), 'On the question of heat transfer to a turbulent flow of fluids in pipes', *Teploenergetika*, No. 4, pp. 63–68.

Petukhov, B. S., Krasnoschekov, E. A., and Protopopov, V. S. (1961), 'An investigation of heat transfer to fluids flowing in pipes under supercritical conditions', *International Developments in Heat Transfer*, Part III, 1961 International Heat Transfer Conference, Boulder, Color., pp. 569–578.

Philip, J. R. (1957), 'Evaporation, and moisture and heat fields in the soil', *J. Meteorol.* **14**, 354–366.

Philip, J. R. (1959), 'The theory of local advection', *J. Meteorol.* **16**, 535–547.

Philip, J. R. (1961), 'The theory of heat flux meters', *J. Geophys. Res.* **66**, 571–579.

Philip, J. R. and De Vries, D. A. (1957), 'Moisture movement in porous materials under temperature gradients', *Trans. Am. Geophys. Un.* **38**, 222–232.

Phillips, D. W. (1978), 'Evaluation of evaporation from Lake Ontario during IFYGL by a modified mass transfer equation', *Water Resour. Res.* **14**, 197–205.

Phillips, D. W. and Rasmusson, E. M. (1978), 'Lake meteorology panel', Proc. IFYGL Wrap-up Workshop Oct. 1977, IFYGL Bull. (Special) No. 22, 13–26, Nat. Ocean. Atmos. Adm., U.S. Dept. Commerce.

Pickett, R. L. (1975), 'Intercomparison of Canadian and U.S. automatic data buoys', *Marine Tech. Soc. J.* **9**, (No. 10), 20–22.

Pierce, F. J. and Gold, D. S. (1977), 'Near wall velocity measurements for wall shear inference in turbulent flows', *U.S. Dept. Inter., Bur. Stand., Spec. Publ.* **484**, (2), 621–648.

Pike, J. G. (1964), 'The estimation of annual run-off from meteorological data in a tropical climate', *J. Hydrol.* **2**, 116–123.

Pinsak, A. P. and Rogers, G. K. (1974), 'Energy balance of Lake Ontario', Proc. IFYGL Symposium, 55th Ann. Meet. Amer. Geophys. Un., 86–101, NOAA, U.S. Dept. Commerce, Rockville, MD.

Plate, E. J. (1971), 'Aerodynamic characteristics of atmospheric boundary layers', AEC Critical Review Series, U.S. Atomic Energy Commission, Div. Tech. Info., 190 pp.

Plate, E. J. and Hidy, G. M. (1967), 'Laboratory study of air flowing over a smooth surface onto small water waves', *J. Geophys. Res.* **72**, 4627–4641.

Pochop, L. O., Shanklin, M. D., and Horner, D. A. (1968), 'Sky cover influence on total hemispheric radiation during daylight hours', *J. Appl. Meteorol.* **7**, 484–489.

Pond, S., Fissel, D. B., and Paulson, C. A. (1974), 'A note on bulk aerodynamic coefficients for sensible heat and moisture fluxes', *Boundary-Layer Meteorol.* **6**, 333–339.

Pond, S., Phelps, G. T., Paquin, J. E., McBean, G. and Stewart, R. W. (1971), 'Measurements of the turbulent fluxes of momentum, moisture and sensible heat over the ocean', *J. Atmos. Sci.* **28**, 901–917.

Prandtl. L. (1904), 'Ueber Flüssigkeitsbewegung bei sehr kleiner Reibung', Verhandl. III, Internat. Math.-Kong., Heidelberg, Teubner, Leipzig pp. 484–491, (1905) (Also in *Gesammelte Abhundlungen*, Vol. 2, Springer-Verlag, Berlin, 1961, pp. 575–584, English in NACA, Tech. Mem. No. 452).

Prandtl, L. (1932), 'Meteorologische Anwendungen der Strömungslehre', *Beitr. Phys. Fr. Atmosph.* **19**, 188–202.

Prandtl, L. und Tollmien, W. (1924), 'Die Windverteilung über dem Erdboden, errechnet aus den Gesetzen der Rohrströmung', *Z. Geophys.* **1**, 47–55.

Prandtl, L. and Wieghardt, K. (1945), 'Ueber ein neues Formelsystem für die ausgebildete Turbulenz', Nachr. Akad. Wissen., Göttingen, Math. Kl. 6–19; See also *Gesammelte Abhandlungen*, Vol. 2, Springer-Verlag, Berlin, 1961 pp. 874–887.

Prescott, J. A. (1940), 'Evaporation from a water surface in relation to solar radiation', *Trans. Roy. Soc. South. Aust.* **64**, 114–125.

Priestley, C. H. B. (1954), 'Convection from a large horizontal surface', *Australian J. Phys.* **6**, 297–290.

Priestley, C. H. B. (1959), *Turbulent Transfer in the Lower Atmosphere*, Univ. Chicago Press, Chicago, Ill., 130 pp.

Priestley, C. H. B. and Taylor, R. J. (1972), 'On the assessment of surface heat flux and evaporation using large-scale parameters', *Monthly Weath. Rev.* **100**, 81–92.

Pruitt, W. O. (1966), 'Empirical method of estimating evapotranspiration using primarily evaporation pans', in *Evapotranspiration and its Role in Water Resources Management*, Am. Soc. Agric. Eng., St. Joseph, Mich., pp. 57–61.

Pruitt, W. O. and Angus, D. E. (1960), 'Large weighing lysimeter for measuring evapotranspiration', *Trans. Am. Soc. Agric. Eng.* **3**, 13–15, 18.

Pruitt, W. O., Morgan, D. L., and Lourence, F. J. (1968), 'Energy, momentum and mass transfers above vegetative surfaces', Tech. Rept. ECOM-0447 (E)-F, Dept. Water Sci. Eng., Univ. Calif., Davis, 49 pp.

Pruitt, W. O., Morgan, D. L., and Lourence, F. J. (1973), 'Momentum and mass transfers in the surface boundary layer', *Quart. J. Roy. Meteorol. Soc.* **99**, 370–386.

Pruitt, W. O., von Oettingen, S., and Morgan, D. L. (1972), 'Central California evapotranspiration frequencies', *J. Irrig. Drain. Div., Proc. ASCE* **98**, 177–184.

Rao, K. S., Wyngaard, J. C., and Coté, O. R. (1974a), 'The structure of the two-dimensional internal boundary layer over a sudden change of surface roughness', *J. Atmos. Sci.* **31**, 738–746.

Rao, K. S., Wyngaard, J. C., and Coté, O. R. (1974b), 'Local advection of momentum, heat, and moisture in micrometeorology', *Boundary-Layer Meteorol.* **7**, 331–348.

Rasmusson, E. M. (1971), 'A study of the hydrology of eastern North America using atmospheric vapor flux data', *Monthly Weath. Rev.* **99**, 119–135.

Rasmusson, E. M. (1977), 'Hydrological application of atmospheric vapor-flux analyses', Operational Hydrol. Rept. No. 11, WMO-No. 476, World Meteorol. Org., 50 pp.

Rasmusson, E. M., Ferguson, H. L., Sullivan, J., and Den Hartog, G. (1974), 'The atmospheric budgets program of IFYGL', Proc. 17th Conf. Great Lakes Res., Internat. Assoc. Great Lakes Res., pp. 751–777.

Raupach, M. R. (1978), 'Infrared fluctuation hygrometry in the atmospheric surface layer', *Quart. J. Roy. Meteorol. Soc.* **104**, 309–322.

Revfeim, K. J. A. and Jordan, R. B. (1976), 'Precision of evaporation measurements using the Bowen ratio', *Boundary-Layer Meteorol.* **10**, 97–111.

Reynolds, O. (1894), 'On the dynamical theory of incompressible viscous fluids and the determination of the criterion', *Phil. Trans. Roy. Soc. London*, **A186** (1895), Part I, 123–161.

Richards, J. M. (1971), 'Simple expression for the saturation vapour pressure of water in the range − 50° to 140°', *Brit. J. Appl. Phys.* **4**, L15–L18.

Richards, L. A. (1931), 'Capillary conduction of liquids through porous mediums', *Physics* **1**, 318–333.

Richards, L. A. (1949), 'Methods of measuring soil moisture tension', *Soil Sci.* **68**, 95–112.

Richardson, L. F. (1920), 'The supply of energy from and to atmospheric eddies', *Proc. Roy. Soc. London* **A97**, 354–373.

Rider, N. E. (1954), 'Eddy diffusion of momentum, water vapour, and heat near the ground', *Phil. Trans. Roy. Soc. London*, **A246**, 481–501.

Rider, N. E. (1957), 'Water losses from various land surfaces', *Quart. J. Roy. Meteorol. Soc.* **83**, 181–193.

Rider, N. E., Philip, J. R., and Bradley, E. F. (1963), 'The horizontal transport of heat and moisture – A micrometeorological study', *Quart. J. Roy. Meteorol. Soc.* **89**, 507–531.

Ritchie, J. T. and Burnett, E. (1968), 'A precision weighing lysimeter for row crop water use studies', *Agron. J.* **60**, 545–549.

Robinson, N. (1966), *Solar Radiation*, Elsevier Publ. Co., New York, 347 pp.

Rohwer, C. (1934), 'Evaporation from different types of pans', *Trans. Am. Soc. Civ. Eng.* **99**, 673–703.

Rosenberg, N. E. and Brown, K. W. (1970), 'Improvements in the Van Bavel-Myers automatic weighing lysimeter', *Water Resour. Res.* **6**, 1227–1229.

Rossby, C.-G. (1932), 'A generalization of the theory of the mixing length with applications to atmospheric and oceanic turbulence', *MIT, Meteorol. Papers, Cambridge, Mass.* **1**, (4), 36 pp.

Rossby, C. G. and Montgomery, R. B. (1935), 'The layers of frictional influence in wind and ocean currents', *MIT, Woods Hole, Pap. Phys. Oceanog. Meteorol.* **3**, (3), 101 pp.

Rossby, C. G., (1940), 'Planetary flow patterns in the atmosphere', *Quart. J. Roy. Meteorol. Soc. Suppl.* **66**, 68–87.

Rotta, J. C. (1951), 'Statistische Theorie nichthomogener Turbulenz', *Z. Phys.*, **129**, 547–572; **131**, 51–77.

Rouse, W. R., Mills, P. F., and Stewart, R. B. (1977), 'Evaporation in high latitudes', *Water Resour. Res.* **13**, 909–914.

Ryan, P. J., Harleman, D. R. F., and Stolzenbach, K. D. (1974), 'Surface heat loss from cooling ponds', *Water Resour. Res.* **10**, 930–938.

Saito, T. (1962), 'On estimation of transpiration and eddy-transfer coefficient within plant communities by energy balance method', *J. Agric. Meteorol.* (*Nogyo Kisho*) *Japan* **17**, 101–105.

Sasamori, T. (1970), 'A numerical study of atmospheric and soil boundary layers', *J. Atmos. Sci.* **27**, 1122–1137.

Satterlund, D. R. (1979), 'An improved equation for estimating long-wave radiation from the atmosphere', *Water Resour. Res.* **15**, 1649–1650.

Satterlund, D. R. and Means, J. E. (1978), 'Estimating solar radiation under variable cloud conditions', *Forest Sci.* **24**, 363–373.

Schlichting, H. (1960), *Boundary Layer Theory* (translated by J. Kestin), 4th edn., McGraw-Hill, N.Y., 647 pp.

Schmidt, W. (1915), 'Strahlung und Verdunstung an freien Wasserflächen; ein Beitrag zum Wärmehaushalt des Weltmeers und zum Wasserhaushalt der Erde', *Ann. d. Hydrogr. u. Mar. Met.* **43**, 111–124.

Schmugge, T. (1978), 'Remote sensing of surface soil moisture', *J. Appl. Meteorol.* **17**, 1549–1557.

Schmugge, T. J., Jackson, T. J., and McKim, H. L. (1980), 'Survey of methods for soil moisture determination', *Water Resour. Res.* **16**, 961–979.

Scholl, D. G. and Hibbert, A. R. (1973), 'Unsaturated flow properties used to predict outflow and evapotranspiration from a sloping lysimeter', *Water Resour. Res.* **9**, 1645–1655.

Schreiber, P. (1904), 'Ueber die Beziehungen zwischen dem Niederschlag und der Wasserführung der Flüsse in Mitteleuropa', *Meteorol. Z.* **21**, 441–452.

Seginer, I. (1974), 'Aerodynamic roughness of vegetated surfaces', *Boundary-Layer Meteorol.* **5**, 383–393.

Seginer, I., Mulhearn, P. J., Bradley, E. F., and Finnigan, J. J. (1976), 'Turbulent flow in a model plant canopy', *Boundary-Layer Meteorol.* **10**, 423–453.

SethuRaman, S. and Raynor, G. S. (1975), 'Surface drag coefficient dependence on the aerodynamic roughness of the sea', *J. Geophys. Res.* **80**, 4983–4988.

Shahane, A. N., Thomas, D., and Bock, P. (1977), 'Spectral analysis of hydrometeorological time series', *Water Resour. Res.* **13**, 41–49.

Shannon, J. W. (1968), 'Use of atmometers in estimating evapotranspiration', *J. Irrig. Drain. Div., Proc. ASCE* **94** (IR3), 309–320.

Shaw, R. H. (1977), 'Secondary wind speed maxima inside plant canopies', *J. Appl. Meteorol.* **16**, 514–521.

Shawcroft, R. W., Lemon, E. R., Allen, L. H. Jr., Stewart, D. W., and Jensen, S. E. (1974), 'The soil-plant-atmosphere model and some of its predictions', *Agric. Met.* **14**, 287–307.

Sheppard, P. A. (1958), 'Transfer across the earth's surface and through the air above', *Quart. J. Roy. Meteorol. Soc.* **84**, 205–224.

Sheppard, P. A., Tribble, D. T., and Garratt, J. R. (1972), 'Studies of turbulence in the surface layer over water (Lough Neagh), Part I Instrumentation, programme, profiles', *Quart. J. Roy. Meteorol. Soc.* **98**, 627–641.

Sheriff, N. and Gumley, P. (1966), 'Heat transfer and friction properties of surfaces with discrete roughnesses', *Int. J. Heat Mass Transfer* **9**, 1297–1319.

Sherman, F. S., Imberger, J., and Corcos, G. M. (1978), 'Turbulence and mixing in stably stratified waters', *Ann. Rev. Fluid Mech.* **10**, 267–288.

Shir, C. C. (1972), 'A numerical computation of air flow over a sudden change of surface roughness', *J. Atmos. Sci.* **29**, 304–310.

Shnitnikov, A. V. (1974), 'Current methods for the study of evaporation from water surfaces and evapotranspiration', *Hydrol. Sci. Bull., Intern. Assoc. Hydrol. Sci.* **19**, 85–97.

Shulyakovskiy, L. G. (1969), 'Formula for computing evaporation with allowance for temperature of free water surface', *Soviet Hydrol. Selec. Papers*, No. 6, 566–573.

Shuttleworth, W. J. and Calder, I. R. (1979), 'Has the Priestley-Talyor equation any relevance to forest evaporation?', *J. Appl. Meteorol.* **18**, 639–646.

Sinclair, T. R., Allen, L. H., and Lemon, E. R. (1975), 'An analysis of errors in the calculation of energy flux densities above vegetation by a Bowen-ratio profile method', *Boundary-Layer Meteorol.* **8**, 129–139.

Slatyer, R. O. and McIlroy, I.C. (1967), *Practical Microclimatology*, CSIRO, Melbourne, Australia, 310 pp.

Smedman-Högström, A.-S. (1973), 'Temperature and humidity spectra in the atmospheric surface layer', *Boundary-Layer Meteorol.* **3**, 329–347.

Smedman-Högström, A.-S and Högström, U. (1973), 'The Marsta micrometeorological field project', *Boundary-Layer Meteorol.* **5**, 259–273.

Smith, S. D. (1974), 'Eddy flux measurements over Lake Ontario', *Boundary-Layer Meteorol.* **6**, 235–255.

Smith, S. D. and Banke, E. G. (1975), 'Variation of the sea surface drag coefficent with wind speed', *Quart. J. Roy. Meteorol. Soc.* **101**, 665–673.

Söderman, D. and Wesantera, J. (1966), 'Some monthly values of evapotranspiration in Finland computed from aerological data', *Geophysica*, **8**, 281–290.

Stanhill, G. (1962), 'The use of the Piche evaporimeter in the calculation of evaporation', *Quart. J. Roy. Meteorol. Soc.* **88**, 80–82.

Stanhill, G. (1969), 'A simple instrument for the field measurement of turbulent diffusion flux', *J. Appl. Meteorol.* **8**, 509–513.

Staple, W. J. (1976), 'Prediction of evaporation from columns of soil during alternate periods of wetting and drying', *Soil Sci. Soc. Am. J.* **40**, 756–761.

Stegen, G. R., Gibson, C. H., and Friehe, C. A. (1973), 'Measurements of momentum and sensible heat fluxes over the open ocean', *J. Phys. Oceanog.* **3**, 86–92.

Stephens, J. C. (1965), 'Discussion of "Estimating evaporation from insolation"', *J. Hydraul. Div., Proc. ASCE* **91**, (HY5), 171–182.

Stephens, J. C. and Stewart, E. H. (1963), 'A comparison of procedures for computing evaporation and evapotranspiration', General Assembly of Berkeley, Int. Assoc. Sci. Hydrology, Publn. No. 62, pp. 123–133.

Stewart, J. B. and Thom, A. S. (1973), 'Energy budgets in pine forest', *Quart. J. Roy. Meteorol. Soc.* **99**, 154–170.

Stewart, R. B. and Rouse, W. R. (1976), 'A simple method for determining the evaporation from shallow lakes and ponds', *Water Resour. Res.* **12**, 623–628.

Stewart, R. B. and Rouse, W. R. (1977), 'Substantiation of the Priesley and Taylor parameter α = 1.26 for potential evaporation in high latitudes', *J. Appl. Meteorol.* **6**, 649–650.

Stricker, H. and Brutsaert, W. (1978), 'Actual evapotranspiration over a summer period in the "Hupsel Catchment"', *J. Hydrol.* **39**, 139–157.

Sutton, O. G. (1934), 'Wind structure and evaporation in a turbulent atmosphere', *Proc. Roy. Soc. London*, **A146**, 701–722.

Sutton, O. G. (1953), *Micrometeorology*, McGraw-Hill Book Co., N.Y., 333 p.

Sutton, W. G. L. (1943), 'On the equation of diffusion in a turbulent medium', *Proc. Roy. Soc. London* **A182**, 48–75.

Sverdrup, H. U. (1935), 'The ablation on Isachsen's Plateau and on the Fourteenth of July glacier in relation to radiation and meteorological conditions', *Geograf. Ann.* (Stockholm) **17**, 145–166.

Sverdrup, H. U. (1937), 'On the evaporation from the oceans', *J. Mar. Res.* **1**, 3–14.

Sverdrup, H. U. (1946), 'The humidity gradient over the sea surface', *J. Meteorol.* **3**, 1–8.

Swinbank. W. C. (1951), 'The measurement of vertical transfer of heat and water vapor by eddies in the lower atmosphere', *J. Meteorol.* **8**, 135–145.

Swinbank, W. C. (1963), 'Long-wave radiation from clear skies', *Quart. J. Roy. Meteorol. Soc.* **89**, 339–348.

Szeicz, G. and Long, I. F. (1969), 'Surface resistance of crop canopies', *Water Resour. Res.* **5**, 622–633.

Takeda, K. (1966), 'On roughness length and zero-plane displacement in the wind profile of the lowest air layer', *J. Meteorol. Soc. Japan, Ser. II* **44**, 101–107.

Talsma, T. (1970), 'Hysteresis of two sands and the independent domain model', *Water Resour. Res.* **6**, 964–970.

Tan, H. S. and Ling, S. C. (1963), 'Quasi-steady micro-meteorological atmospheric boundary layer over a wheatfield', in E. R. Lemon (ed.), *The Energy Budget at the Earth's Surface*, ARS, USDA

Production Res. Rept. No. 72, Washington, DC, pp. 7–25.

Tanner, C. B. (1960), 'Energy balance approach to evapotranspiration from crops', *Soil Sci. Soc. Am. Proc.* **24**, 1–9.

Tanner, C. B. and Jury, W. A. (1976), 'Estimating evaporation and transpiration from a row crop during incomplete cover', *Agron. J.* **68**, 239–242.

Tanner, C. B. and Pelton, W. L. (1960), 'Potential evapotranspiration estimates by the approximate energy balance method of Penman', *J. Geophys. Res.* **65**, 3391–3413.

Taylor, G. I. (1935), 'Statistical theory of turbulence', *Proc. Roy. Soc. London* **A151**, 421–478.

Taylor, G. I. (1938), 'The spectrum of turbulence', *Proc. Roy. Soc. London* **A164**, 476–490.

Taylor, P. A. (1969a), 'On wind and shear stress profiles above a change in surface roughness', *Quart. J. Roy. Meteorol. Soc.* **95**, 77–91.

Taylor, P. A. (1969b), 'On planetary boundary layer flow under conditions of neutral thermal stability', *J. Atmos. Sci.* **26**, 427–431.

Taylor, P. A. (1969c), 'The planetary boundary layer above a change in surface roughness', *J. Atmos. Sci.* **26**, 432–440.

Taylor, P. A. (1970), 'A model of airflow above changes in surface heat flux, temperature and roughness for neutral and unstable conditions', *Boundary-Layer Meteorol.* **1**, 18–39.

Taylor, P. A. (1971), 'Airflow above changes in surface heat flux, temperature and roughness; an extension to include the stable case', *Boundary-Layer Meteorol.* **1**, 474–497.

Taylor, R. J. (1960), 'Similarity theory in the relation between fluxes and gradients in the lower atmosphere', *Quart. J. Roy. Meteorol. Soc.* **86**, 67–87.

Taylor, R. J. (1961), 'A new approach to the measurement of turbulent fluxes in the lower atmosphere', *J. Fluid Mech.* **10**, 449–458.

Tennekes, H. (1970), 'Free convection in the turbulent Ekman layer of the atmosphere', *J. Atmos. Sci.* **27**, 1027–1034.

Tennekes, H. (1973), 'The logarithmic wind profile', *J. Atmos. Sci.* **30**, 234–238.

Tennekes, H. and Lumley, J. L. (1972), *A First Course in Turbulence*, The MIT Press, Cambridge, Mass., 300 pp.

Thom, A. S. (1971), 'Momentum absorption by vegetation', *Quart. J. Roy. Meteorol. Soc.* **97**, 414–428.

Thom, A. S. (1972), 'Momentum, mass and heat exchange of vegetation', *Quart. J. Roy. Meteorol. Soc.* **98**, 124–134.

Thom, A. S. (1975), 'Momentum, mass and heat exchange of plant communities', in J. L. Monteith, (ed.) *Vegetation and the Atmosphere*, Vol. I. Principles, Academic Press, London, pp. 57–109.

Thom, A. S. and Oliver, H. R. (1977), 'On Penman's equation for estimating regional evaporation', *Quart. J. Roy. Meteorol. Soc.* **103**, 345–357.

Thom, A. S., Stewart, J. B., Oliver, H. R., and Gash, J. H. C. (1975), 'Comparison of aerodynamic and energy budget estimates of fluxes over a pine forest', *Quart. J. Roy. Meteorol. Soc.* **101**, 93–105.

Thornthwaite, C. W. (1948), 'An approach toward a rational classification of climate', *Geograph. Rev.* **38**, 55–94.

Thornthwaite, C. W. and Holzman, B. (1939), 'The determination of evaporation from land and water surfaces', *Monthly Weath. Rev.* **67**, 4–11.

Timofeev, M. P. (1954), 'Change in the meteorological regime on irrigation', *Izv. Akad. Nauk, S.S.S.R., Ser. Geogaf.* No. 2, 108–113.

Townsend, A. A. (1965a) 'Self-preserving flow inside a turbulent boundary layer', *J. Fluid Mech.* **22**, 773–797.

Townsend, A. A. (1965b), 'The response of a turbulent boundary layer to abrupt changes in surface conditions', *J. Fluid Mech.* **22**, 799–822.

Townsend, A. A. (1966), 'The flow in a turbulent boundary layer after a change in surface roughness', *J. Fluid Mech.* **26**, 255–266.

Tschinkel, H. M. (1963), 'Short-term fluctuation in streamflow as related to evaporation and transpiration', *J. Geophys. Res.* **68**, 6459–6469.

Tsukamoto, O., Hayashi, T., Monji, N., and Mitsuta, Y. (1975), 'Transfer coefficients and turbulence-flux relationship as directly observed over the ocean during the AMTEX '74', Scient. Report, 4th AMTEX Study Confer., (Tokyo, Sept. 1975), pp. 109–112.

Tsvang, L. R., Koprov, B. M., Zubkonskii, S. L., Dyer, A. J., Hicks, B. B., Miyake, M., Stewart, R. W., and McDonald, J. W. (1973), 'A comparison of turbulence measurements by different instruments: Tsimlyansk field experiment', *Boundary-Layer Meteorol.* **3**, 499–521.

Turc, L. (1954, 1955), 'Le bilan d'eau des sols: relations entre les précipitations, l'évaporation et l'écoulement', *Ann. Agron.* **5**, 491–595; **6**, 5–131.

Uchijima, Z. (1962), 'Studies on the micro-climate within plant communities' (1): 'On the turbulent transfer coefficient within plant layer', *J. Agric. Meteorol.* (*Nogyo Kisho*) *Japan* **18**, 1–9.

Uchijima, Z., Udagawa, T., Horie, T. and Kobayashi, K. (1970), 'Studies of energy and gas exchange within crop canopies' (8): 'Turbulent transfer coefficient and foliage exchange velocity within a corn canopy', *J. Agric. Meteorol.* (*Nogyo Kisho*) *Japan* **25**, 215–227.

Uchijima, Z. and Wright, J. L. (1964), 'An experimental study of air flow in a corn plant-air layer', *Bull. Nation. Inst. Agric. Sci. Japan* (*Nogyo Gijutsu Kenkyusho Hokoku*) **A11**, 19–66.

Unsworth, M. H. and Monteith, J. L. (1975), 'Long-wave radiation at the ground', *Quart. J. Roy. Meteorol. Soc.* **101**, 13–24.

Van Bavel, C. H. M. (1966), 'Potential evaporation: the combination concept and its experimental verification', *Water Resour. Res.* **2**, 455–467.

Van Bavel, C. H. M. (1961), 'Lysimetric measurements of evapotranspiration rates in the eastern United States', *Proc. Soil Sci. Soc. Am.* **25**, 138–141.

Van Bavel, C. H. M. (1967), 'Changes in canopy resistance to water loss from alfalfa induced by soil water depletion', *Agric. Meteorol.* **4**, 165–176.

Van Bavel, C. H. M. and Myers, L. E. (1962), 'An automatic weighing lysimeter', *Agric. Eng.* **43**, 580–583, 587–588.

Van Bavel, C. H. M. and Reginato, R. J. (1965), 'Precision lysimetry for direct measurement of evaporative flux', Internat. Sympos. Methodol. of Plant Eco-Physiol., Montpellier, France, 1962, pp. 129–135.

Van Hylckama, T. E. A. (1968), 'Water level fluctuation in evapotranspirometers', *Water Resour. Res.* **4**, 761–768.

Van Wijk, W. R. (1963), 'General temperature variations in a homogeneous soil', in W. R. Van Wijk, (ed.), *Physics of Plant Environment*, North Holland Pub. Co., Amsterdam, pp. 144–170.

Van Wijk, W. R. and De Vries, D. A. (1963), 'Periodic temperature variations in a homogeneous soil', in W. R. Van Wijk, (ed.), *Physics of Plant Environment*, North Holland Publ. Co., Amsterdam, pp. 102–143.

Van Wijk, W. R. and Scholte Ubing, D. W. (1963), 'Radiation', in W. R. Van Wijk (ed.) *Physics of Plant Environment*, North Holland Pub. Co., Amsterdam, pp. 62–101.

Verma, S. B., Rosenberg, N. J., and Blad, B. L. (1978), 'Turbulent exchange coefficients for sensible heat and water vapor under advective conditions', *J. Appl. Meteorol.* **17**, 330–338.

Warhaft, Z. (1976), 'Heat and moisture flux in the stratified boundary layer', *Quart. J. Roy. Meteorol. Soc.* **102**, 703–707.

Webb, E. K. (1960), 'On estimating evaporation with fluctuating Bowen ratio', *J. Geophys. Res.* **65**, 3415–3417.

Webb, E. K. (1964), 'Further note on evaporation with fluctuating Bowen ratio', *J. Geophys. Res.* **69**, 2649–2650.

Webb, E. K. (1966), 'A pan-lake evaporation relationship', *J. Hydrology*, **4**, 1–11.

Webb, E. K. (1970), 'Profile relationships: The log-linear range, and extension to strong stability', *Quart. J. Roy. Meteorol. Soc.* **96**, 67–90.

Weisman, R. N. (1975), 'Comparison of warm water evaporation equations', *J. Hydraul. Div., Proc. ASCE* **101** (HY10), 1303–1313.

Weisman, R. N. and Brutsaert, W. (1973), 'Evaporation and cooling of a lake under unstable atmospheric conditions', *Water Resour. Res.* **9**, 1242–1257.

Wesely, M. L., Eastman, J. A., Cook, D. R., and Hicks, B. B. (1978), 'Daytime variations of ozone eddy fluxes to maize', *Boundary-Layer Meteorol.* **15**, 361–373.

Wesely, M. L. and Hicks, B. B. (1975), 'Comments on "Limitations of an eddy-correlation technique for the determination of the carbon-dioxide and sensible heat fluxes"', *Boundary-Layer Meteorol.* **9**, 363–367.

Wieringa, J. (1972), 'Tilt errors and precipitation effects in trivane measurements of turbulent fluxes over open water', *Boundary-Layer Meteorol.* **2**, 406–426.

Wieringa, J. (1974), 'Comparison of three methods for determining strong wind stress over Lake Flevo', *Boundary-Layer Meteorol.* **7**, 3–19.

Williams, R. J., Broersma, K., and Van Ryswyk, A. L. (1978), 'Equilibrium and actual evapotranspiration from a very dry vegetated surface', *J. App. Meteorol.* **17**, 1827–1832.

Williamson, R. E. (1963), 'The management of soil salinity in lysimeters', *Soil Sci. Soc. Am. Proc.* **27**, 580–583.

Willis, W. O. (1960), 'Evaporation from layered soils in the presence of a water table', *Soil Sci. Soc. Am. Proc.* **24**, 239–242.

Wilson, R. G. and Rouse, W. R. (1972), 'Moisture and temperature limits of the equilibrium evapotranspiration model', *J. Appl. Meteorol.* **11**, 436–442.

Wooding, R. A., Bradley, E. F., and Marshall, J. K. (1973), 'Drag due to regular arrays of roughness elements of varying geometry', *Boundary-Layer Meteorol.* **5**, 285–308.

Wright, J. L. and Brown, K. W. (1967), 'Comparison of momentum and energy balance methods of computing vertical transfer within a crop', *Agron. J.* **59**, 427–432.

Wüst, G. (1937), 'Temperatur- und Dampfdruckgefälle in den untersten Metern über der Meeresoberfläche', *Meteorol. Z.* **54**, 4–9.

Wyngaard, J. C. (1973), 'On surface layer turbulence', in D. A. Haugen (ed.), *Workshop on Micrometeorology*, Amer. Meteorol. Soc., pp. 101–149.

Wyngaard, J. C., Arya, S. P. S., and Coté, O. R. (1974), 'Some aspects of the structure of convective planetary boundary layers', *J. Atmos. Sci.* **31**, 747–754.

Wyngaard, J. C. and Coté, O. R. (1971), 'The budgets of turbulent kinetic energy and temperature variance in the atmospheric surface layer', *J. Atmos. Sci.* **28**, 190–201.

Yadav, B. R. (1965), 'Total solar radiation in relation to duration of sunshine', *Indian J. Meteorol. Geophys.* **16**, 261–266.

Yaglom, A. M. (1972), 'Turbulent diffusion in the surface layer of the atmosphere', *Izv. Acad. Sci. U.S.S.R., Atmos. Oceanic Phys.* **8**, 579–593 (English edn. AGU 333–340.)

Yaglom, A. M. (1976), 'Semi-empirical equations of turbulent diffusion in boundary layers', *Fluid Dynamics Trans. (Polish Acad. Sci., Inst. Fund. Technol. Res., Warsaw)* **7** (II), 99–144.

Yaglom, A. M. (1977), 'Comments on wind and temperature flux-profile relationships', *Boundary-Layer Meteorol.* **11**, 89–102.

Yaglom, A. M. and Kader, B. A. (1974), 'Heat and mass transfer between a rough wall and turbulent fluid flow at high Reynolds and Péclet numbers', *J. Fluid Mech.* **62**, 601–623.

Yamada, T. (1976), 'On the similarity functions A, B and C of the planetary boundary layer', *J. Atmos. Sci.* **33**, 781–793.

Yamamoto, G. (1950), 'On nocturnal radiation', *Sci. Rep. Tohoku Univ. (Sendai, Japan), Ser. 5, Geophys.* **2**, 27–43.

Yamamoto, G. and Shimanuki, A. (1964), 'Profiles of wind and temperature in the lowest 250 meters in Tokyo', *Sci. Rep. Tohoku Univ. (Sendai, Japan), Ser. 5, Geophys.* **15**, 111–114.

Yamamoto, G. and Miura, A. (1950), 'Evaporation by natural convection', *Sci. Rep. Tohoku Univ. (Sendai, Japan), Ser. 5, Geophys.* **2**, 48–50.

Yap, D., Black, T. A., and Oke, T. R. (1974), 'Calibration and tests of a yaw sphere-thermometer system for sensible heat flux measurements', *J. Appl. Meteorol.* **13**, 40–45.

Yasuda, N. (1975), 'The heat balance at the sea surface observed in the East China Sea', *Sci. Rep. Tohoku Univ. (Sendai, Japan), Ser. 5, Geophys.* **22**, 87–105.

Yeh, G. T. and Brutsaert, W. (1970), 'Pertubation solution of an equation of atmospheric turbulent diffusion', *J. Geophys. Res.* **75**, 5173–5178.

Yeh, G. T. and Brutsaert, W. (1971a), 'A numerical solution of the two dimensional steady-state turbulent transfer equation', *Monthly Weath. Rev.* **99**, 494–500.

Yeh, G. T. and Brutsaert, W. (1971b), 'Sensitivity of the solution for heat flux or evaporation to off-diagonal turbulent diffusivities', *Water Resour. Res.* **7**, 734–735.

Yeh, G. T. and Brutsaert, W. (1971c), 'A solution for simultaneous turbulent heat and vapor transfer between a water surface and the atmosphere', *Boundary-Layer Meteorol.* **2**, 64–82.

Yih, C.-S. (1952), 'On a differential equation of atmospheric diffusion', *Trans. Am. Geophys. Un.* **33**, 8–12.

Yotsukura, N., Jackman, A. P., and Faust, C. R. (1973), 'Approximation of heat exchange at the air-water interface', *Water Resour. Res.* **9**, 118–128.

Yu, S. L. and Brutsaert, W. (1967), 'Evaporation from very shallow pans', *J. Appl. Meteorol.* **6**, 265–271.

Yu, S. L. and Brutsaert, W. (1969a), 'The generation of an evaporation time series for Lake Ontario', *Water Resour. Res.* **5**, 785–796.

Yu, S. L. and Brutsaert, W. (1969b), 'Stochastic aspects of Lake Ontario evaporation', *Water Resour. Res.* **5**, 1256–1266.

Yordanov, D. and Wippermann, F. (1972), 'The parameterization of the turbulent fluxes of momentum, heat and moisture at the ground in a baroclinic planetary boundary layer', *Beitr. Phys. Atmos., Contr. Atmos. Phys.* **45**, 58–65.

Zilitinkevich, S. S. (1969), 'On the computation of the basic parameters of the interaction between the atmosphere and the ocean', *Tellus* **21**, 17–24.

Zilitinkevich, S. S. and Deardorff, J. W. (1974), 'Similarity theory for the planetary boundary layer of time-dependent height', *J. Atoms. Sci.* **31**, 1449–1452.

Zilitinkevich, S. S., Laikhtman, D. L., and Monin, A. S. (1967), 'Dynamics of the atmospheric boundary layer', *Izv. Acad. Sci. U.S.S.R., Atmos. Oceanic Phys.* **3**, 297–333 (English edn. AGU, 170–191).

Index